T0251195

Near-Earth Laser Communications

Optical Science and Engineering
Founding Editor: Brian J. Thompson, University of Rochester
Rochester, New York

For more information about this series, please visit: https://www.crcpress.com/Optical-Science-and-Engineering/book-series/CRCOPTSCIENG

Near-Earth Laser Communications

Second Edition

Edited by

Hamid Hemmati

CRC Press
Taylor & Francis Group
Boca Raton London New York

CRC Press is an imprint of the
Taylor & Francis Group, an **informa** business

Second edition published 2021
by CRC Press
6000 Broken Sound Parkway NW, Suite 300, Boca Raton, FL 33487-2742

and by CRC Press
4 Park Square, Milton Park, Abingdon, Oxon, OX14 4RN

First issued in paperback 2023

© 2021 Taylor & Francis Group, LLC
CRC Press is an imprint of Taylor & Francis Group, an Informa business

Reasonable efforts have been made to publish reliable data and information, but the author and publisher cannot assume responsibility for the validity of all materials or the consequences of their use. The authors and publishers have attempted to trace the copyright holders of all material reproduced in this publication and apologize to copyright holders if permission to publish in this form has not been obtained. If any copyright material has not been acknowledged please write and let us know so we may rectify in any future reprint.

Except as permitted under U.S. Copyright Law, no part of this book may be reprinted, reproduced, transmitted, or utilized in any form by any electronic, mechanical, or other means, now known or hereafter invented, including photocopying, microfilming, and recording, or in any information storage or retrieval system, without written permission from the publishers.

For permission to photocopy or use material electronically from this work, access www.copyright.com or contact the Copyright Clearance Center, Inc. (CCC), 222 Rosewood Drive, Danvers, MA 01923, 978-750-8400. For works that are not available on CCC please contact mpkbookspermissions@tandf. co.uk

Trademark notice: Product or corporate names may be trademarks or registered trademarks, and are used only for identification and explanation without intent to infringe.

Publisher's Note
The publisher has gone to great lengths to ensure the quality of this reprint but points out that some imperfections in the original copies may be apparent.

ISBN 13: 978-1-4987-7740-7 (hbk)
ISBN 13: 978-1-03-265254-2 (pbk)
ISBN 13: 978-0-429-18672-1 (ebk)

DOI: 10.1201/9780429186721

Typeset in Times LT Std
by Cenveo® Publisher Services

Contents

Preface

The second edition of *Near-Earth Laser Communications* is an updated version of the book first published in 2009. Most chapters focus on devices and hardware design and development methodology, while a few chapters emphasize both the background theoretical treatment and the hardware needed to make a specific flight or ground subsystem function properly.

This book provides an updated and comprehensive view of the developments in laser communications technology and where it is making impact. It covers all major sub-assemblies of a laser communication transceiver from the flight subsystem to the ground subsystem. The objective of the new edition is the same as that of the first edition, i.e., to emphasize the hardware aspects of a free-space laser communications system.

The technology of free-space optical communications, also known as laser communications or lasercom, is no longer a technology under basic development; rather, it is ready and the EDRS satellites use optical links operationally. This technology can now deliver from space and high-altitude platforms (HAPs) data rates that compare with that of fiber-optics technology as it mostly uses the components that were developed by the fiber-optics industry. Data-rates on the order of 100s of Gbps and over 1 Tbps are now possible and subject of upcoming demonstrations. Perceived risks associated with the lasercom technology have been largely alleviated owing to a number of successful optical link demonstrations from HAPS and the Earth orbit. This technology is likely to find applications at a much larger scale than ever before in civilian and non-civilian, and Earth and space science observations.

The field of lasercom deployment is growing rapidly as it offers lower system size, mass, and power consumption, along with as much as two orders of magnitude higher bandwidth delivery when compared to its nearest competitors. Key advantages are brought about by the higher frequency, which results in the narrowing of the optical beam's divergence. However, with the ever-increasing demand for bandwidth and higher data-rates lasercom is nowhere near its technology peak, and much remains to be explored.

This book is intended to be a general reference and a textbook for researchers and senior and graduate-level university students working in the field of telecommunications and electro-optical systems. It may also be of use to engineers or project leaders who would like to gain a deep understanding of principles behind the field of free-space optical communications.

This book is an amalgamation of the vast experiences of its authors. It is designed to acquaint the reader with the basics of laser satellite communications. The emphasis is on device technology, implementation techniques, and system trades. The detailed theory behind laser communications has been covered in detail by a few books published earlier.

Chapter 1 is an introduction to near-Earth laser communications technology. Justification for higher bandwidth and recent successful demonstrations are described here.

Chapter 2 discusses design drivers, design trades, link budgets for acquisition, tracking and pointing, and communications links.

Chapter 3 details approaches for laser beam pointing, acquisition, and tracking, with an emphasis on device technologies and implementation strategies.

Chapter 4 elaborates on the flight laser transmitters for coherent and direct detection.

Chapter 5 discusses the flight optomechanical assembly.

Chapter 6 discusses channel coding techniques, describing applicable modulation for free-space optical communications.

Chapter 7 describes the photodetectors and receivers as applicable to both the flight and ground transceivers.

Chapter 8 discusses numerous implications of the atmospheric channel on laser beam propagation for both downlinks and uplinks.

Chapter 9 describes the ground terminal and provides examples from recent successful flight links.

Chapter 10 discusses methodologies for flight qualification of components and subsystems, including both optical and optoelectronic assemblies.

Chapter 11 looks at the approaches and configurations for cross-links and optical networking.

Chapter 12 concludes the book with a brief summary of ongoing activities and those planned for the near future.

I am indebted to all the contributing authors from the United States, Europe, and Japan for taking time off from their busy schedules to share their knowledge. There are many others who have contributed directly or indirectly to the development of this book. They include members of JPL, Facebook, MIT Lincoln Lab, Tesat-Spacecom, and NICT's lasercom teams who have contributed immensely to the advancement of the free-space laser communications technology. Finally, I would like to express my deepest gratitude to my wife Azita, my daughter Elita and my son Dr. David Hemmati for their continued and unfailing love and support.

Hamid Hemmati
Los Angeles, California
hemmati@ieee.org

List of Contributors

Adolfo Comerón
Department of Signal Theory and
    Communications
Technical University of Catalonia
Barcelona, Spain

Marcos Reyes García-Talavera
Technology Division
Instituto de Astrofisica de Canarias
Tenerife, Spain

Jon Hamkins
Jet Propulsion Laboratory
California Institute of Technology
Pasadena, California

Hamid Hemmati
Facebook Inc.
Menlo Park, California

Nikos Karafolas
European Space Agency
European Space Research and
    Technology Centre
Noordwijk, The Netherlands

Werner Klaus
National Institute of Information
    and Communications Technology
    (NICT)
Tokyo, Japan

Klaus Kudielka
Sr. Systems Engineer
Synopta GmbH
Zurich, Switzerland

Robert G. Marshalek
Ball Aerospace
Boulder, Colorado

Bruce Moision
GoogleX
Santa Clara, California

Sabino Piazzolla
Jet Propulsion Laboratory
National Aeronautics and Space
    Administration
California Institute of Technology
Pasadena, California

Klaus Pribil
NewComp
Freienstein, Switzerland

Zoran Sodnik
European Space Agency
European Space Research and
    Technology Centre
Noordwijk, The Netherlands

Morio Toyoshima
Space Communication Group
National Institute of Information
    and Communications Technology
    (NICT)
Tokyo, Japan

Peter J. Winzer
Nokia Bell Labs
Holmdel, New Jersey

Editor Biography

 **Hamid Hemmati** received his master's degree in physics from the University of Southern California, and a PhD degree in physics from Colorado State University in 1981. He is an Engineering Director at Facebook Inc. Prior to that he was with the JPL-NASA-Caltech for 28 years, working primarily on the satellite laser communications technology. From 1983 to 1986 he worked at the NASA Goddard Space Flight Center on the Cosmic Background Explorer (COBE) spacecraft and on free-space laser communications. As a post-doctoral fellow at NIST, University of Colorado (Boulder, 1981–1983), Dr. Hemmati worked on ultra-stable atomic clocks based on laser-cooled trapped ions.

He is the editor and author of two books, *Deep Space Optical Communications* and *Near-Earth Laser Communications*, and the author of seven other book chapters. He is the recipient of NASA's Exceptional Service Medal, NASA Space Act Board Awards, and 36 NASA certificates of appreciation. He is a Fellow member of OSA (Optical Society of America) and the SPIE (Society of Optical Engineers).

Dr. Hemmati's research interests include providing global Internet connectivity, specifically for the unconnected or under-connected population; advancing laser and millimeter-wave communications technologies for satellite, airborne, and terrestrial applications; fiber-optic installation cost reduction; and data center transceiver optics.

1 Near-Earth Laser Communications

Hamid Hemmati

CONTENTS

1.1   INTRODUCTION

Today, the fiberoptics technology is managing much of the terrestrial access and backbone networks at tens of tera-bit-per-second (Tbps) collective capacity [1]. However, due to steep cost or geological obstacles, fiberoptic networks have been impractical in certain rural and isolated areas. Mobile networking is another application where wired communications is not applicable [2]. Telecommunications via satellites and high-altitude platforms (HAPs) may bridge this gap and in particular deliver backhaul service coverage to sparsely populated areas.

Link reliability and availability aside, current communications capacity via airborne and spaceborne platforms now constitutes only a small fraction of those provided by the fiberoptics networks. However, such platforms equipped with a multitude of Earth-observing sensors or broadband communications systems are experiencing an exponential growth in telecom data volumes.

Conventional air and space platforms use radio the frequency (RF), microwave (MW), and millimeter-wave (MMW) spectrum for communications. Geosynchronous orbit (GEO) satellites deliver the bulk of these links [3]. By 2022, the ViSat-3 GEO satellites with a total capacity of >1 Tbps per satellite is expected to establish the highest throughput satellite (HTS). Downlinking of hundreds of beams each operating at 1 to 2 Gbps and with capacity demand dependent beam-hopping capability to activate and deactivate the beams makes this sizeable link capacity possible [4]. In early 2020s, several constellations of LEO satellite are also planning to create multi-Tbps collective capacity in space [5–7]. Owing to the huge growth forecasted in the volume of information linked globally, entirely new approaches for data communications are required for future spacecraft. RF and MMW satellite telecom technology are hindered by the available spectrum bandwidth limitations and regulations. Their system technologies are now pushing the state-of-the-art limits. MMW links at slightly higher frequencies are promising limited relief to these obstacles [8].

The laser (optical) communications (lasercom) technology with currently unregulated spectrum and with multi-THz of telecom bandwidth can potentially augment the conventional RF/MW/MMW communications technology with orders of magnitude capacity enhancement [9]. As an example, uplink to GEO satellite at rates of hundreds of Gbps to multi-Tbps beamed from spatially diverse ground stations would majorly enhance the current capacity limits. Inter-satellite crosslinks at the rate of tens of Gbps would also enable large capacity LEO satellite constellations with much fewer numbers of ground-based gateway stations [10]. Relative to RF/MW communications, for nearly the same flight mass and input power, the lasercom technology provides tens of dB link margin enhancement. The additional margin may be traded for substantial increase in data-rate or significant reductions in aperture (telescope) diameter, weight, and power-consumption. Since the year 2000, a number of highly successful lasercom demonstrations from airplane, the Earth orbit and the Moon have reaffirmed that this technology is ready for operational use.

This chapter provides a high-level overview of the status of near-Earth laser communications technology developments and future research opportunities. Subsequent chapters describe each subsystem in greater detail. By near-Earth we mean links with airborne, Earth orbiting, and the lunar platforms. A number of publications describe excellent performances of recent lasercom demonstrations, follow-up technology development efforts, and lasercom theoretical treatments [11–12]. Here, we emphasize critical requirements and design drivers, the status of current subsystem technologies, and pathways to achieve the immense potential of laser communications. This chapter summarizes state of the art in subsystem hardware technologies, underlying principles and its technology trends.

1.1.1 WHY LASER COMMUNICATIONS?

A. Lower link losses due to lower beam divergence

Dominated by the laws of diffraction, the width of the high frequency (short wavelength) optical/laser communications beams are three to four orders of magnitude narrower than the (lower frequency) RF communications beams. Consequently, the transmitted signal can be delivered to the intended receiver in a less lossy manner. The resulting benefits include data-rate delivery comparable to that of fiberoptics communications, reduced SWaP (size, weight, and power consumption), and precision navigation, tracking, and ranging that could be conducted along with the telecom function [13]. Examples of benefits include frequency reuse, i.e., using the same wavelength for multiple links, improved channel security, reduced mass, reduced power consumption, reduced size, ability to track and communicate with the sun within the field of view, multifunctionality with other electro-optic instruments, and precision ranging [14].

B. Abundant unregulated spectrum

Unlike RF communication systems with restricted spectrum usage, optical links are not subject to frequency regulation. This is an attractive prospect for high-bandwidth applications. Frequency reuse represents an additional advantage, made possible due to the small divergence angle of the communications laser beam.

C. Leverages advancements in fiberoptics industry

Lasercom links efficiently leverage the huge and continuous investment that has gone into the fiberoptics industry to support the huge demand posed by the exponential growth of the Internet globally.

D. Benefits tactical applications

Relative to RF systems, lasercom systems are difficult to intercept and jam.

1.1.2 WHY NOT LASER COMMUNICATIONS?

Despite a multitude of potential advantages, the perceived risks associated with lasercom has often been the primary impediment to its prevalent use in space. The lasercom technology provides a powerful tool in the telecommunications toolbox and not an approach that could make the conventional telecom technologies obsolete. Some of the potential lasercom drawbacks include:

1. The adverse effects of the atmosphere on the laser beam including attenuation (e.g., clouds) and index of refraction fluctuations (e.g., turbulence and scintillation) resulting in deep fades and beam breakup, among examples. The outcome is lower link availability compared with the RF band for through-atmosphere links. For applications such as Internet delivery requiring high link availability ($\geq 99.9\%$), use of lasercom for through-atmosphere downlink from and uplink to the satellite may be impractical. But applications where timely delivery of data is not critical, such as telecom with Earth-observing satellites can be used effectively.
2. Higher energy per bit at optical vs. RF communications frequencies since optical photons are more energetic than radio-, micro-, and millimeter waves, resulting in additional quantum-noise in the optical receiver [15]. That is, optical detectors (detecting the signal's intensity) are quantum noise limited with power detection of ~5E-19 W, while RF receivers (detecting the signal's field) are thermal noise limited with power detection capability of ~1E-21 W [16]. Given this limitation, unlike RF communications, wide-beam-divergence (especially omni-directional transmitters) transmission of optical beams is very power inefficient. However, energetic optical photons allow for a high and even noiseless single-photon detection probability; a phenomenon that is not possible with RF communications, making possible unique communications architectures such as photon-starved regimes.
3. The requirement for line of sight in most outdoor links (>10s of meters in range) often becomes an impediment; and
4. Lack of off-the-shelf subsystems qualified for use in space and lack of extensive ground infrastructure drive costs and development time.

Undoubtedly, by implementing the lasercom technology industry would usher a new era of global telecom for transmitting high volumes of both sensors and broadband data.

1.1.3 LASER COMMUNICATIONS MATURITY

Today, laser communications from HAPs and the Earth orbit is no longer a maturing technology. Rather, this telecommunication toll is now mature enough where flight transceivers are available commercially from multiple sources and are being deployed for operational use [17, 18]. Table 1.1 summarizes historic milestones in laser communications technology demonstrations and most recently operational deployments.

TABLE 1.1

Spacecraft Lasercom Links Successfully Performed

Year	Project	Link	Max. Data Rate (Gbps)	Organization and Notes
1981	AFTS [19]	Airplane		McDonnell Douglas
1991	TALC [20]	Plane-submarine		GTE
1992	GOPEX [21]	Ground to deep space	N/A	Laser beam pointing from ground to a satellite in deep space
1996	RME [22]	Space relay		Ball Aerospace
1995	LCE [23]	GEO-Ground	0.001	National Institute of Information and Comm Tech (NICT, Japan), Jet Propulsion Laboratory JPL/NASA, Duplex links.
2001	GeoLITE	GEO-Ground	>1	Lincoln Lab (USA). Duplex
2001	SILEX [24]	LEO-GEO	0.05	European Space Agency. Duplex
2002	ALEX	GEO-Air	>1	Lincoln Lab. Duplex links to GeoLITE
2005	LUCE [25]	LEO-GEO LEO-Ground	0.05	JAXA (Japan), OICETS spacecraft
2006	LOLA [28]	Air-GEO	0.05	France; Duplex links to SILEX
2008	LCTSX [29]	LEO-LEO LEO-Ground	5.5	DLR/TESAT-Spacecom (Germany) Coherent detection
2013	Alphasat [30]	LEO-GEO	1.8	European Space Agency (ESA)
2013	LLCD [31]	Moon-Earth	0.622	NASA/Lincoln Lab/JPL
2014	Sentinel-A [32]	LEO-GEO	1.8	Operational use from satellites
2014	OPALS [33]	LEO-Earth	0.175	NASA/JPL
2014	SOTA [34]	LEO-Earth	0.01	NICT
2016	EDRS-A [35]	LEO-GEO	1.8	ESA
2016	OSIRISV2 [36]	LEO-Earth	1	DLR
2017	Sentinel-2 [37]	LEO-GEO	1.8	ESA
2020 (plan)	LCRD [38]	GEO-Earth Earth-GEO	1.244	NASA Goddard Space Flight Center

With innovations in lasercom optical assemblies along with advancements in the fiberoptics technology that are directly applicable to lasercom, lower mass, power, size, and cost flight systems are expected in the near future. Cost of the flight transceiver is driven primarily by those required for flight qualification. Volume manufacturing of flight qualified lasercom systems remains a challenge that will likely be facilitated through automated integration and test. Lower cost ground stations with diameter on the order of ≈ 0.4 to 1 m are also of prime interest. Toward this goal, commercial entitities are aiming at providing ground lasercom ground station networks commercially throughout the globe [39].

1.1.4   USE CASES

Point-to-point optical communication links include satellite (or aircraft)-to-satellite (or aircraft), satellite (or aircraft)-to-airborne platforms (aircraft or balloon) and satellite (or aircraft)-to-ground-based stations. Both inter-orbit (e.g., LEO-LEO) and intra-orbit (e.g., LEO-GEO) links are of interest. With the exception of very short-range links where beam broadening is possible, distance between the two ends of the link is a good metric for the magnitude of the link difficulty since the required pointing accuracy, output power of transmitters, size, mass, power consumption, and deployment cost all increase with the link range.

While future Earth optical receivers may be located on airborne or Earth-orbiting platforms to circumvent the atmospheric issues, operational uses of the lasercom would likely to rely on lower risk and less costly ground-based receivers.

Depending on the ground station's location, link availability over a year time for laser beam propagation through clouds and other atmospheric constituents could range from <30% to >70%. Satellite crosslinks (e.g., LEO-LEO, LEO-GEO, and GEO-GEO) or airplane crosslinks above the clouds and much of the atmosphere (e.g., air-to-air, and air-to-satellite) links are immune to the adverse effects of the atmosphere. As a result, applications such as broadband Internet delivery that require high link availability may not be able to rely on the lasercom technology. However, transmission of delay-tolerant data, such as those from sensor payloads, could well take advantage of lasercom technology's high data capacity.

Alternatively, data may be sent from air or low-Earth-orbit satellite to MEO or GEO satellite followed by RF or MMW downlink to the Earth. Burst mode of operation is another approach to transmit many gigabytes of data to Earth in a short period of time.

A number of LEO satellite constellations for global telecommunications are presently under development and have focused attention on the need for inter-satellite (LEO-LEO) links (ISLs). Optical ISLs are naturally suited to this application due to tens of Gbps data-rate capability and an atmosphere-free environment [40]. Current challenges are delivery of ≥10 Gbps over the communications range of ≈6000 km and minimum of a 7-year lifetime. Cost of each lasercom transceiver in quantity (>1000) needs to be reduced by one to two orders of magnitude relative to previous flight laser communications system. Volume manufacturing of flight qualified units are required at the rate of approximately a few transceivers per day. A number of manageable paradigm shifts have to be implemented relative to the state of the art to achieve these requirements.

Past laser communications technology demonstrations primarily focused on the success of the project at the expense of larger mass, power, volume, and cost that could have been possible. Much of the system designs were baselined around the year 2000. Since then, significant advancements in the development of components, particularly those of fiberoptics components, directly applicable to free-space communications transceivers, have occurred. Reduced complexity and cost of volume-manufacturable flight transceivers would facilitate its prevalent use.

1.2 SUBSYSTEM TECHNOLOGIES

1.2.1 DESIGN DRIVERS

Key design drivers for laser communications from airplanes and satellites include managing the space loss due to the vast distance between the two terminals by adequate laser power and antenna gains at each end of the link, precision laser beam pointing, the sun angles, and propagation through the atmosphere that often limits the link availability.

The requirement for sub-microradian to several-microradian beam-pointing drives the optomechanical design segment of a spaceborne and airborne transceiver. Such a stringent requirement on the precision of beam pointing places challenging to achieve tolerances on the optomechanical and thermomechanical stability. Efficient, lightweight and long-life laser transmitters, optimal coding and modulation, and efficient and low-noise signal detection and component properties that make them amenable to air and space qualification are among the other design drivers.

Figure 1.1 illustrates a block diagram of lasercom link signal flow including interface between the flight transceiver on an air or space platform and the Earth-based station and the optical channel in between. Any instability of either the air- or the space-vehicle, and or the environment surrounding the ground station might adversely affect the link performance. Examples include excessive vibrations in the lasercom payload hosting airplane, unmitigated spacecraft random vibrations and background radiation surrounding the satellite. Turbidity and scintillation above the ground station could cause a number of deleterious effects on the laser beam traversing through the atmosphere.

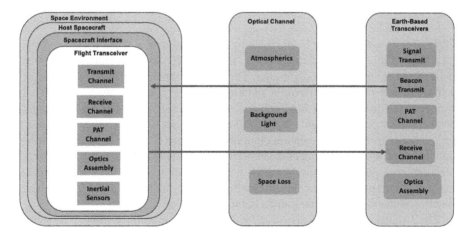

FIGURE 1.1 Schematic of interaction between the air- to spaceborne optical transceiver and a ground station illustrating a variety of link-influencing parameters on the satellite-side, the optical-channel, and the ground-side of the communications system.

1.2.2 SPATIAL ACQUISITION, TRACKING, AND LASER BEAM POINTING AND STABILIZATION

Near-infrared laser wavelength's narrow beam divergence leads to the requirements for precision beam pointing by both ends of the link, often on the order of a few microradians. This requirement makes the laser beam pointing to or from a moving platform one of the most challenging aspects of free-space laser communications. Often a dedicated pointing, acquisition, and tracking (PAT) subsystem tasked with stabilizing the line-of-sight is incorporated with the optics assembly to satisfy this highly stringent requirement.

The problem of lasercom beam pointing can in general be decomposed into (1) optical line-of-sight stabilization and (2) providing the appropriate pointing reference to the receiver location. The latter is generally achieved by providing a beacon from the receiver location, and the former, by using a high bandwidth control loop to sense and correct for the platform jitter. For a near-Earth lasercom system, a strong beacon from the remote transceiver, or use of the communications beam itself as a beacon, usually serves both purposes. General steps leading to precise laser beam pointing between two satellites involves the following phases:

Acquisition: Compensation of the initial laser beam pointing errors arising from spatial acquisition; e.g., spacecraft attitude ephemeris and knowledge error of its location.

Tracking: Tracking out the dynamic local angular disturbances imparted onto the transceiver by the external (e.g., the host spacecraft) and internal active elements with accuracy amounting to fraction of a beamwidth; often at a few microradian level.

Beam pointing: Once the two transceivers have locked onto each other, this subsystem compensates for the relative cross velocities and motions (including vibration) of the two platforms.

Certain design drivers for beam pointing and stabilization include tracking the receiving station with an accuracy on the order of 1/10 the beamwidth with microradian level residual line-of-sight jitter. This level of pointing accuracy needs to be achieved while the host platform is vibrating, jittering, creeping, the beacon signal experience frequent fades at small sun angles introduce significant background light. A spacecraft's platform angular micro-vibrations can stem from guidance and control activities including momentum dumping and retrorocket firings. To achieve the tight beam pointing requirements some form of a cooperative beacon that accurately identifies the opposite terminal location is always required. The acquisition, tracking, and pointing process may be characterized by the pointing budget, link budget, timeline/probability allocations and reliability allocations.

The 3-sigma total pointing accuracy is defined as 3 times jitter + bias. The angular width of a 1500-nm laser beam transmitted through a 15-cm-diameter diffraction-limited telescope is ~10 μrad. A pointing loss factor is typically allocated to the power link margin to account for statistical pointing-induced fades (PIF) and mispoint angles. For instance, a 2-dB pointing loss allocation with a 1% PIF, assumed for the above system, entails a total pointing accuracy of at least 5.4 μrad. The 2 dB allocation is then divided into bias and jitter mis-point errors [41]. To effectively deliver the signal to the ground station, the lasercom transmitter must be capable of tracking the receiving station, such that jitter error is less than approximately 10% of the transmit beamwidth (in this case approximately 1 μrad jitter error).

The residual jitter can be calculated given the knowledge of the power spectral density (PSD) of the pointing drift and jitter (N_f), and the control system's disturbance rejection (R_f) via [42]: $\Theta^2_{rms} = \int |R(f)|^2 N(f) \cdot df$.

To establish a robust link, beamwidth of the transmitter and receiver's field of view have to be adequately larger than Θ_{rms}. Unless the exact value of the N_f is known at the system engineering design stage, it is advisable to assume a conservative value that can be modified with time as better data becomes available.

It is imperative to understand the spectrum and amplitude of vibrational disturbances of the host platform to which the optical transceiver will be mounted. Very often, these values are not thoroughly identified until the platform's assembly is underway, or after measurements are made on an existing platform. Platform micro-vibration frequencies are in the range of 0.1 to 100s of Hz with amplitudes typically much larger than the transmit laser beamwidth. Passive isolators and inertially referenced stabilized platforms utilizing angle sensors and accelerometers can adequately reject frequencies >10s of Hz [36]. Inertial reference unit (IRU) options include AC-coupled devices like angle sensors such as the magneto-hydrodynamic type, accelerometers and DC-coupled sensors like fiber-coupled gyros.

The control loop bandwidth needs to be greater than ten times higher than the maximum frequency disturbance one wants to compensate. One way to reduce the loop bandwidth is to limit the input disturbance spectrum. A reliable beacon signal, e.g., a dedicated beacon source, emanating from the target, or use of star-trackers in conjunction with dedicated pointing control subsystems can help to reject lower frequencies and to maintain accurate beam pointing [43].

It is also feasible to circumvent the need for a fast-tracking subsystem by mounting a compact lasercom transceiver on a platform that can effectively isolate the transceiver from the host platform. Disturbance-free platform (DFP) is an example of one such device [44]. Another example is the miniaturized active vibration isolation system (MVIS II) by Honeywell for isolation of flight optical systems, it has an expected in-flight capability of over 20 dB RMS and active jitter suppression over the 1 to 200 Hz range [45, 46]. Some of the advantages of this architecture are that (a) by reducing the amount of line-of-sight jitter coupled into the optical element, the focal plane array (FPA) can be integrated over a much longer period of time, enabling the use of faint beacon as pointing references. Due to the long integration time, the pointing concept is insensitive to the scintillation-induced fades of the uplink beacon, (b) in addition to providing mechanical isolation, the DFP can also provide coarse and fine pointing adjustment and thereby relaxing the flight transceiver pointing requirement, and (c) by eliminating the need for high bandwidth line-of-sight control, the optical package can be significantly simplified, resulting in mass and power savings.

Errors in position determination, boresight calibration, and residual tracking not compensated by the pointing subsystem control loop, comprise the total pointing error for a lasercom transceiver. Inadequate compensation of host platform vibrations by the pointing control loop results in residual tracking error (and therefore the residual pointing error), which can be minimized by utilizing a high bandwidth control loop. Beacon signal centroiding update rate as high as several kHz may be required if a reliable inertial navigation system with comparable update rates has not incorporated with the system. Typically, a quadrant detector is adequate for laser beam tracking when the beacon signal level is high, and the field of regard is small, as in inter-satellite links. However, achieving sub-microradian accuracy laser pointing necessitates a larger number of pixels on the detector array with longer integration times to increase the signal-to-noise ratio (SNR). Subarray of a focal-plane-array sensor (e.g., a CCD or CMOS) and fiber nutating elements are examples of other approaches that, depending on the communication scenario, can assist in achieving higher pointing accuracy.

Figure 1.2 shows a simplified block diagram of the PAT subsystem and its key components that have to work together in a nested loop to precisely point the laser transmitter's beam to the direction of the receive terminal. Options are a separate transmit and receive optical apertures or a single aperture for both transmit and receive. The latter has been shown repeatedly to be the best approach when precise

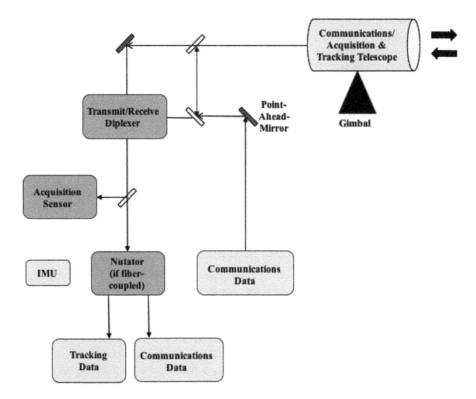

FIGURE 1.2 An example of a PAT subsystem block diagram representing critical components. A common path transmit/receive aperture is shown. The laser transmitter's beam bounces off a fast-steering (fine-pointing) mirror before it exits the telescope. The acquisition sensor tracks the direction of the outgoing beam.

(microradian level) beam pointing is required since it largely minimizes the possibility of drift between separate transmit and receive apertures [47]. Other key components of the PAT channel include a coarse-pointing assembly (CPA), a fine-beam-pointing or fast-steering mirror assembly (FSM), a quadrant detector or a multi-pixel focal-plane array to track the direction of the incoming and outgoing beams, a point-ahead mechanism [48]. Additionally, as required passive optical elements to route and isolate beams.

In cases where the laser beam pointing angles are large, such as LEO satellite with ground or LEO-LEO satellite communications, a CPA becomes an essential subsystem of the lasercom terminal. Since the required beam-pointing angles may vary widely, beam-pointing control will typically need to be applied over a large angular span while achieving high accuracy jitter compensation. In practice, this is often accomplished with a combination of coarse and fine-beam steering controls. Coarse beam pointing is an essential subsystem of a mobile laser communication terminal. This subsystem is a low-bandwidth mechanism with significant angular motion engaged for initial open loop pointing, raster or spiral scanning, acquisition searching, and variable rate slewing.

Options for the coarse-pointing of the lasercom terminal include a gimbaled telescope and gimbaled mirror, each offering specific advantages. A gimbaled flat mirror is an actuated mirror located in front of the telescope aperture to manipulate laser transmitter's beam. A gimbaled flat is particularly suitable for GEO-ground communications due to a relatively narrow angular range from the satellite point of view.

Non-mechanical CPAs and FSMs suitable for flight use is maturing gradually and could become viable devices in near future [49]. The technology of non-mechanical coarse-pointing and fine-beam steering devices based on optical phased-array technology is advancing rapidly and gradually becoming mature enough for field implementation [50, 51].

Chapter 3 of this book provides a comprehensive description of issues and technologies for laser beam pointing, acquisition, and tracking.

1.2.3 ROBUST FLIGHT OPTOMECHANICAL ASSEMBLY (OMA)

Precision (microradian level) beam pointing between aerial or spaceborne terminal and Earth, the communication range between transceivers, the sun angles, environmental conditions, and the spacecraft launch loads are among the multiple design challenges of a flight optics assembly. Accommodating these requirements necessitating high temporal and thermal stability for the OMA's structure. Other driving requirements are a few to tens of centimeter diameter telescope while minimizing its mass and volume and at the same time sufficient structural stability, support multiple wavelengths, and sustantial degree of the background and spurious light rejection.

An example of optical beam paths within one particular configuration of a flight lasercom transceiver is shown in Figure 1.3. The optical train includes a common transmit/receive aperture, followed by transmit, receive, and boresight channels, shown in red, green, and blue beam paths, respectively. The transmit laser signal reflects from a two-axis fine-pointing mirror before traveling out through the telescope. The beacon signal from a remote receiver is imaged onto a focal plane

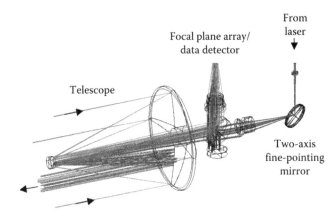

FIGURE 1.3 Transceiver optical train example showing the three main critical optical channels.

detector array for spatial acquisition and tracking. The receive path may accommodate both data and beacon signals. The beacon signal is used to aid with acquisition, tracking, and precise transmit laser beam pointing. A small portion of the laser transmitter's signal is also imaged onto the focal plane (through the boresight channel) for reference.

Other key requirements include minimum mass and swept volume, a structure with high fundamental frequency, ability to support multiple optical wavelengths (for wavelength division multiplexed operation), and excellent background light rejection. Reflective mirrors and the telescope structure may be electroformed or fabricated entirely from aluminum, or from a variety of silicon carbide compounds. These materials, along with beryllium and low thermal expansion glass, are excellent candidates for mirror substrates [52–54].

For optical designs with a single aperture for both transmit and receive, due to the need for detection of very faint receive (bacon and data) signals in the presence of strong transmit laser light that does get scattered within the optics assembly, often transmit/receive signal isolation on the order of 100 to 140 dB is required. This is an optical design challenge, but such combined systems are less susceptible to angular drift between transmit and receive channels. For typical spaceborne links, the optics assembly has to provide adequate stability between the transmit and receive signals, ensuring that maximum free-space offset between the two beams is <0.1 μrad. The optomechanical system's structural design has to withstand the expected launch load without deterioration.

An alternative implementation approach uses one to several strands of optical fiber in the aft optics portion of the optical assembly to route the transmitted or the received signals [55, 56]. This approach has potential to significantly simplify the optomechanical system development, and to minimize the required tolerances on the optomechanical assembly [57].

Stray-light requirements are often a driving factor in optical terminal design. Examples are uniform distribution of stray light over the tracking detector field of view and internal stray-light levels limited to 1 μW/nm·sr (measured relative to telescope pupil). An optical system of class 300 or better cleanliness is frequently employed to minimize the scattered light that can find its way to any one of the system sensors.

A solar background light-rejecting window, with low wavefront distortion at the telescope aperture, minimizes the heat load from sunlight entering the telescope. Nevertheless, external heaters are required in conjunction with the optical system to maintain a uniform (non-gradient) temperature on the order of 300 ± 1 °K.

Spaceborne applications, in particular, are often severely mass-constrained or operate in challenging thermal environments. Depending on the application, mass constraints and thermal load conditions, a variety of materials including aluminum, beryllium, SiC, and AlBeMet have been employed to form the OMA's mechanical structure and the mirror substrates [58]. Chapter 5 of this book provides additional discussion on the optomechanical assembly for the flight lasercom transceivers.

1.2.4 GROUND TELESCOPE

Currently, there is a limited number of 0.6 to >1.0 m diameter optical ground stations, adequate for communications with LEO satellites. These are located primarily in Japan, Europe, and the United States [59–60]. Bridgewave Inc. is now attempting to establish a global network of optical ground stations [39]. Lasercom ground station specific requirements that distinguish it from the astronomy ground telescopes set demanding characteristics on the telescope optics and opto-mechanical assembly. These include operations at both daytime and nighttime, operating at small sun angles without damaging the optics, tracking relatively fast-moving planes and LEO satellites with adequate precision and the ability to compensate for the atmospheric turbulence effects. Figure 1.4 shows a picture of NASA-JPL's 1-m-diameter optical communications, and the 20-cm-diameter adaptive-optics-equipped optical communications ground station.

FIGURE 1.4 (Left) NASA-JPL's 1-m-diameter lasercom-dedicated telescope (called OCTL) capable of tracking LEO and GEO spacecraft and fast-moving aircraft. This telescope is capable of operating at daytime to within 10° of the sun, and with an add-on filter, to <3° of the sun. (Right) a 20-cm-diameter adaptive-optics-equipped optical communications ground station in Southern California developed by Facebook Inc.

Telescope's primary mirror and the high precision two-axis gimbal are among the cost drivers for a lasercom ground station. Active compensation of a thin mirror, spin-casted mirrors, and molded composite materials with post-polishing for high Strehl are among the technologies developed to reduce the cost of the primary mirror [61]. Background light filtering both at the entrance aperture and at the aft optics will be essential to optimize the SNR while operating at small sun angles [62]. This includes spectral, spatial, and temporal filtering with the requirement that the spectral bandpass of the filter has to be larger than the spectral width of the transmit laser. Tunable spectral filters might be required to track the Doppler shift of the communication's wavelength. Following safety procedures to avoid eye-damage and aerial to space vehicle damage is an essential part of any lasercom ground station. Laser interlocks, radars, cameras, and human spotters are often implemented for safe operations.

1.2.5   OPTOMECHANICAL ASSEMBLY FOR GROUND TRANSCEIVER

For the near-Earth links, a variety of telescope configurations with near-diffraction limited monolithic or segmented primary mirrors with diameters on the order of 0.3 to 1.5 m are adequate. Driving requirements include operations during both daytime and nighttime, signal reception within a few degrees of the solar disk, operations at zenith angles up to 80°, partial compensation of atmospheric effects, and telescope two-axis gimbaling with arc-second level pointing accuracy. Note that most astronomy telescopes are designed for nighttime operations only and daytime operation generates a new set of opto-mechanical design challenges.

Deployment of a large number of telescopes would require per telescope cost reduction. Current research efforts include development of potentially low-cost composite material, membrane, and electroformed primary mirrors. Other related research includes actively corrected mirrors for mirror-aberration or image-motion compensation and high-accuracy stabilization, in which activations are applied either directly to the telescope mirror, or a deformable mirror or a spatial light modulator positioned in the aft optics of the telescope [63, 64]. The two-axis telescope gimbal, the dome/structure that holds the telescope(s), and the primary mirror constitute major cost drivers for an optical station.

Proper mitigation of background and stray light will often be required. Filtering schemes applicable to lasercom include spatial filtering, use of multiple detectors, temporal filtering, use of polarized light, modulation/coding, solar filtering at the entrance to the telescope, specially designed telescopes to prevent sunlight from entering the focal plane, and spectral filtering. Even though as narrow a filter bandwidth as possible is desired, the allowable spectral filter bandwidth will be limited by the laser linewidth and data rate as well as the Doppler shift.

To avoid operations outages and negative consequences arising from the Sun at or near the telescope field of view, a lasercom ground receiver has to operate at small (a few degrees) Sun angles. The deleterious effects of sunlight entering the telescope include damage to optics or baffles, thermally induced aberrations and misalignment, poor seeing, and excessive stray light due to scattered light. Development of multimeter diameter optical filters, to be placed at the entrance aperture of the

FIGURE 1.5 A 1-m-diameter band-pass filter at the entrance aperture of a telescope (OCTL).

ground telescope (or in the dome shutter opening) as a prefilter, is now feasible. Figure 1.5 shows a 1-m-diameter band-pass filter centered at 1064 nm wavelength, utilizing a membrane substrate [65]. The additional wavefront aberration introduced by the filter was on the order of a fraction of a wave (at 1000 nm), acceptable for most lasercom applications.

Laser beam propagation from the ground to air or space must follow safety-related requirements and guidelines established by various government agencies. These include the OSHA (for work force protection), the FAA (in the United States) to protect pilots and aircraft, and the Laser Clearinghouse of Space Command (US operations) responsible for protecting space assets. JPL has developed a multitiered safety system to meet the coordination, monitoring, and reporting functions required by the relevant regulatory government bodies to ensure safe laser beam propagation from the OCTL facility [66]. Tier-0 relates to laser operators meeting OSHA safety standards for protection. For example, access to the high-power laser areas is restricted and interlocked. As illustrated in Figure 1.6, Tier-1 identifies aircraft in the FAA-controlled airspace with a long-wave infrared (8–13 µm) camera and spans the range from the telescope dome to 3.4 km. Tier-2 detects at-risk aircraft by radar and spans the range from 3.4 km to the aircraft service ceiling in FAA airspace. Tier-3 covers the beam propagation into space and will require coordination with the Laser Clearinghouse [67]. The operational procedures that comply with the agency's safe laser beam propagation guidelines should facilitate the desire to operate the ground stations autonomously.

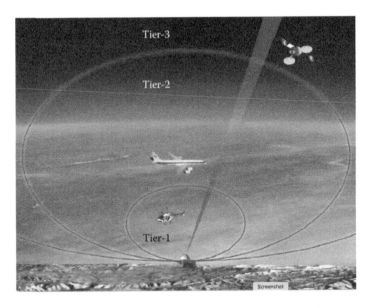

FIGURE 1.6 Three JPL defined safety tiers for ground to space laser beam propagation from the OCTL.

1.2.6 SIGNAL RECEPTION

The signal detection channel critically assesses the optical channel performance, and all associated aspects of the channel and the system, such as channel models and capacity, detection, modulation and error correcting codes.

The block diagram shown in Figure 1.7 is one of many approaches of the data communications chain where the data to be transmitted is modulated and coded prior to being digitally applied to the laser beam. The laser's output is amplified to the level specified by the link budget and physically beam-shaped via a multitude of passive optics prior to propagation into medium. Upon reception of the

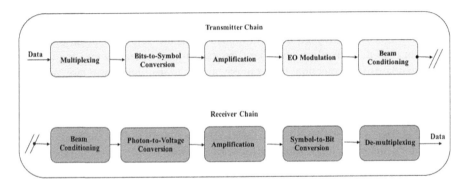

FIGURE 1.7 Example of a signal transmission and reception chain.

optical signal by the opposite terminal, optical-to-electrical signal is generated by the receiver followed by amplification and decoding in order to recover the originally transmitted signal.

Available implementation options include detection using a discrete detector or fiber-coupled detectors. Advantages of fiber coupling are (a) allowing use of pre-detection amplifiers for higher receiver sensitivity (b) more effective implementation of wavelength division multiplexed (WDM) schemes, and (c) single-mode receiver more effectively reject background. Downside is the more complex implementation of telescope (adaptive optics), and fiber coupling.

In the receiver chain, photon-to-electron conversion is often accomplished via PIN (positive-intrinsic-negative) detectors with no internal gain, and APD (avalanche-photo-diode) detectors with gain that could be >100. Detector's quantum-efficiency, signal responsivity, spectral responsivity and a variety of internal and inherent noise parameters characterize these devices. Photon-counting Geiger-mode detectors and super-conducting arrays constitute some of the most efficient detectors, but due to a variety of reasons, such as deadtime, they are typical not suited for high-rate (multi-Gbps) communications [68].

Key driving receiver requirement includes high receiver sensitivity to minimize the required signal photons/bit to achieve enhanced link performance at a certain given bit-error rate. Fundamental modes of lasercom operation are coherent and incoherent – also known as direct detection. Conventionally, RF, microwave, and millimeter-wave communications use coherent detection. A fundamental difference between RF and optical signal detection is that the RF receiver measures the field of the signal, whereas optical detectors measure its intensity (field square). Coherent detection provides as much as 4 to 6 dB of link margin improvement over incoherent detection and enables excellent rejection of background light, but its implementation is more complex than incoherent detection. Also, correction of the phase-front of the received optical beam that has traveled through the atmosphere is often required for coherent detection. Preamplified coherent receiver performance can approach the theoretical limits [69, 70]. Incoherent detection along with extremely low-noise single-photon-sensitive (photon-counting) detectors can have a performance that approaches that of coherent detection [71].

Besides the atmospheric effects on coherent communications, in the past, realization of this technology for wireless links suffered from implementation difficulties. However, thanks to significant advances in the coherent fiberoptics technology, high-rate modems, on the order of 100 to 400 Gbps and on the way to >1 Tbps in in near future, have advanced significantly and are being used extensively in fiberoptic networks. The same technology is directly applicable to free-space optical links. Coherent communication benefits include effective mitigation of the background light and high detection efficiency. Modern coherent transceivers achieve a sensitivity on the order of a few photons per bit, several dBs from the capacity limit of 1.44 bits/photon for a coherent homodyne receiver [72].

Chapter 7 of this book discusses signal detectors and detection in significant detail.

1.2.7 MODULATION AND CODING

Forward error correction (FEC) is an effective tool for introducing redundancy at the transmitter and to correct receiver's detection errors [73]. At data rates on the order of tens of Gbps, the link margin enhancement as a result of the sensitivity gain could be as much as 10 dB. FECs do typically introduce implementation complexity due to the requirement for additional processing and overhead on the order of <10% [74]. Commercially available FECs at 40 Gbps have relatively low overhead and apply Reed-Solomon codes. The more advanced FECs rely on iterative and soft-decision-decoded low-density parity check (LDPC) or turbo codes to achieve significantly higher receiver sensitivity [75].

Laser modulators superimpose the data to be transmitted onto the transmit laser beam with a particular modulation format designed for the link by the communications systems engineers. To correct the signal transmission faults, error correcting codes are often implemented simultaneously with modulation. Both modulation and decoding for wireless optical communications are now advanced and mature technologies. Serially concatenated pulse position modulation (SCPPM) and LDPC codes are among the examples of codes that have brought the state of the art to within a fraction of 1 dB of the theoretical Shannon limit (channel capacity) [74]. Pulse position modulation (PPM) allows multiple bits per photon detection and is ideal for photon-starved channels. Binary differential phase-shift keying (DPSK) is another high-efficiency code applied in conjunction with optically preamplified receivers to high-rate laser communications.

Shannon's classical treatment of information theory sets capacity limits on the highest data communications reliability for a given receiver, while Holevo's quantum-mechanical treatment of information theory optimizes link capacity over the best possible receiver [75]. Holevo's theorem may be applied to predict the performance of an optical receiver optimized to achieve quantum-limited detection (not proven experimentally yet).

Some of the popular data transmission formats in lasercom include on-off keying (OOK), PPM, and phase-shift keying (PSK). Prior to transmit signal amplification, the encoder output often feeds into an easily modulatable low-power, high-beam quality laser transmitter (oscillator) in a master oscillator power amplifier (MOPA) configuration. Even though more remains to be done, modulation and coding is a relatively mature subsystem of a lasercom transceiver. This is illustrated in Figure 1.8 where bit error rate (BER) is plotted as a function of E_b/N_0, the bit energy-to-noise density for different modulation and coding schemes. The waterfall behavior of the newly developed codes reflects their high efficiency. A high code rate (the ratio of data bits to total bits transmitted in the code words) refers to high information content and low coding overhead. Chapter 6 provides an overview of channel coding techniques for airborne and spaceborne laser communications.

1.2.8 FLIGHT LASER TRANSMITTER

Highly efficient flight laser transmitters capable of multi-Gbps to multi-Tbps bandwidth modulation at moderate average power levels and with high spatial beam

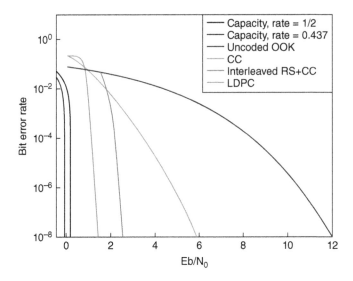

FIGURE 1.8 Performance of different modulation schemes, indicating that the recently developed codes for optical detection are within <1 dB of capacity limit. (OOK, on-off keying; CC, convolutional code; RS, Reed-Solomon; LDPC, low-density parity check) (see Chapter 6).

quality are required for near-Earth lasercom in the direct detection mode. Certain fiber amplifiers, solid-state amplifiers, and semiconductor laser diodes lend themselves well to this application.

Moderate output power up to a few watts, in high spatial beam quality (for direct detection) and both high spatial and longitudinal beam quality (for coherent detection); high wall-plug electrical-to-optical efficiency, ease of output beam modulation to tens and hundreds of Gbps, lifetime on the order of 7 to 15 years, respectively, for LEO and GEO orbit under the satellite orbit's radiation environment and launch loads, and operations in vacuum are among the specific requirements for satellite and under the satellite or airplane launch loads are among the laser transmitter requirements.

Today's off-the-shelf laser transmitters meet the output power and beam quality requirements of near-Earth laser communication links. Except for radiation tolerance and operation in a vacuum, Telcorida-qualified commercial off-the-shelf lasers meet nearly all of the demanding requirements of airborne and spaceborne operations. Desired improvement areas include higher overall efficiency (including laser drivers and thermal management), and improved life span for missions lasting 10 to 15 years.

Several laser wavelengths are prominent in lasercom. Among them, the 1550 nm wavelength is most popular since these devices are readily available commercially for the fiber optics industry, and the atmospheric attenuation and eye hazards are less for 1550 nm than for 800 or 1064 nm. MOPA configurations such as erbium-doped fiber amplifier (EDFA) the mainstay of the fiberoptics communications networks are among the most suitable laser transmitters for free-space communications.

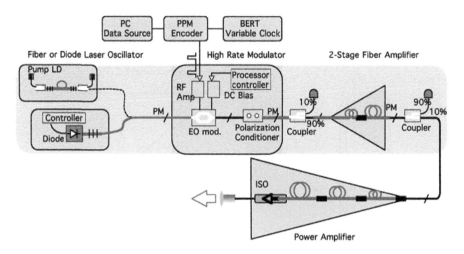

FIGURE 1.9 Schematic of a multistage amplifier to generate tens of dBm of modulated laser power. Output power is scalable by the number of amplifier stages.

Such amplifiers preserve the beam and modulation quality at hundreds of Gbps while boosting the input power of a few milliwatts from a master oscillator to multi-watts. The EDFA's 1545 ± 20 nm approximate wavelength is also about two orders of magnitude more eye safe than the shorter wavelength lasers in the 800 nm region [76].

Figure 1.9 schematically illustrates at high level one of several approaches for a multistage oscillator-amplifier. Gain, gain coefficient, gain ripple, noise figures, spectral and modulation bandwidth, maximum output power, center wavelength, and nonlinear effects at higher powers constitute some of the key characteristics of the laser amplifier with ramifications on the overall communications link budget.

Chapter 4 of this book has a detailed discussion on laser transmitters for coherent communications (4a) and laser transmitters for direct-detection communications (4b).

1.3 ATMOSPHERIC CHANNEL

Laser beams propagating from air or space through the Earth's atmosphere and toward a ground-based optical station may experience received signal disturbances due to atmospheric attenuation (absorption and scattering) and scintillation and turbulence. These effects, which may occur independently or simultaneously, depend on the atmospheric conditions when communications occur. Absorption and scattering resulting from cloud cover and fog are, by far, the primary source of atmospheric attenuation since the constituent particle sizes are comparable in size to the laser's wavelength. A significant portion of the atmospheric-induced optical link degradation can be mitigated through proper site selection and site configuration. Scintillation may result in signal fades at the receiver, beam spreading at the telescope focal plane, wavefront tip/tilt, background light due to direct sunlight and sky radiance, and other but similar undesirable phenomena.

1.3.1 Atmospheric Background Light

Elevated noise levels may occur at the system's photodetector due to the atmospheric background light entering the telescope during both daytime and nighttime. Throughout the day, levels of the sky radiance may fluctuate by one to two orders of magnitude. The variations depend on the receiver's altitude, line-of-sight sun angle to the receiver, cloud density, atmospheric aerosol concentration and the optical systems' straylight rejection characteristics.

1.3.2 Atmospheric Attenuation

Cloud and fog attenuation may vary in magnitude anywhere from 1 to >200 dB, depending on their density. In cases of severe attenuation, enhancement of the transmit laser power would not be able to compensate the attenuation losses as every doubling of the laser power just adds another 3 dB to the link margin. Currently, ground station diversity where the ground telescopes are located in truly independent weather cells is the only practical method to circumvent atmospheric attenuation resulting from cloud cover.

Decorrelation of the weather condition has been measured over approximately three weather cells, where each cell is approximately 400 km across demonstrating that multiple optical receiving stations separated by a few hundred kilometers should provide close to full station weather availability [77].

1.3.3 Atmospheric Turbulence/Scintillation

The sunlight-induced heating of the Earth's surface transfer's its stored thermal energy to the surrounding atmosphere. The effect results in an index of refraction variation of the atmosphere where the laser beam traverses towards aerial and space platforms. Highly dynamic lens and prism-like atmospheric zones with differing size and index are created within the atmosphere with narrower zones more adjacent to the Earth and midday correlated with the peak temperature of the day and wider zones as one gets away from the Earth and at moderate temperatures of the day. The magnitude of the turbulence effects may vary by as much as an order of magnitude throughout the day. The severity of turbulence effects often subsides just prior to dawn and dusk [78]. Ground station altitude, atmospheric pressure, and wind speed are among other factors that affect the atmospheric region's behavior. The movement rate of the zones varies in the sub-kHz to multi-kHz range. The net effect is intensifying the adverse effects on the laser beam particularly on the uplink beam and less so in downlink beams. Deep fades with major intensity fluctuations may last a few milliseconds [79].

Additional undesired turbulence-induced phenomena on the beam that propagates within the optical medium includes phase fluctuations, spatial and temporal intensity fluctuations, and angle-of-arrival fluctuations. The observed phenomena include beam steering and focal image dancing, beam wander, focusing, defocusing, and speckle. The concern is that the received optical beam at times or entirely misses the small photodetector active area and the resulting communication link fading.

Fluctuations of laser intensity at the receiver are partially caused by atmospheric refractive index fluctuations (turbulence). The Hufnagel-Valley theorem models these fluctuations for different altitudes and estimated values might be characterized by the so-called refractive index structure constant C^2_n [80]. The phenomenon of aperture averaging reduces the overall deep fade probability. Both the amplitude and phase of a laser beam can be affected by atmospheric turbulence. For weak turbulence, C_n^2 at the ground level is on the order of 6×10^{-17} m$^{-2/3}$, C_n^2 for medium turbulence is on the order of 2×10^{-15} m$^{-2/3}$, and at strong turbulence $C_n^2 \approx 2 \times 10^{-13}$ m$^{-2/3}$ [81]. Lognormal statistics approximately characterizes weak scintillation while exponential statistics approximately characterizes strong turbulence conditions [82].

The Fried parameter, known as r_0, characterizes the atmospheric coherence length and is related to the C_n^2 constant with $\lambda^{6/5}$ proportionality to wavelength of the beam. That is, r_0 (roughly the diameter of the atmospheric blubs referred to earlier to as the atmospheric zones) increases as the beam's wavelength increases while the corresponding turbulence effect decrease. When the turbulence effects are most severe, value of r_0 may be as low as 1.5 cm while under weak turbulence conditions value of r_0 would be on the order of tens of cm (at wavelength of 500 nm).

A number of instruments are commercially available to quantitatively characterize the atmosphere including photometers, daytime optical and infrared all-sky cameras, nighttime differential image motion monitor (DIMM), and a multi-aperture scintillation sensor (MASS).

1.3.4 MITIGATION OF ATMOSPHERIC EFFECTS

Besides allocating adequate link margin to compensate for moderate fades, to typically avoid deep fades and to be able to couple the received optical beam at the ground station into a single-mode fiber (5–10 μm diameter core), much of the atmospheric turbulence effects need to be compensated effectively.

Some of the effective fade-mitigation techniques applied to the downlink beam include adaptive optics; telescope aperture averaging; automatic repeat request; delayed signal diversity, spatial transmitter diversity, wavelength diversity and use of long interleaver codes [83]. Techniques for uplink turbulence mitigation include multi-beam transmission and uplink adaptive optics which is currently in the early stages of investigation.

1.3.5 APERTURE AVERAGING EFFECTS

The aperture averaging effect becomes significant once the received aperture size is equal to or greater than the atmospheric coherence length, r_0 [84]. Starting with aperture diameters (D) greater than approximately 0.3 m, the aperture averaging effect can mitigate amplitude scintillation effects, but not frequency scintillation. When $D/r_0 > 1$, a single-spatial-mode receiver's performance in coherent detection schemes, or single-mode fiber coupling scheme for preamplified direct detection, can be severely degraded by the corrupted phase error in the wavefront. The Adaptive Optics (AO) technique can partially correct the beam's wavefront, compensating for residual phase errors that can introduce losses to the lasercom link budget. Kaufmann calculates the single-mode receiver fade statistics for imperfect phase correction and identifies conditions under which communications link losses are minimized [85].

1.3.6 Downlink Adaptive Optics

The effects of atmospheric turbulence need to be addressed anytime the receiver is located on the ground. The AO system enables a reduction of the angular area of sky encompassed by the communications detector. This effect improves the free-space communications link performance by directing a larger portion of the received power onto the detector (Figure 1.10) [86]. Also, diminishing the amount of sky background light being seen by the detector, resulting in improved SNR. The extent of improvement with AO compensation depends on the modulation format, gains >6 dB has been experimentally validated under laboratory-emulated atmospheric conditions involving turbulence and background light [87, 88]. The fade statistics of the turbulent atmospheric channel have also been analyzed with and without AO correction [89]. For large diameter telescope (with diameter larger than the minimum atmospheric r_0 value), the adaptive optics hardware will be often essential to couple the received signal into a single-mode fiber. Chapter 9 of this book provides a comprehensive discussion of the requirements and implementation strategies for the ground transceiver.

An AO system is equipped with a wave-front sensor (WFS) and a pixelated deformable mirror (DM). WFS sensor analyzes a portion of the received signal for deviations from a perfect wavefront. The error signals are then sent to the DM to dynamically correct the wave-front of the arriving beams [90]. Prior to incorporating an AO system with the receiving telescope, only a minute fraction (only one mode)

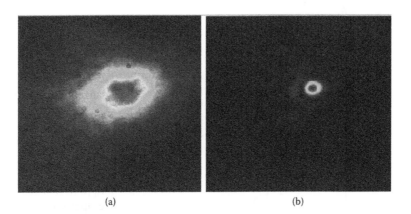

(a)   (b)

FIGURE 1.10 Focal spot sizes before (a) and with (b) AO compensation in the presence of turbulence. For (a) BER = 0.2 and spot size 1.3 mm and (b) BER = 0.8×10^{-4} and spot size of 0.25 mm. The saturated r_0 was approximately 9 mm for a 7-cm beam in the laboratory testbed (corresponding to r_0 = 13 cm at 1064 nm on the sky for a 1-m-diameter telescope). The above images are scaled by the automatic gain control of the camera and are not scaled relative to each other.

of the received may be coupled into a single-mode fiber. Effective coupling of the received signal into a single-mode fiber allows us to take advantage of a large number of components that have been well developed for the fiberoptics industry, such as the high-rate coherent fiberoptic transceivers equipped with modems at hundreds of Gbps [91].

1.3.7 ENCODING, DECODING, AND PROTOCOLS

Mitigation of transient outages, caused by scintillation fades and transient obscurant through the implementation of protocols and error control methods remains the focus of research [92]. The aim here is to ensure a certain quality of service compatible with the required data rate. The adapted protocol is a function of acceptable data rates, architecture, and acceptable minimum data rates. Depending on the architecture in a networked environment, some of the common protocol trades include a fixed path or dynamic rerouting, FEC or retransmission, delay and retransmit or delay and discard.

Atmospheric fades are caused by scintillation and generally have a burst nature. Major burst errors can last from a few milli-seconds to tens of milli-seconds long. The long outage period implies that FEC (and in general coding) should spread the data over a period much longer than the expected fade period. Long interleaver codes spread the data stream burst errors by reformatting the bits to be transmitted, followed by reestablishing the original pattern upon data reception. For this technique to work, the interleaver depth has to be much greater than the longest anticipated duration fades. This may result in long delays (as much as 200 ms) and the requirement for extensive memory use onboard the spacecraft [93, 94]. Figure 1.11 shows

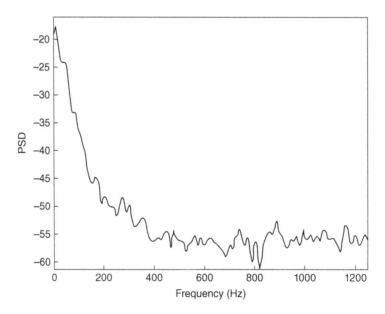

FIGURE 1.11   PSD frequency for the case of strong turbulence.

the power spectral density (PSD) of the fading time series for the case of strong turbulence as function of frequency. The coherence time of the fading process is inversely proportional to the bandwidth of the PSD.

Explicitly designed data relay circuits can group the data into frames, maintain synchronous connections and provide for Automatic Retransmission reQuest (ARQ) of lost or erred frames. In one rendition of several available schemes to identify frames for the ARQ process, the network processor configures the data into appropriately sized frames with headers and frame sequences. The frames are forwarded to an adaptor for formatting the data and adding error check bits. Frame acquisition and synchronization is followed by transmission to the encoder and hand-off to the laser source. At the receiving station, an interface adapter and the network processor check for errors and frames out of sequence. For each received frame, the appropriate ARQ control frame is generated and forwarded to the transmitter for notifying the other transceiver.

1.3.8 MULTI-BEAM UPLINK

Spatial transmitter diversity or multi (4–16) beam propagation of mutually incoherent beams can help mitigate amplitude scintillation only. This phenomenon relies on sufficient separation of two or more parallel beam transmitters that results in signal transmission through statistically independent air masses [95, 96]. Wavelength diversity relies on laser beams of unequal wavelengths experiencing different disturbances as they pass through air masses with varying indices of refraction [97].

In the multi-beaming scheme, a number of mutually incoherent and independent laser beams carrying the same data stream is transmitted from the ground station to the aerial or space platforms. This technique is shown to effectively reduce the probability of uplink fades [98]. The transmit beams are wavelength separated and distance separated by greater than the value of r_0 for the ground station's location to prevent interference. At up to four beams ($N = 4$), signal variance decreases approximately as $1/N$, for $N = 5$ to 16, signal variance is reduced by $1/N^{0.5}$. Above 16 beams, further benefits of multi-beaming are diminished. Even with the benefit of multi-beaming, the optical channel remains susceptible to occasional fades lasting up to a few tens of ms. A normalized received signal variance as recorded by the detector on the spacecraft and for a number of different uplink beams is presented in Figure 1.12.

1.3.9 UPLINK ADAPTIVE OPTICS

Scintillation effects on the uplink beam are much more prominent than on the downlink beam. However, the technique of uplink AO is still in the early stages and not nearly as well developed as the downlink AO. Different implementation approaches, such as a laser guide star, would be required to effectively accomplish uplink AO. Also, owing to the point-ahead effect, the channel is not reciprocal. Experiments are now being conducted to test out the effectiveness of the uplink AO system [99, 100].

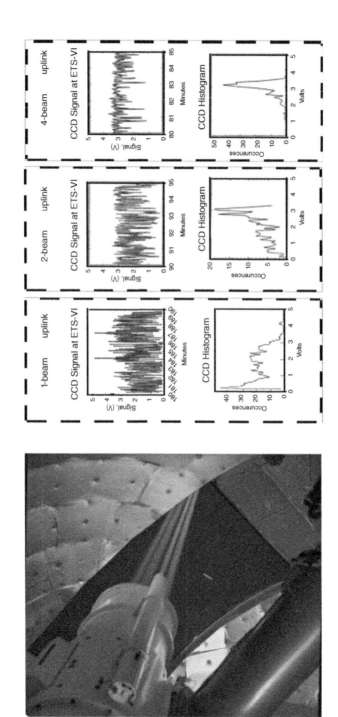

FIGURE 1.12 Multi-beam laser uplink mitigating the effects of atmospheric scintillation on beam wander. Up to four beam uplink form a 60-cm-diameter telescope at the Table Mountain Observatory in California (the GOLD demonstration by JPL and NICT). Beams that were incoherent relative to each other were propagated through independent atmospheric coherent cells. Each laser beam is delayed relative to the other by greater than laser's coherent length.

1.4 LASERCOM APPLICATIONS

Free space laser communications technology has made a number of unique applications. A few examples are outlined below.

1.4.1 POINT-TO-MULTIPOINT LINKS

Considering that the primary advantage of lasercom stems from its narrow beam, point-to-multipoint optical communications using a highly divergent transmit laser beam would be highly inefficient. An array of point-to-point links where each beam is pointed to different directions, could partially satisfy point-to-multipoint communications requirements. Agile beam steering using mechanical or non-mechanical beam scanner, optical phased arrays and wavelength-division multiple-access technologies could be used in this implementation. Optical communications technology may prove itself applicable to networking circumstances where RF frequency reuse might not be possible and suffer from limited channel capacity. A comprehensive description of multiple access optical links and optical networking is provided by Chan [101, 102].

Multiplatform high-performance satellite communications, multiplatform multistatic sensing, interoperable satellite communications, extendable data-volume satellite communications, shared spaceborne processing and reconstitution of severed terrestrial networks are among the unique networking applications of laser communications.

1.4.2 NETWORKS AND MULTIPLE ACCESS LINKS

Since RF technology is limited in data-rate growth potential beyond a few Gbps per beam, lasercom technology can become exceedingly useful, particularly in a network environment where frequency reuse may not be possible [103, 104].

Use of transceivers designed for point-to-point communications to simultaneously link with a number of other transceivers can easily result in unacceptably high mass, power consumption, size, and cost for a given station. Multiaccess links relying on advanced beam multiplication systems, such as wavelength-division multiple access, and agile beam steerers such as optical phased arrays or micro-electromechanical systems, can solve this problem when these technologies have advanced to high enough technology readiness levels [105]. Many point-to-multipoint designs replicate a standard fiber-optic distribution network. Chapter 11 of this book provides a broad overview of the laser-communications networking architecture and scenarios.

1.4.3 IMPROVED NAVIGATION AND RANGING

Navigation and precision ranging data are additionally provided by the spacecraft's telecommunications subsystem. Higher-quality navigation and precision ranging to the satellite is possible with lasercom relative to their RF frequencies owing to the short wavelength of the laser beams [106]. Time-of-flight measurements and inference from coding frame boundaries can make sub-cm precision ranging to the satellite possible. Doppler information is gleaned from the recovered PPM slot clock or by the symbol rate of other modulation techniques. Analysis and simulations indicate that, depending on how the PPM slot clock is phase locked, Doppler and range data 2 to 10 times better than Ka-band can be obtained using PPM and direct detection.

1.4.4 MULTIFUNCTIONAL TRANSCEIVERS

A lasercom instrument comprises telescope(s), lasers, imager(s), detectors, and beam directors. Passive and active electro-optic systems, e.g., imagers, spectrometers, laser ranging, and laser remote sensing, typically include two or more of the same subsystems. For small areal or satellite platforms where size, mass, and power consumption are key design drivers, this feature provides a unique opportunity for the telecommunications transceiver to be combined with other electro-optic instruments. Field-of-view requirements for each instrument or the type and quality of the laser used may result in compromises within the instrument design. A proof-of-concept multifunctional instrument combining the functions of lasercom and visual imager within the same instrument has been developed [107].

1.4.5 RETRO-MODULATOR LINKS

Modulating retro-reflectors (MRR) might be viewed as an active barcode (vs. a typical fixed, one-time write, barcode). An MRR incorporates a passive retro-reflector (corner cube) with an active element as one of its surfaces that acts like a very fast shutter. For free-space links, the data to be transmitted is modulated onto the remotely located interrogating laser beam via the shutter. The small cross section of the retroreflector, bandwidth of the shutter, and the $1/R^4$ effect limit the communications bandwidth to tens of Mbps and the link distance to a few tens of kilometers. However, due to achievable ±50° corner-cube field of view, the transceiver beam pointing, a major challenge for conventional lasercom systems, is highly simplified and constrained only to coarse pointing of the interrogator beam. Also, low mass, compact size, and power consumption of the retro-modulator make certain niche applications feasible [108, 109]. Low weight, low power consumption, and highly compact size are among the attractive aspects of retro-modulators. Accomplishments include data rates on the order of tens of Mbps and link distance up to a few tens of kilometers [110].

1.5 TECHNOLOGY VALIDATIONS

Past technology validations in space emphasized successful experimental demonstration without much regard for system mass and power consumption. Moreover, some of the earlier optical communication demonstrator designs were constrained by the status of subsystem technology. To become practical for operational use, mass, power, size, and cost must be reduced significantly and simultaneously. Recent technological advancements including efficient high-power laser transmitters (e.g., fiber amplifiers), efficient modulation and coding, and high-sensitivity photodetectors (e.g., photon counters), should all greatly help in relaxing flight system parameters that burden the airborne or spaceborne host platforms.

1.5.1 RECENT VALIDATIONS

Since the early days of laser development, both terrestrial and space lasercom systems have been under development in the United States, Europe and Japan. Only in the past decade were major lasercom systems from Earth-orbiting spacecraft demonstrated successfully.

1.5.1.1 Lasercom Demonstrations from LEO, GEO, and the Moon

1.5.1.1.1 LEO-LEO and LEO-GEO

TerraSAR-X, developed by the TESAT Spacecom for LEO and GEO orbit satellites is capable of bidirectional multi-Gbps links and nearly 10 years of life in space. Applications include LEO-LEO, LEO-GEO, and both LEO and GEO through-atmosphere links with ground and with HAPs [111, 112]. Figure 1.13 pictorially illustrates the TESAT lasercom system used for both satellite crosslinks and links between satellite and Earth.

The European Data Relay System (EDRS) series of satellites make use of these lasercom payloads for relaying data from the Earth observation satellite sensor instruments in LEO to the GEO orbit. Data gathered at LEO is relayed to GEO satellite for downlink to Earth via Ka-band [35]. This established the first commercial and operational use of lasercom from space by the customers of the satellite, including the European Space Agency.

The geostationary EDRS system provides data relay services to Sentinel-1 and -2. Each Sentinel will communicate with an EDRS satellite via a laser communications

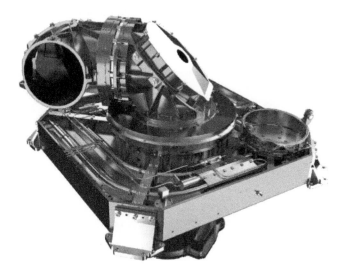

FIGURE 1.13 TESAT laser communication flight transceiver. (Courtesy of TESAT Spacecom.)

link, which can relay the large volume of data gathered by the satellite. The EDRS satellite will relay the data to the ground via a Ka-band link. With EDRS, Sentinel instrument data are directly down-linked via data relay to processing and archiving centers, while routine spacecraft telemetry continues to be received at ESA's Kiruna station, via X-band radio.

1.5.1.1.2 Lunar Link

The 2013 Lunar Laser Communications Demonstration (LLCD) mission, a NASA-funded project at the MIT Lincoln Laboratory demonstrated lasercom downlink to the Earth at 622 Mbps and uplink to the Moon at the rate of 20 Mbps. The flight transceiver had a mass of approximately 31 kg and consumed 90 W of DC input power [113].

The LLCD demonstrated fast acquisition followed by precise closed-loop tracking of the uplink and downlink beams. The pulse-position-modulation (PPM) scheme, developed earlier for free-space lasercom links with deep space, was applied to enhance the link margin. This modulation approach necessities higher peak-to-average power from the laser than the conventional modulation schemes [114]. To improve signal detectivity, the downlink detection scheme applied superconducting single-photon-sensitive detectors. These detectors are particularly suitable to the photon-starved regime of the planetary links. Centimeter-class ranging to the satellite was simultaneously conducted with the communications links. The ground stations were located in New Mexico (developed by the Lincoln Laboratory), the Southern California (by JPL), and at the Canary Islands (by ESA).

Figure 1.14 schematically illustrates LLCD's flight and ground transceivers.

This highly successful demonstration led to the Laser-Communications-Relay-Demonstration (LCRD) for bi-directional links with a GEO satellite. Slated for launch in 2019, the LCRD mission intends to conduct GEO-and-Earth at 2 Mbps up to 1.244 Gb/s via a differential DPSK, burst-mode modulated beam at 1550 nm.

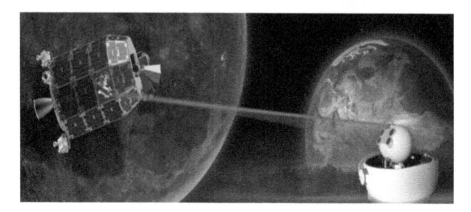

FIGURE 1.14 The LLCD pictorial of 622 Mbps data transmission from the Moon to Earth via laser communications.

1.6   FUTURE DIRECTIONS, DEMONSTRATIONS, AND OPERATIONAL SYSTEMS

As the demand for data capacity increases exponentially, the conventional RF communication technology may extend its limits of data delivery and the uniqueness of lasercom for delivering Tbps level data rate, commensurate with the ever-increasing fiberoptics communications will emerge. Further improvements are anticipated via both component developments and more simplified architectures which are often made possible owing to technology advancements in associated areas.

1.6.1   INTER-SATELLITES LINKS

At least seven constellations of LEO satellites with as many as thousands of spacecrafts networked together through inter-satellite crosslinks (ISLs) and or through ground stations are under development at the time of this writing [115]. As the channel is vacuum, a number of LEO constellation ventures and optical ISL manufacturers have baselined lasercom inter-satellite links at multi-Gbps links over ranges of up to 6000km [116]. At four optical transceivers per satellite for in-plane and out-of-plane communications between satellites, a number of challenges arise. These include cost of flight-qualified transceivers with 7 to 10-year lifetime, and volume manufacturing where a few thousand units must be developed in a short time.

1.6.2   COMPONENT TECHNOLOGIES

Component technologies used in lasercom systems are continuously evolving and a number of developments, such as photonics integrated circuits (PICs) [117]. PICs, improved non-mechanical beam steerers, and wafer thin (flat) metalenses are among the technologies that can substantially reduce cost, size, mass, and power (CSWaP) of future lasercom systems [118, 119]. Moreover, simplified lasercom architectures and continued leveraging of fiber optics and components developed by the fiberoptics industry can help to further minimize the CSWaP of upcoming lasercom systems [120].

The coherent transceivers developed by the fiberoptic industry for communications at rates approaching 1 Tbps in an ever-decreasing size, mass, power, and cost remains an extremely powerful tool for the free-space optical communications technology [121]. Direct-detection systems can leverage only the signal's intensity, i.e., only one degree-of-freedom of the optical channel. On the other hand, coherent detection leverages two quadratures in two polarizations, i.e., four degrees of freedom of the signal-mode-fiber channel. By mixing the faint incoming signal with a strong local oscillator, as much as 20-dB link margin may be gained via coherent detection.

In the next two decades, the lasercom technology is projected to transform future space communications system architectures by a staggering increase in capacity while maintaining an affordable mass, power, size, and cost. Standardization of telecom interoperability protocols is estimated to minimize the operational costs. Availability of one or more commercial ground station networks servicing future lasercom-equipped satellites is also anticipated to drive down the cost of operational systems. For near-Earth satellite communications, the laser communications technology is no longer an emerging technology, rather a technology that is ready for reliable deployment. Future technology development would serve to improve performance and increase affordability. As examples, significant technology improvements are foreseen, through the following developments on the spacecraft side:

Laser: Wall-plug efficiency comparable to RF SSPA and TWTAs
Amplifiers: Lower inherent noise and higher electrical efficiency
Optics: Lighter-weight, more compact, and higher throughput
Detectors: Higher detection efficiency, lower noise
Receivers: Larger effective aperture and improved throughput and
                    sensitivity
Spurious signal: Improved mitigation techniques
Reliability: Ruggedized components with greater lifetime

1.7   EVOLVING NON-OPTICAL TECHNOLOGIES

While licensed Ku and Ka-band communication bands (in the 10–30 GHz region) with high atmospheric link availability, but limited spectrum bandwidth availability, the achievable capacity may reach a plateau in the next decades. In such a case, higher frequencies, e.g., Q/V-band, E-band, G-band in the 200 GHz regime and sub-THz to THz frequencies will find specific and important use for a variety of applications. At these frequencies, the atmospheric absorption losses are higher than lower frequencies and spatial diversity will be required to improve overall link availability. However, the broader spectrum available at these frequencies can enable higher rate mm-wave communications links feasible. Concentrated technology development efforts would make them highly viable choices for specific applications and could become strong competitors for certain applications of the lasercom technology.

Examples of using higher frequencies for telecom is the recent demonstration by the Facebook Connectivity of a 40 Gbps bidirectional link at E-band frequency (71–76 and 81–86 GHz) over a terrestrial range of 13 km (limited by the test site), and in links with fast moving aircraft at 20 km range, and E-band links by Northrop/DARPA at 100 Gbps over tens of km [122, 123]. At these frequencies, depending on location link availability varies between ~40 and >90% per year (limited by heavy rain), competing well with terrestrial or through-cloud optical communications that is limited by atmospheric attenuation (cloud, fog, dust, rain, and snow) and atmospheric turbulence.

1.8   CONCLUSION

In summary, the lasercom technology is a powerful tool in the telecommunications toolbox but would not make the conventional RF/MW communications obsolete. This technology will enable certain links that cannot be accomplished otherwise, particularly as the links' data rate approached high Gbps to Tbps regimes. Despite a multitude of potential advantages, the perceived risks associated with lasercom have often been the primary impediment to its prevalent use in space.

Free-space laser communications with aircrafts and spacecrafts has reached a level of maturity adequate for stand-alone operational or mission enhancing deployment without placing additional attitude control and stability requirements on the host platform. Atmospheric attenuation, communications within a few degrees of the sun angle, and lack of a large aperture telescope infrastructure for lasercom data reception remain engineering or programmatic challenges. We conclude that ground station diversity can adequately mitigate cloud cover. The advent of affordable, large diameter (1–2 m), non-diffraction-limited ground-based telescopes should make the ground infrastructure development realizable. Robotics integration and test of flight transceivers should make possible large quantity delivery for the large constellation of satellites.

REFERENCES

1. Cisco, "Cisco global cloud index: Forecast and Methodology, 2015-2020," White paper (2016).
2. K.-C. Chen, T. Zhang, R. D. Gitlin, and G. Fettweis, "Ultra-low latency mobile networking," IEEE Network, V. 33(2), pp. 181–187 (2019).
3. E. Lutz, "Towards the terabit/s satellite—Interference issues in the user link," Inter. J. of Satellite Comm. and Networking, June 17 (2015).
4. C. Miller, "How and why commercial high-capacity satellites offer superior performance and survivability in the future space threat continuum," 32nd Space Symposium, Technical Track, Colorado Springs, Colorado, United States of America. Presented on April 11–12 (2016).
5. A. Karasuwa, J. Eastman, and I. Otung, "Design considerations for high-throughput satellite communication system," Proc. 21st Ka and Broadband Satellite Conf. (2015).
6. J. Poliak, D. Giggenbach, F. Moll, F. Rein, C. Fuchs, and R. Mata Calvo, "Terabit-throughput GEO satellite optical feeder link testbed," 13th Inter. Conf. on Telecom. (ConTel), July 13–15 (2015).
7. H. Hemmati, "Lower frequency bands emerging as valid alternatives to free-space laser communications in terrestrial, aerial, and satellite links," Proc. SPIE, V. 10910 (2019).
8. C. K. Chong, D. A. Layman, W. L. McGeary, W. Menninger, M. Ramay, and X. Zhai, "Q/V-band high-power uplink helix TWT for future high-data-rate communications, IEEE Trans. on Electron Devices, V. 65 (2018).
9. H. Hemmati, *Near-Earth Laser Communications*, CRC Press, Rochester NY (2009).
10. C. Fuchs, S. Poulenard, N. Perlot, J. Riedi, and J. Perdigues, "Optimization and throughput estimation of optical ground networks for LEO-downlinks, GEO feeder links and GEO-relays," Proc. SPIE, V. 10096, (2017).
11. R. M. Gagliardi and S. Karp, *Optical Communications*, 2nd ed., Wiley, New York (1995).
12. H. Hemmati and D. Caplan, "Optical satellite communications," in *Optical Fiber Telecommunications Volume VIB*, 6th ed. A. E. Wilner, T. Li, and I. Kaminow, eds. Academic Press, Waltham, MA, Chapter 4 (2013).
13. A. Rietdorf, C. Daub, and P. Loef, "Precise positioning in real-time using navigation satellite and telecommunication," Proc. of the 3rd Workshop on Positioning, Nav. and Comm. W C' 6 .173/948 (2006).
14. H. Hemmati, *Deep Space Optical Communications*, Wiley, Hoboken, NJ (2006).

15. B. A. Bash, S. Guha, D. Goeckle, and D. Towsley, "Quantum noise limited optical communication with low probability of detection," IEEE Inter. Symp. on Information Theory, July (2013).

16. S. B. Alexander, *Optical Communication Receiver Design*, SPIE Optical Engineering Press (1997).

17. M. Gregory, F. Heine, H. Kämpfner, R. Meyer, R. Fields, and C. Lunde, "TESAT laser communication terminal performance results on 5.6 Gbit coherent inter satellite and satellite to ground links," Inter. Conf. on Space Optics (ICSO) (2010).

18. F. Heine, D. Troendle, C. Rochow, K. Saucke, M. Motzigemba, R. Meyer, M. Lutzer, E. Benzi, and H. Hauschildt, "Progressing towards an operational optical data relay service," Proc. SPIE, V. 10096 (2017).

19. J. Maynard and M Ross, "Airborne Flight Test System (AFTS) Final Report," USAF Contract #F33615-76-C-1002, 26 October (1981).

20. L. B. Stotts, "Strategic laser communications program," Proc. of the NAVY/DARPA Fourth Technical Interchange Meeting, NOSC Technical Document, TD 352, V. I (1979).

21. K. E. Wilson, J. R. Lesh, and T.-Y. Yan, "GOPEX: A laser uplink to the Galileo space-craft on its way to Jupiter," Proc. SPIE, V. 1866 (1993).

22. D. L. Bagley, "Relay mirror experiment," Proc. SPIE, V. 2699 (1996).

23. Y. Arimoto, M. Toyoshima, M. Toyoda, T. Takahashi, M. Shikatani, and K. Araki, "Preliminary result on laser communication experiment using ETS-VI," Proc. SPIE, V. 2381 (1995).

24. T. T. Nielsen and G. Oppenhaeuser, "In-orbit test result of an operational optical inter-satellite link between ARTEMIS and SPOT4, SILEX," Proc. SPIE, V. 4635 (2002).

25. K. Nakagawa and A. Yamamoto, "Preliminary design of laser utilizing communica-tions equipment (LUCE) installed on optical inter-orbit communications engineering test satellite (OICETS)," Proc. SPIE, V. 2381 (1995).

26. G. Planche, V. Chorvilli, and L. Le Hors, "Optical communications between an aircraft and a GEO relay satellite: Design & flight results of the LOLA demonstrator," Proc. of the 7th Inter. Conf. on Space Optics (ICSO), ESTEC (2008).

27. R. Lange, B. Smutny, B. Wandermoth, R. Czichy, and D. Giggenbach, "142 km, 5.625 Gbps free-space optical link on homodyne BPSK modulation," Proc. SPIE, V. 6105 (2006).

28. G. Muehlnikel, H. Kämpfner, F. Heine, H. Zech, and D. Troendle, "The Alphasat GEO laser communication terminal flight acceptance tests," Proc. ICSOS (2012).

29. D. M. Boroson, J. J. Scozzafava, D. V. Murphy, and B. S. Robinson, "The lunar laser communications demonstration (LLCD)," Third IEEE Inter. Conf. on Space Mission Challenges for Information Tech. (2009).

30. H. Zech, F. Heine, D. Troendle, P. M. Pimentel, K. Panzlaff, M. Motzigemba, R. Meyer, S. Phillip-May, R. Fields, and C. Lunde, "LCTS on ALPHASAT and Sentinel 1a: in orbit status of the LEO TO GEO data relay system," Inter. Conf. on Space Optics (ICSO), Tenerife, Canary Islands (2014).

31. M. W. Wright, J. Kovalik, J. Morris, M. Abrahamson, and A. Biswas, "LEO-to-ground optical communications link using adaptive optics correction on the OPALS down-link," Proc. SPIE, V. 9739, (2016).

32. A. Carrasco-Casado, H. Kunimori, H. Takenaka, T. Kubo-Oka, M. Akioka, T. Fuse, Y. Koyama, D. Kolev, Y. Munemasa, and M. Toyoshima, "LEO-to-ground polariza-tion measurements aiming for space QKD using Small Optical TrAnsponder (SOTA)," Optics Express, V. 24, pp. 12254–12266 (2016).

33. K. Bohmer, M. Gregory, F. Heine, H. Kämpfner, R. Lange, M. Lutzer, and R. Meyer, "Laser communication terminals for the European data relay system," Proc. SPIE, V. 8246 (2012).

34. C. Schmidt and C. Fuchs, "The OSIRIS Program—First results and outlook," Proc. IEEE Inter. Conf. on Space Optical Systems and Applications, ICSOS (2017).

35. M. Gregory, F. F. Heine, H. Kämpfner, R. Lange, M. Lutzer, and R. Meyer, "Commercial optical inter-satellite communication at high data rates," Optical Engineering, V. 51 (2012).

36. D. J. Israel, B. Edwards, and J. Staren, "Laser communications relay demonstration (LCRD) update and the path towards optical relay operations," IEEE Aerospace Conf. (2017).

37. E. Herz, C. Dahn, E. Carney, and J. Campagna, "BridgeSat laser communication scheduling: A case study," 2018 SpaceOps Conf. (2018).

38. S. Kaur, "Analysis of inter-satellite free-space optical link performance considering different system parameters," Opto-Electronics Review, V. 27(1), pp. 10–13 (2019).

39. S. Lee, J. W. Alexander, and M. Jeganathan, "Pointing and tracking subsystem design for optical communication link between the International Space Station and ground," Proc. SPIE, V. 3932 (2000).

40. B. Mathason, M. Albert, D. Engin, H. Cao, K. G. Petrillo, J. Hwang, K. N. Le, et al., "CubeSat lasercom optical terminals for near-earth to deep space communications," Proc. SPIE, V. 10910 (2019).

41. G. Ortiz, A. Portillo, S. Lee, and J. Ceniceros, "Functional demonstration of accelerometer-assisted beacon tracking," Proc. SPIE, V. 4772, pp. 112–117 (2001).

42. C. C. Chen, H. Hemmati, A. Biswas, G. Ortiz, W. Farr, and N. Pedrerio, "Simplified lasercom system architecture using a disturbance-free platform," Proc. SPIE, V. 6105, 610505 (2006).

43. W. Liu, Y. Gao, W. Dong, and Z. Li, "Flight test results of the microgravity action vibration isolation system in China's Tianzhou mission," Microgravity Science and Tech., V. 30(6), pp. 995–1009 (2018).

44. M. B. McMickell, T. Kreider, E. Hansen, T. Davis, and M. Gonzalez, "Optical payload isolation using the miniature vibration isolation system," Proc. SPIE, V. 6527 (2007).

45. G. Zhao, W. Deng, H. Xia, and M. Zhang, "Research on common aperture device for high-energy laser system," Proc. SPIE, V. 11170 (2019).

46. E. D. Miller, "An inverse-kinematic approach to dual-stage servo control for an optical pointing system," Proc. SPIE, V. 10524 (2018).

47. P. McManamon, "An overview of optical phased array technology and status," Proc. SPIE, V. 5947 (2005).

48. X. Wang, L. Wu, X. He, X. Huang, and Q. Tan, "Theoretical analysis on power stability and switch time of the non-mechanical beam steering using liquid crystal optical phased array," J. Liquid Crystals, V. 45 (2018).

49. M. Ziemkiewicz, S. Davis, S. D. Rommel, D. Gann, B. Luey, J. D. gamble, and M. Anderson, "Laser-based satellite communication system by non-mechanical electro-optic scanners," Proc. SPIE, V. 828 (2016).

50. A. C. Foster, M. Kossey, N. Macfarlane, C. G. Rozk, T. Lott, R. Osiander, and N. Mosavi, "Chip-scale optical phased arrays for inter-spacecraft communications," Defense and commercial sensing, Computer Science Engineering (2019).

51. CIEEE J. of Selected Topics in Quantum Electronics, V.25 (2019). V. Poulton, M. J. Byrd, P. Russo, E. Timurdogan, M. Khandaker, D. Vermeulen, and M. R. Watts, "Long-range LIDAR and free-space data communication with high-performance optical phased arrays," V.25 (2019).

52. Z. Dong, A. Jiang, Y. Daim, and J. Xue, "Space-qualified fast steering mirror for image stabilization system for space astronomical telescope," Applied Optics, V. 57, pp. 9307–9315 (2018).

53. C. Fuchs, H. Henniger, and D. Giggenbach, "Fiberbundle receiver: A new concept for high-speed and high-sensitivity tracking in Optical Transceivers," Proc. SPIE, V. 6709 (2007).

54. K. Takahashi and Y. Arimoto, "Compact optical antennas using free-form surface optics for ultrahigh-speed laser communication systems," Optical Eng., V. 47 (2008).

55. E. A. Swanson and R. S. Bondurant, "Using fiber optics to simplify free-space lasercom systems," Proc. SPIE, V. 1218 (1990).

56. K. Habib and E. Lewis (eds.), "Frontier research and innovation in optoelectronic technology and industry," Proc. the 11th Inter. Symposium on Photonics and Optoelectronics (SOPO 2018), Kunming, China (2018).

57. M. R. García-Talavera, J. Sánchez-Capuchino, F. Tenegi, A. Alonso, N. Vego, Y. Martín, C. Rivera, and M. Stumpf, "Design of a ground terminal for deep-space optical communications," Proc. IEEE Intern. Conf. on Space Optical Systems and Applications (2017).

58. K. E. Wilson, M. Britcliffe, and N. Golshan, "Progress in design and construction of the Optical Communications Telescope Laboratory (OCTL)," Proc. SPIE, V. 3932, pp. 112–116 (2000).

59. C. C. Chen and J. R. Lesh, "Overview of the optical communications demonstrator," Proc. SPIE, V. 2123 (1994).

60. A. Biswas, F. Khatri, and D. Boroson, "Near-sun free-space optical communications from space," IEEE Aerospace Conf. (2006).

61. H. Hemmati and Y. Chen "Active optical compensation of low-quality optical system aberrations," Optics Lett., V. 31, p. 1630 (2006).

62. R. C. Romero, A. B. Meinel, M. P. Meinel, and P. C. Chen, "Ultralightweight and hyperthin rollable primary mirror for space telescopes," Proc. SPIE, V. 4013 (2000).

63. W. T. Roberts, "Optical membrane technology for deep space optical communications filters," IEEE Aerospace Conf. (2005).

64. K. Wilson, W. T. Roberts, V. Garkanian, F. Battle, R. Lablanc, H. Hemmati, and P. Robles, "Plan for safe laser beam propagation from the optical communications telescope laboratory," JPL IPN Progress Report, pp. 42–152 (2003).

65. K. E. Wilson, N. Page, J. Wu, and M. Srinivasan, "The JPL optical communications telescope laboratory (OCTL) test bed for the future optical Deep Space Network," Inter. Space Conf. of Pacific-basin Societies, Nagoya Japan (2003).

66. B. S. Robinson, D. O. Caplan, M. L. Stevens, R. J. Barron, E. A. Dauler, and S. A. Hamilton, "1.5-photons/bit photon-counting optical communications using Geiger-mode avalanche photodiodes," IEEE LEOS Summer Topical Meetings (2005).

67. A. O. Aladeloba, A. J. Phillips, and M. S. Woolfson, "Improved bit error rate evaluation for optically pre-amplified free-space optical communication systems in turbulent atmosphere," IET Optoelectronics, V. 6, pp. 26–33 (2012).

68. M. S. Erkilinc, D. Lavery, K. Shi, B. Thomsen, R. Killey, S. Savory, and P. Bayvel, "Comparison of low-complexity coherent receivers for UDWDM-PONs (l-to-the-user)," J. of Lightwave Tech., V. 36 (2018).

69. J. A. Mendenhall, P. I. Hopman, D. M. Boroson, C. J. Digenis, R. C. Shoup, L. M. Candell, G. Zogbi, D. O. Caplan, and D. R. Hearn, "Design of an optical photon counting array receiver system for deep space communications," Proc. IEEE, V. 95(10), pp. 2059–2069 (2007).

70. J. R. Pierce, "Optical channels: Practical limits with photon counting," IEEE Trans. Comm., COM-26, pp. 1819–1821 (1978).

71. S. Poulenard, B.Gadat, J. F. Chouteau, T. anfray, C. Poulliat, C. Jego, O. Hartmann, G. Artaud, and H. Meric"Forward error corrfecting code for high data rate LEO satellite optical downlinks," ICSO 2018, Chania, Greece (2018).

72. J. Zhang, Z. Yu, J. Zhang, B. Bai, C. Chen, and F. Huang "Speeding up LDPC decoder for inter-frame pipeline for wireless laser communications,"2019 IEEE/CIC International Conference on Communications in China (ICCC).

73. B. Moision and J. Hamkins, "Coding and modulation for free-space optical communications," in *Near-Earth Laser Communications (this book)* (1st and 2nd eds.), CRC Press, Boco Raton, FL, Chapter 6 (2009).

74. S. K. Chung, G. D. Fomey, T. J. Richardson, and R. Urbanke "On the design of low-density parity-check codes with 0.0045 dB of Shannon limit," IEEE Comm. Lett., V. 5, pp. 58–60 (2001).

75. B. Erkmen, B. Moision, and S. Dolinar, "On approaching the ultimate limits to resource efficiency in photonics communication," Proc. SPIE, V. 8246 (2012).

76. N. W. Spellmeyer, J. C. Gottschalk, D. O. Caplan, and M. L. Stevens, "High-sensitivity 40 Gb/s RZ-DPSK with forward error correction," IEEE Photon. Tech. Lett., 16(6), pp. 1579–1581 (2004).

77. C. Fuchs and D. Moll, "Ground station network optimization for space-to-ground optical communication link," IEEE/OSA J. of Optical Comm. and Networking, V. 7 (2015).

78. V. Kornilov, A. Tokovinin, N. Shatsky, O. Voziakoval, S. Potanin, and B. Safonov," Combined MASS-DIMM instrument for atmospheric turbulence studies," Mon. Not. R. Astron. Soc., V. 382(3), pp. 1268–1278 (2007).

79. S. Trisno, I. I. Smolyaninov, S. D. Milner, C. C. Davis, and K. W. Billman, "Delayed diversity for fade resistance in optical wireless communications through turbulent media," Proc. SPIE, V. 5596, 385–394 (2004).

80. D. Gulich, G. Funes, D. Perez, and L. Zunino, "Estimation of Cn2 based on scintillation of fixed targets imaged through atmospheric turbulence," Optics lett., V. 40, pp. 5642–5645 (2015).

81. J. L. Bufton, "Comparison of vertical profile turbulence structure with stellar observations," Applied Optics, V. 12, pp. 1785–2793 (1973).

82. L. C. Andrews and R. L. Phillips, *Laser Beam Propagation through Random Media*, 2nd ed. SPIE Press (2005).

83. R. Pernice, A. Ando, A. Parisi, A. C. Cino, and A. C. Busacca, "Moderate-to-strong turbulence generation in a laboratory indoor free space optics link and error mitigation via RaptorQ codes," 18th IEEE Inter. Conf. on Transparent Optical Networks (ICTON), pp. 10–14, July (2016).

84. D. L. Fried, "Statistics of a geometric representation of wavefront distortion," J. Optical Soc. America, V. 55, pp. 1427–1435 (1965).

85. J. Kauffmann, "Performance limits of high-rate space-to-ground optical communications through the turbulent atmospheric channel," Proc. SPIE, V. 2381, pp. 171–182 (1995).

86. R. Tyson, "Bit-error rate for free-space adaptive optics laser communications," J. Optical Soc. America, V. 19, pp. 753–758 (2002).

87. N. Martinez, L. F. Rodriguez-Ramos, and Z. Sodnik, "Toward the uplink correction: Application of adaptive optics techniques on free-space optical communications through the atmosphere," Optical Engineering, V. 57 (2018).

88. J. C. Juarez, A. J. Goers, J. E. Maloicki, R. J. Dimeo, and V. Bedi, "Evaluation of curvature adaptive optics for airborne laser communication systems," Proc. SPIE, V. 10770 (2018).

89. B. M. Levine, E. A. Martinsen, A. Wirth, A. Jankevics, M. Toledo-Quinones, F. Landers, and T. L. Bruno, "Horizontal line-of-sight turbulence over near-ground paths and implications for adaptive optics corrections in laser communications," Applied Optics, V. 37, pp. 4553–4560 (1988).

90. M. W. Wright, J. Roberts, W. Farr, and K. Wilson, "Improved Optical Communications Performance using adaptive optics and pulse position modulation," IPN Prog. Report 42–161 (2005).

91. T. Weyrauch, M. A. Vorontsov, J. Gownes, and T. Bifano, "Fiber coupling with adaptive optics for free-space optical communications," Proc. SPIR, V. 4498 (2002).

92. S. Yamamoto, H. Takahira, and M. Tanaka "5 Gbit/s optical transmission terminal equipment using forward error correcting code and optical amplifier," Electronics lett., V. 30, pp. 254–255 (1994).

93. M. Yu, J. Li, and J. Ricklin, "Efficient forward error correction coding for free-space optical communications," Proc. SPIE, V. 5550 (2004).

94. D. Zhang, S. Hao, Q. Zhao, L. Wang, Q. Zhao, and X. Wan, "High-throughput inter-leaving scheme in free space optical communication system," 17th Inter. Conf. on Comm. Tech. (ICCT), October (2017).

95. I. I. Kim, H. Hakakha, P. Adhikari, E. Korevar, and A. Majumdar, "Scintillation reduction using multiple transmitters," Proc. SPIE, V. 2990, pp. 102–112 (1997).

96. A. Biswas and S. Piazzolla, "Multi-beam laser beacon propagation over lunar distance: Comparison of predictions and measurements," SPIR Proc. V. 10096 (2017).

97. D. Giggenbach, B. Wilkerson, L. Brandon, H. Henninger, and N. Perlot, "Wavelength-diversity transmission for fading mitigation in the atmospheric optical communication channel," Proc. SPIE, V. 6304 (2006).

98. L. Zhou, Y. Tian, R. Wang, T. Wang, T. Sun, C. Wang, and X. Yang, "Mitigating effect on turbulent scintillation using non-coherent multi-beam overlapped illumination," Optics and Laser Tech., V. 91, pp. 97–105 (2017).

99. C. J. Pugh, J.-F. Lavigne, J.-P. Bourgoin, B. L. Higgins, and T. Jennewein, "Adaptive optics benefits for quantum key distribution from ground to satellite," https://arxiv.org/abs/1906.04193v1 (2019).

100. N. Martinez, L. F. Rodriguez-Ramos, and Z. Sodnik, "Toward the uplink correction: Application of adaptive optics techniques on free-space optical communications through the atmosphere," Optical Engineering, V. 57 (2018).

101. V. W. S. Chan, "Free space optical communications," J. Lightwave Tech., V. 24, p. 4762 (2006).

102. V. W. S. Chan, "Optical satellite networks," J. Lightwave Tech., V. 21, pp. 2811–2827 (2003).

103. H. Hauschildt, C. Elia, H. L. Moeller, and J. M. Perigues Armengol, "HydRON: High throughput optical network," Proc. SPIE, V. 10910 (2019).

104. C. Chen, A. Grier, M. Malfa, E. Booen, H. Harding, C. Xia, and M. Hunwardsen, et al., "Demonstration of bi-directional coherent air-to-ground optical link," Proc. SPIE, V. 10524 (2018).

105. B. W. Segura and W. Mathlouthi, "Multi-point free space optical system," US patent # 10,009,107 (2018).

106. Y. Chen, K. Birnbaum, and H. Hemmati, "Active laser ranging over planetary distances with millimeter accuracy," Appl. Phys. Lett., V. 102, p. 241107 (2013).

107. H. Hemmati, "Combined laser communications and laser ranging for Moon and Mars," Proc. SPIE, V. 7199 (2009).

108. W. S. Rabinovich, R. Mahon, P. Goetz., E. Waluschka, D. S. Katzer, S. Binari, and G. C. Gilbreath, "A cat's eye multiple quantum well modulating retro-reflector," IEEE Photonics Tech. Lett., V. 15, pp. 461–463 (2003).

109. H. Hemmati, C. Esproles, W. Farr, W. Liu, P. Estabrook, G. C. Gilbreath, and W. S. Rabinovich, "Retro-modulator links with a mini-rover," Proc. SPIE, V. 5338, pp. 50–55 (2004).

110. J. Ohgren, F. Kullander, L. Sjoqvist, Q. Wang, S. Junique, S. Almqvist, and B. Noharet, "A high-speed modulated retro-reflector communication link with a transmissive modulator in a cat's eye optics arrangement," Proc. SPIE, V. 6736 (2007).

111. H. Zech, F. Heine, D. trundle, S. Seel, M. Motzigemba, and S. Phillip-May, "LCT for EDRS: LEO to GEO optical communications at 1.8 Gbps between Alphasat and Sentinel 1a," Proc. SPIE, V. 9647 (2015).

112. F. Heine, H. Kämpfner, R. Czichy, R. Meyer, and M. Lutzer, "Optical inter-satellite communication operational," MILCOM Comm. Conf. (2017).

113. D. V. Murphy, J. E. Kansky, M. E. Grein, R. T. Schulein, M. M. Willis, and R. E. Lafon, "LLCD operations using the Lunar Lasercom Ground Terminal," Proc. SPIE, V. 8971 (2014).

114. D. O. Caplan, B. S. Robinson, R. J. Murphy, and M. Stevens, "Demonstration of 2.5-Gslot/s optically-preamplified M-PPM with 4 photons/bit receiver sensitivity," Optical Fiber Comm. Conf. (OFC), Anaheim (2005).

115. "Telesat LEO – What makes it work?," https://www.telesat.com/services/leo/what-makes-it-work

116. M. Knapek, A. Al-Mudhafar, S. Muncheberg, K. Shortt, and M. Soutullo, "Development of a laser communication terminal for large LEO constellations," IEEE Inter. Conf. on Space Optical Systems and Applications (ICSOS), November (2017).

117. J. Klamkin, H. Zhao, B. Song, Y. Liu, B. Isaac, S. Pinna, F. Sang, and L. Coldren, "Indium phosphide photonic integrated circuits; technology and applications," 2018 IEEE BiCMOS and Compound Semiconductor Integrated Circuits and Tech. Symp., October 15–17 (2018).

118. J. Yang, I. Ghimire, P. C. Wu, S. Gurung, C. Arndt, D. P. Tsai, H. Wai, and H. Lee, "Photonic crystal fiber metalens," Nanophotonics, V. 8, pp. 443–449 (2019).

119. J. Fridlander, V. Rosborough, F. Sang, S. Pinna, S. Estrella, L. Johansson, and J. Klamkin, "Photonic integrated transmitter for space optical communications," Proc. SOIR, V. 10910 (2019).

120. R. Saathof, S. Kuiper, W. Crowcombe, H. De Man, D. de Lange, N. van der Valk, L. Kramer, and E. Fritz, "Opto-mechatronics system development for future intersatellite laser communications," Proc. SPIE, V. 10910 (2019).

121. R. J. Aniceto, R. Milanowski, S. McClure, A. Aguilar, S. Moro, E. D. Miller, K. Cahoy, N. Nicholson, and D. Greene, "Heavy ion radiation assessment of 100G/200G commercial optical coherent DSP ASIC," Proc. SPIE, V. 10910 (2019).

122. Q. Tang, A. Tiwari, I. del Portillo, M. Reed, H. Zhou, D. Shmueli, G. Ristroph, S. Cashion, D. Zhang, J. Stewart, P. Bondalapati, Q. Qu, Y. Yan, B. Proctor, and H. Hemmati, "Demonstration of 40Gbps bi-directional air-to-ground millimeter wave communication link," IEEE International Microwave Symposium (IMS), June 2019

123. https://www.microwavejournal.com/articles/30932-ngc-darpa-set-new-standard-for-wireless-transmission-speed

2 Systems Engineering and Design Drivers

Morio Toyoshima

CONTENTS

2.1 INTRODUCTION

An end-to-end optical communication systems design is driven by the desired link characteristics and environmental/channel constraints placed on the flight and receive transceivers. There are several major design drivers, including tracking and pointing link margin allocation, communication link margin, data rate and bit error rate (BER) over a given communication range, limitations imposed by the affordable mass and DC power at the spacecraft on the flight transceiver aperture size, and the available ground transceiver aperture size. The acquisition and tracking link margin is perhaps the single most important portion of an optical communication link design, and deserves the most attention. Some of the major link characteristics are described below.

Delivery of high data rates necessitates narrow beamwidths transmitted from a given flight transceiver. This implies diffraction-limited or near-diffraction-limited optical systems. The typical requirement of beam pointing (approximately 1/10 of the beamwidth) necessitates precision beam pointing by the flight transceiver. As will be discussed in Chapter 3, establishing an efficient link requires maintenance of the beam pointing losses to a minimum. Beam pointing becomes progressively challenging as the transceiver's aperture size increases (i.e., the diffraction-limited beamwidth decreases). Therefore, typical near-Earth transceivers are designed with aperture diameters on the order of 0.05 to 0.25 m.

A robust link imposes the need for adequate allocation of acquisition, tracking, the pointing link margin, and the overall communication link margin.

Bidirectionality of the link supports many vital spacecraft functions, including uplink command, onboard software upgrades, assistance for recovery during emergencies (e.g., spacecraft safing mode), navigation and ranging data information, and (possibly) a beacon for precise downlink beam pointing.

Ranging and navigation data extraction may be accomplished through round-trip time of flight measurements, as part of the telecommunication overhead in the transmitted data packets. Range measurements yield Doppler measurements that provide useful data for ranging and navigation.

Operation at a small Sun angle may result in solar conjunction (or opposition) outages when Sun angles at the receivers (on each end of the link) are small. This outage arises from increased solar background noise at the receiver, and at the spacecraft's pointing and tracking detector. Stray light (light scattered into the receivers) is the primary culprit here. Optical filtering schemes (prefilter, post-filter, Lyot, and spatial filters) and control of optical surface quality and contamination will assist in rejecting or minimizing background light seen by the receivers.

2.1.1 RELIABILITY

Thermal management under solar illumination conditions can be a complicated issue. Because thermal blanketing of the telescope does not help with the entering sunlight, the telescope structure and optics face an increase in temperature. Optical filtering at the transmit and receive laser wavelengths can help mitigate this problem.

Radiation tolerance of optics (mirrors, lenses, filters, etc.) and electronics (processors, microchips, and sensors) are required for certain orbits around the Earth. Use of reflective optics, hard coatings, and shielding of sensitive components can help protect these components.

Component and subsystem reliability proportionate to a mission's lifetime is required for all missions. This implies implementation of adequate reliability and redundancy of critical and/or less reliable components (with sacrificing system performance).

Proper functioning over mission phases, from shortly after the launch to the end of the mission, is often required. Some system performance degradation is always anticipated, and built into the link budget for operation at the mission's end.

Atmospheric availability, caused primarily by signal attenuation due to cloud cover, is a limitation for optical communication links. Some implications of cloud-related attenuation include the following:

1. Weather statistics indicate that at some of the best observatory sites, a single station's visibility will be no better than approximately 65%.[1] The link availability value decreases further if significant spacecraft reacquisition time is required, or a given station is nonoperational due to technical difficulties. Multiple ground stations within the field of regard of the flight transceiver are needed to provide weather diversity reception to improve the coverage probability. Use of data buffers (memory) onboard the spacecraft, in conjunction with data communication protocols, can also improve link availability.
2. For an acquisition, tracking, and pointing (PAT) subsystem design that relies on an uplink laser beacon (emanated from the receiver station), the occasional cloud outage can cause the downlink to wander off the desired pointing location. In this case, the link availability must account for both uplink and downlink outages.

2.1.2 ROLE OF RF BACKUP

Weather-dependent availability can impose a significant operational limitation on an optical communications link. This may become particularly significant for critical maneuver periods (e.g., trajectory correction and orbit insertion) and spacecraft safe-mode operation, where it is in a standby mode attempting to recover from an emergency. Due to the optical terminal's narrow beam, it is more difficult to recover from emergencies by solely utilizing the optical telecommunication subsystem.

Significantly higher link margins for the link than typically assumed and (particularly) spatial diversity of ground receivers, locating them in different weather patterns, can help mitigate the weather availability issues for optical communications. Alternatively, addition of a low-capability Radio Frequency (RF) backup system to complement the optical link under adverse link conditions should greatly improve link availability.

2.1.3 REQUIREMENTS

Establishment of requirements is the first step in the design process. For a given mission, the more precise the transceiver's objectives and requirements are, the more precise the design will be. The flow-down of requirements from well-defined Level 1 to 6 (or finer) is essential in the systems engineering process, leading to successful development of a flight transceiver.

The following procedures help in defining system requirements:

1. Identify desired characteristics of the subsystem.
2. Identify technology readiness level and constraints for each subsystem.

3. Generate a baseline set of assumptions for the spacecraft environment (thermal, radiation levels, platform attitude control, platform vibration frequency spectrum, and mechanical interface specifications).
4. Generate a baseline set of assumptions for the receiving station.
5. Identify mission's operational requirements.
6. Identify design constraints and design drivers, including:
   a. Limitations on the host spacecraft (available mass, size, and power).
   b. Limitations on the receiver station (telescope diameter, quantity of stations, locations, performance of the optoelectronic receiver, and availability).
   c. Establishing outages due to Sun interference as determined by the spacecraft trajectory.
   d. Requirement of minimum data rates and bit error rates.
   e. Requirement of system reliability requirements, such as determined by the desired mission lifetime.

In general, it is prudent to shift the burden from the flight transceiver to the ground transceiver. The major subsystems for an optical communication transceiver and their relevant design drivers are listed below:

2.1.3.1 Flight Transceiver

Subsystem	Major Design Drivers
Laser transmitter	Wavelength, efficiency, output power, modulation rate, format and extinction ratio, and reliability
Acquisition, tracking, and pointing (ATP) fine-pointing mirror	Bandwidth, pointing accuracy, angular range, repeatability and reliability
Focal plane array	Device type, array size, pixel size, noise, spectral band, field of view, dynamic range, sensitivity, update rate, and reliability.
Received beacon signal	Beacon motion
Mechanical and thermal interface with the host platform	Vibration environment, deadband cycle, thermal effects of sun exposure
Electronics and software	Temperature environment, special voltages, processor speed, buffering, timing

2.1.3.2 Ground Receiver

Subsystem	Major Design Drivers
Atmospheric effects compensation	Visibility, attenuation, elevation angle, Sun angle, solar loading, turbulence, scattering
Optics	Aperture size, sunlight rejection, temperature control
Uplink laser beacon	Spatial mode quality, output power, quantity of lasers, atmospheric effect mitigation
Receiver	Detector type, amplification type, noise, detection efficiency, field of view
Signal processing	Signal-to-noise ratio, synchronization

2.2   BEAM POINTING BUDGET

A detailed discussion of PAT is provided in Chapter 3. A brief description of the ATP budget is given here. Precision pointing and tracking controls are required between laser communication transceivers in satellite-borne optical communication systems. The pointing subsystem's overall pointing error includes a *random error* component that has a short time constant (e.g., may vary from frame to frame), and a *static pointing error* with a long time constant. Random error stems from sensor noise and control system error. The static error stems from ephemeris errors, alignment errors, and algorithm errors.

The optical beamwidth (on the order of microradians) is a few orders of magnitude narrower than the attitude stability of the host satellite, generally assumed to be about 0.1° (1.75 mrad). This leads to drastic reductions in antenna gain, unless special care is taken to reduce the uncertainty, which is the most serious technological problem of satellite-borne optical communication systems.

The optical system relies upon precision pointing and tracking to reduce this uncertainty. Detailed knowledge of satellite micro-vibrational noise is needed for the proper design and engineering of pointing and tracking systems.

2.2.1   EXAMPLE OF POWER SPECTRAL DENSITY

Sudey and Sculman[2] and Wittig and coworkers,[3, 4] respectively, reported the power spectral density (PSD) on the LANDSAT-4 satellite, and the power spectra of a transient caused by a thruster firing on the OLYMPUS satellite. The European Space Agency (ESA) applied as a specification for the optical communication payload, Semiconductor Laser Intersatellite Link Experiment (SILEX), the following PSD for the angular base motion[4]:

$$S(f) = \frac{160}{1+f^2}, \quad [\mu rad^2/Hz] \qquad (2.1)$$

with an root mean square (rms) value of approximately $\Theta_{rms} = \sqrt{\int S(f)df} = 16 \mu rad$.

Similarly, the Japan Aerospace Exploration Agency (JAXA, formerly NASDA) specified the PSD that covered the transient caused by thruster firing

$$S(f) = \frac{160}{1+(f/10)^2}. \quad [\mu rad^2/Hz] \qquad (2.2)$$

The rms value of micro-vibration is $\Theta_{rms} = 50 \mu rad$ in this case. Other vibration signatures were obtained with a ground-based CO_2 laser radar while the retractable boom on a satellite was moving.[5, 6] The micro-vibrational environment was measured by using Laser Communications Equipment (LCE) onboard the ETS-VI satellite, the radial rms value of which was $\Theta_{rms} = 16.3 \mu rad$.[7] The PSDs for antenna tracking and slewing for the Optical Inter-orbit Communications Engineering Test Satellite (OICETS) are plotted during the in-orbit experiments, and the rms angular

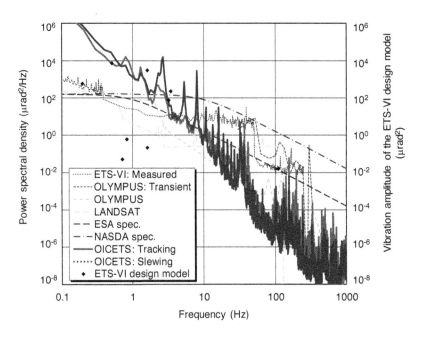

FIGURE 2.1 Power spectral density of the satellite micro-vibrations measured in space for several space demonstrations.

disturbances from 1 Hz to 1024 Hz were estimated to be 23.8 μrad rms during tracking and 43.8 μrad rms during slewing as the worst cases, respectively.[8] These PSDs are plotted in Figure 2.1.

2.2.2 POINTING ERROR STRUCTURE

The pointing error in optical space communications is caused by different sources. The structure of pointing errors is depicted in Figure 2.2. The pointing error consists of two main sources of tracking and point-ahead (or point-behind) errors. The tracking and point-ahead errors consist of two-axis components. Each one-axis error is added as the root sum square (RSS). The tracking error σ_{track} is a random error, and the point-ahead error σ_{point} becomes a bias error. The total pointing jitter error σ is given as the linear sum of those by

$$\sigma = \sigma_{track} + \sigma_{point}, \tag{2.3}$$

where

$$\sigma_{track} = \sqrt{\sigma_{microvib}^2 + \sigma_{attitude}^2 + \sigma_{coarse}^2 + \sigma_{fine}^2}, \tag{2.4}$$

$$\sigma_{point} = \sqrt{\sigma_{ephemeris}^2 + \sigma_{logic}^2 + \sigma_{command}^2 + \sigma_{sensor}^2 + \sigma_{calibration}^2 + \sigma_{deform}^2}, \tag{2.5}$$

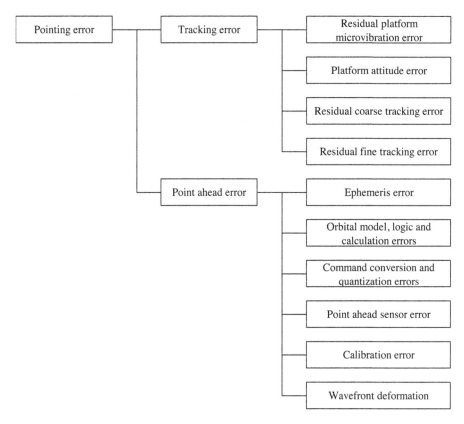

FIGURE 2.2 Pointing error structure for optical space communications.

Where

$\sigma_{microvib}$ is the residual platform micro-vibration,
$\sigma_{attitude}$ is the platform attitude error,
σ_{coarse} is the residual coarse error,
σ_{fine} is the fine tracking error,
$\sigma_{ephemeris}$ is the ephemeris error,
σ_{logic} is the orbital model, logic, and calculation error,
$\sigma_{command}$ is the command conversion and quantization error,
σ_{sensor} is the point-ahead sensor error,
$\sigma_{calibration}$ is the calibration error, and
σ_{deform} is the error due to wavefront deformation in the optics system.

These errors can be added as RSS when the errors are uncorrelated with each other.

Transmitter pointing has difficulties if the transmission distances are such that the propagation time is significant and relative motion is involved. In this case, the transmitter must actually point the optical beam ahead of the current receiver position to allow reception. This pointing procedure is called "point ahead." Let z be the distance from the transmitter to the satellite and let τ_c be the round-trip propagation time for light. Then, τ_c is equal to $2z/c$, where c is the speed of light. The distance the satellite moves along its orbit during τ_c is $\tau_c v$, where v is the satellite tangential velocity. Assuming a small angle, the necessary point-ahead angle θ_L in radians can be approximated as[9]

$$\theta_L \cong \frac{\tau_c v}{z} = \frac{2v}{c}. \tag{2.6}$$

As typical cases, the point-ahead angles correspond to several tenths of microradians between two satellites, and ~17 μrad between a ground station and a geostationary earth orbit (GEO) satellite. The computer controls (based on orbital calculations) are used to program these point-ahead angles.

2.2.3 RESIDUAL TRACKING ERROR

Generally, a high bandwidth feedback control is used to stabilize the beam-steering elements of a tracking system. Figure 2.3 shows a block diagram of such a system with existing closed-loop feedback vibration compensation. In this case, the disturbance is modeled as being additive to the output. The output $E_{out}(s)$ is given by[10]

$$E_{out}(s) = \frac{G(s)P_{ref}(s) + S(s)}{1 + G(s)}, \tag{2.7}$$

where $P_{ref}(s)$ is the reference signal at the positioning system and s stands for the Laplace variable. From a classical control standpoint, the disturbance $S(s)$ effects may be minimized by making the gain $G(s)$ as large as possible. In this case, the functional block $G(s)$ represents the closed-loop fine-pointing mechanism (FPM). The drawback of closed-loop control for high-frequency disturbance spectra is that, as the vibration waveform affects plant $G(s)$, no compensation is made until such time as some error is observed in the laser beam direction $E_{out}(t)$, which becomes the residual tracking error.

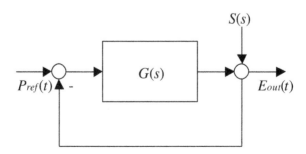

FIGURE 2.3   A standard feedback compensator.

2.2.4 RECEIVED POWER AND BER IN THE ABSENCE OF JITTER

For definiteness, it is considered a Gaussian-shaped laser beam of $1/e^2$ half-width, divergence angle w_0, and transmitted power P_t.[11, 12] In the absence of jitter, the irradiance at a wavelength λ and at a propagation distance z in the far field, with the offset angle α from the center of the beam, can be expressed as[13]

$$I^0(\alpha,z) = \frac{P_t \tau_t}{z^2} \frac{2}{\pi w_0^2} \exp\left(-2\frac{\alpha^2}{w_0^2}\right), \tag{2.8}$$

where τ_t is the optical loss of the transmitter. The received optical power at the receiver is given by

$$P_r = I^0(\alpha,z) A_r \tau_r, \tag{2.9}$$

where

A_r is the area of the receiver aperture and
τ_r is the optical loss of the receiver.

The received optical power stimulates the photonic current in the photo detector for which the signal-to-noise ratio (SNR) can be represented by the parameter Q. The BER of the optical receiver for the intensity-modulation-direct-detection system with nonreturn-to-zero is given by[14]

$$\mathrm{BER}(Q) = \frac{1}{2}\mathrm{erfc}\left(\frac{Q}{\sqrt{2}}\right), \tag{2.10}$$

where $\mathrm{erfc}\left(Q/\sqrt{2}\right)$ is the complementary error function. As the SNR can be considered constant in the absence of jitter, the average BER is given directly by Eq. (2.10).

2.2.5 RECEIVED POWER AND AVERAGE BER IN THE PRESENCE OF JITTER

The pointing errors have a significant effect on link performance and the bit error probability.[12] The pointing probability density function (PDF) with random jitter of σ in rms becomes the Nakagami-Rice distribution and is given by[15–17]

$$p_j(\alpha,\rho) = \frac{\alpha}{\sigma^2}\exp\left(-\frac{\alpha^2+\rho^2}{2\sigma^2}\right)I_0\left(\frac{\alpha\rho}{\sigma^2}\right), \tag{2.11}$$

where

α is the random pointing error angle,
ρ is the bias error angle from the center of the Gaussian beam, and
$I_0(x)$ is the modified Bessel function of order zero.

It is important that random jitter σ depends on the time duration that is considered. This pointing error is estimated from Eq. (2.3).[18] The random jitter that includes the worst case should be assumed in order to evaluate long-term stability of systems (such as those used in commercial satellites).[19, 20] The tracking control loop should be designed to reduce micro-vibration disturbances shown in Subsection 2.2.3, and to meet the requirement of residual random jitter. The bias error angle can even be regarded as the zero mean random value because the optical axis calibration of the transmitted laser beam will be carried out periodically by operation of the optical system during such a long time span. It is assumed that the bias error angle can be regarded as zero, and then the pointing PDF reduces to the Rayleigh distribution:

$$p_j(\alpha, 0) = \frac{\alpha}{\sigma^2} \exp\left(-\frac{\alpha^2}{2\sigma^2}\right). \tag{2.12}$$

Therefore, the PDF of the received optical intensity $p(I)$ becomes the beta distribution and is given by[15, 17, 21, 22]

$$p(I) = \beta I^{\beta-1} \text{ for } 0 \leq I \leq 1, \tag{2.13}$$

$$\bar{I} = \frac{\beta}{\beta+1}, \tag{2.14}$$

where

I is the normalized intensity,
$\bar{I}$ is its average value, and
$\beta = w_0^2/4\sigma^2$.

Therefore, the unconditional BER in the presence of jitter should be averaged with respect to the PDF of the received optical intensity and is given by

$$\overline{\text{BER}}(Q_r) = \int_0^1 p(I)\,\text{BER}\left(IQ_r\,\frac{\beta+1}{\beta}\right)dI,$$

$$= \frac{Q_r(\beta+1)}{2} \int_0^1 I^{\beta-1}\,\text{erfc}\left(\frac{IQ_r}{\sqrt{2}}\,\frac{\beta+1}{\beta}\right)dI. \tag{2.15}$$

where Q_r is the required SNR parameter for the desired average BER that equals a_{BER}. The mathematical expression in Eq. (2.15) is based on the fact that the frequency bandwidth of $p(I)$ is up to several kHz and (on the other hand) that Q_r corresponds to the data rate of the communication signal of which the frequency bandwidth is much higher than that of $p(I)$. Figure 2.4 shows degradation of the BER performance as a function of Q_r, and the ratio of the beam divergence angle w_0 to the random jitter σ. The desired average BER is obtained by solving $\overline{\text{BER}}(Q_r) = a_{BER}$, which the

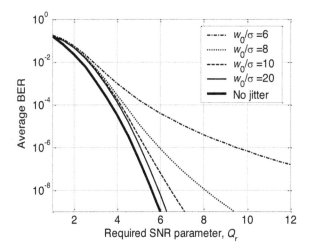

FIGURE 2.4 Average BER characteristics as a function of the required SNR parameter and the ratio of the beam divergence angle to the random jitter.

designers can choose as a free parameter. Therefore, the power penalty for the average BER that results from variation of the received optical signal is defined as

$$L_j = \left(\frac{Q \big|_{\mathrm{BER}(Q)=a_{BER}}}{Q_r \big|_{\overline{\mathrm{BER}}(Q_r)=a_{BER}}} \right). \tag{2.16}$$

Figure 2.5 shows the power penalty L_j as a function of w_0/σ obtained by numerical calculation. The average BER is gradually degraded owing to the variation of signal when $(w_0/\sigma) < 12$.

2.3 INTER-SATELLITE OPTICAL LINK DESIGN

2.3.1 LINK PERFORMANCE MARGIN

A link control table (LCT) helps us analyze and estimate the required telecommunication link signal power (upon providing all relevant aspects of the link condition). This includes the required BER and the transmitter and receiver characteristics at each side of the link. Link margin, which represents the difference between the available (detected) signal power and required power, is an important output of the LCT. Large values of the link margin (approximately >3dB) are typically desired.

2.3.2 LINK EQUATION FOR OPTICAL COMMUNICATION CHANNEL

The average received optical signal power in the presence of jitter can be expressed by:

$$P_r = P_t \tau_t G_t L_r G_r \tau_r \tau_j L_j, \tag{2.17}$$

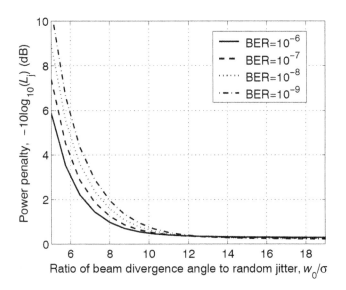

FIGURE 2.5 Power penalty for the desired average BER as a function of the ratio of the beam divergence angle to the random jitter.

where

P_r is the total average power measured at the input to the receiver,
P_t is the average transmit power,
τ_t is the optical loss of the transmitter, and
G_t is the peak transmit antenna gain, expressed by the formula:

$$G_t = \frac{8}{w_0^2},$$ (2.18)

L_r is the space loss, represented by:

$$L_r = \left(\frac{\lambda}{4\pi z}\right)^2,$$ (2.19)

G_r is the receiving antenna gain, measured from the receiver, and expressed as:

$$G_r = \frac{4\pi A_r}{\lambda^2},$$ (2.20)

τ_r is the optical loss of the receiver,
τ_j is the average pointing-loss due to random-pointing jitter, represented by:

$$\tau_j = \frac{w_0^2}{4\sigma^2 + w_0^2},$$ (2.21)

L_j is the power penalty of the optical receiver, given by:

$$L_j = \left(\frac{Q\big|_{\mathrm{BER}(Q)=a_{BER}}}{Q_r\big|_{\overline{\mathrm{BER}}(Q_r)=a_{BER}}} \right). \tag{2.22}$$

One can find additional loss parameters of τ_j and L_j by comparison with the conventional link equation. Eq. (2.17) estimates the received optical power for the average BER. It is useful to design the optimum system parameters (such as the beam divergence angle and random jitter), as shown in Subsection 2.3.5, and the power penalty for average BER in Figure 2.5.[23, 24]

2.3.3 LINK EQUATION FOR OPTICAL TRACKING CHANNEL

The fade statistics is needed for the optical tracking channel.[25–28] As shown in Eq. (2.13), the PDF of the received signal becomes the beta distribution. The average received optical power at the tracking sensor is given by

$$P_{rs} = P_t \tau_t G_t L_r G_r \tau_r \tau_j \tau_{rs}, \tag{2.23}$$

where τ_{rs} is the transmission loss of the optical path to the tracking sensor. The fade level F_T from the average level is defined by

$$F_T = \frac{\beta+1}{\beta} P_F^{\frac{1}{\beta}}, \tag{2.24}$$

where

$$P_F = \int_0^{F_T \bar{I}} p(I)\, dI = \left(F_T \bar{I} \right)^{\beta}, \text{ and } \bar{I} = \frac{\beta}{\beta+1}. \tag{2.25}$$

P_F is the allowable fade probability at the tracking sensor. The surge level S_T is similarly defined by

$$S_T = \frac{\beta+1}{\beta}(1 - P_S)^{\frac{1}{\beta}}, \tag{2.26}$$

Where

$$P_S = \int_{S_T \bar{I}}^{1} p(I)\, dI = 1 - \left(S_T \bar{I} \right)^{\beta}, \text{ and } \bar{I} = \frac{\beta}{\beta+1}. \tag{2.27}$$

P_S is the allowable surge probability at the tracking sensor. The dynamic range required for the tracking sensor to cope with only random angular jitter is

$$D_{jitter} = \frac{S_T}{F_T}. \tag{2.28}$$

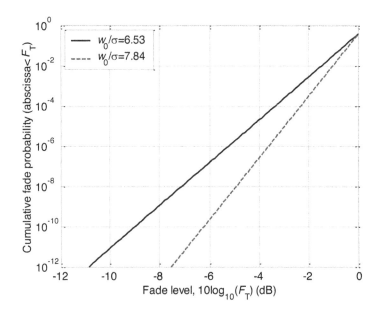

FIGURE 2.6 Cumulative fade probability as a function of the fade level when $(w_0/\sigma)_{opt} =$ 6.53 and 7.84.

The cumulative fade and surge probabilities are also plotted as a function of the fade and surge levels in Figures 2.6 and 2.7.

In practice, the variation of distance (due to the orbital motion of satellites) should also be taken into account in the calculation of this dynamic range. As the range variation influences on the received power (a mean value) and the fade and surge levels in this chapter are relative parameters, the total dynamic range (including range variation) is given by

$$D_{total} = \frac{S_T}{F_T}\left(\frac{z_{max}}{z_{min}}\right)^2,\qquad(2.29)$$

where z_{max} and z_{min} are the maximum and minimum distances (respectively) between two satellites in their orbital motions. The dynamic range for the tracking sensor should usually cover D_{total}. However, transmitted laser power or optical attenuation should be adjustable according to the range condition, such that the dynamic range for the tracking sensor is less than D_{total}.

2.3.4 EXAMPLE OF A SPACE-TO-SPACE OPTIMUM LINK DESIGN

Reflecting results in the concrete optical system, it is the most expedient way to show optical link budget estimates of the communication and tracking channels. Table 2.1 shows an example of the optimum link design in the presence of jitter. The methodology for estimating the optimum link is as follows: (1) determine the achievable

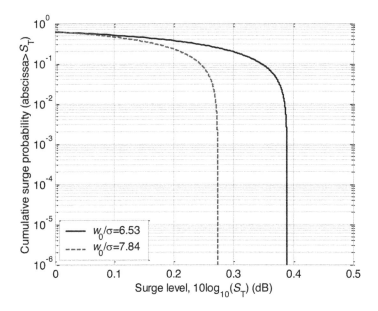

FIGURE 2.7 Cumulative surge probability as a function of the surge level when $(w_0/\sigma)_{opt} = 6.53$ and 7.84.

pointing error of the optical communication system by modeling or making the engineering structural model, (2) estimate the optimum beam divergence angle w_0 by using the approximate version of Eqs. (2.32) and (2.33) in Subsection 2.3.5, which gives the beam waist w_{waist} of the Gaussian beam from $w_0 = \lambda/\pi w_{waist}$, then, the transmitting antenna diameter D can be determined from $D = 2\alpha_t W_0$ with the truncation ratio α_t and the beam radius W_0 at the telescope aperture, (3) read the power penalty for average BER from Figure 2.5, and then (4) adjust the transmitting optical power in order to meet the required power for the optical receiver. Once the beam divergence angle and pointing jitter error are determined, fade and surge levels at the allowable fade and surge probabilities can be estimated for the optical tracking channel from Eqs. (2.25) and (2.27). With range variation, the dynamic range for sensors should be satisfied using Eq. (2.29) during the communication phase.

2.3.5 Optimum Ratio of w_0/σ Versus Average BER

In practice, the power penalty is limited by not only the BER degradation that results from jitter, but also the dependence of the received optical power on the laser beams' divergence angle. The average on-axis irradiance for the receiver in the presence of jitter is derived as:

$$\langle I(0,z)\rangle = \frac{P_t\tau_t}{z^2}\frac{2}{\pi w_0^2}\frac{\beta}{\beta+1}L_j,$$

$$= I^0(0,z)\tau_j L_j,$$

(2.30)

TABLE 2.1

Example of the Optimum Link Budget Estimated for Space-to-Space Optical Tracking and Communication Channels

ITEM	UNIT	FORWARD/RETURN LINKS	
		TRACKING	COMM.
TX POWER, P_t	W	1.0	1.0
	dBm	30.0	30.0
TX BEAM WAIST $(1/e^2)$	cm	4.8	4.8
$(w_0/\sigma)_{opt}$ (@BER=10^{-9})	-	7.84	7.84
POINTING JITTER, σ	μrad(rms)	2.60	2.60
TX BEAM DIVERGENCE, $2w_0$	μrad	40.8	40.8
TX OPTICS LOSS, τ_t	dB	-2.0	-2.0
WAVELENGTH, λ	m	1.55E-06	1.55E-06
AVERAGE POINTING LOSS, τ_j	dB	-0.1	-0.1
TX GAIN, G_t	dB	102.8	102.8
DISTANCE, L	m	5.00E+06	5.00E+06
SPACE LOSS, L_r	dB	-272.2	-272.2
RX ANTENNA DIAMETER	cm	10.0	10.0
RX GAIN, G_r	dB	106.1	106.1
RX OPTICS LOSS, τ_r or τ_{rs}	dB	-4.0	-2.0
TRACKING SENSOR POWER, P_{rs}	dBm	-39.2	-
FADE LEVEL, F_T(P_F=10^{-2})	dB	-1.0	-
SURGE LEVEL, S_T	dB	0.3	-
RANGE VARIATION, $(R_{max}/R_{min})^2$	dB	10.0	-
DYNAMIC RANGE, D_{TOTAL}	dB	11.3	-
RX POWER, P_r	dBm	-	-37.2
POWER PENALTY FOR AVERAGE BER, L_j	dB	-	-2.0
DATA RATE	bps	-	2.488E+09
SENSITIVITY (@BER=10^{-9})	photons/bit	-	90
	dBm	-	-45.4
AVERAGE MARGIN	dB	-	6.2

where $\tau_j = \beta/(\beta + 1)$ is the average pointing loss due to jitter. The combined total power penalty in the average received optical power is given by

$$L_p = -10\log\left[\frac{\langle I(0,z)\rangle}{\langle I(0,z)\rangle|_{BER=10^{-6},max}}\right],$$

$$\propto -10\log P_t + 20\log z + 10\log\left(w_0^2 + 4\sigma^2\right) - 10\log L_j.$$

(2.31)

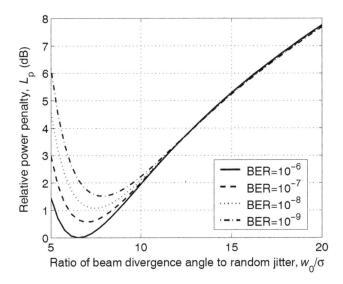

FIGURE 2.8 Total power penalty for the optical link as a function of the ratio of the beam divergence angle to the random jitter for different values of average BER when the transmitted power, the distance, and the random jitter are assumed to be constant against the beam divergence angle.

Although this equation cannot be analytically solved to get the minimum power penalty by using the condition $dL_p/dw_0 = 0$, numerical results are readily obtained and plotted in Figure 2.8 as a function of w_0/σ and the desired average BER, when the transmitted power P_t, the distance z, and random jitter σ are assumed to be constant versus the beam divergence angle w_0. One can see that there is an optimum value that minimizes the required transmitter power. The numerical calculation was done and the obtained results of the optimum ratio are tabulated in Table 2.2. Being of simple use to designers, an analytic approximation for $(w_0/\sigma)_{opt}$ for $10^{-12} \leq \overline{\text{BER}} \leq 10^{-2}$, as obtained by a least squares fit to the calculated values, is given by

$$\left(\frac{w_0}{\sigma}\right)_{opt} = a_6x^6 + a_5x^5 + a_4x^4 + a_3x^3 + a_2x^2 + a_1x + a_0, \tag{2.32}$$

$$x = \log_{10}\left(\overline{\text{BER}}\right), \tag{2.33}$$

where the coefficients a_i $(i = 0, 1, ..., 6)$ are listed in Table 2.3. The optimum ratio $(w_0/\sigma)_{opt}$ as a function of the desired average BER is easily given by Eqs. (2.32) and (2.33). For example, $(w_0/\sigma)_{opt} = 6.53$ and $(w_0/\sigma)_{opt} = 7.84$ can be obtained for desired average BERs of 10^{-6} and 10^{-9}, respectively. Even if there is any range variation in z, the ratio of the beam divergence angle to random jitter obtained from Eqs. (2.32) and (2.33) is optimum with respect to the desired average BER.

TABLE 2.2
Optimum Ratio of the Beam
Divergence Angle to Random
Jitter for Each Desired BER

BER	$(w_0/\sigma)_{opt}$
10^{-2}	4.12431
10^{-3}	4.85198
10^{-4}	5.47376
10^{-5}	6.02693
10^{-6}	6.53065
10^{-7}	6.99640
10^{-8}	7.43177
10^{-9}	7.84207
10^{-10}	8.23126
10^{-11}	8.60228
10^{-12}	8.95751

2.4   GROUND-TO-SPACE OPTICAL LINK DESIGN

2.4.1   AVERAGE BER

The PDF of the received optical intensity $p_{total}(I)$ in the presence of atmospheric turbulence under weak turbulence conditions becomes the lognormal distribution, and the PDF due to random pointing jitter is the beta distribution. The joint PDF in the presence of atmospheric turbulence and random jitter is given by[15, 17]

$$p_{total}(I) = \frac{1}{2\bar{I}} erfc\left[\frac{\ln\left(I/\bar{I}\right)+\sigma_I^2\left(\beta+1/2\right)}{\sqrt{2}\sigma_I}\right] \times \beta \exp\left[\frac{\sigma_I^2}{2}\beta(\beta+1)\right]\left(\frac{I}{\bar{I}}\right)^{\beta-1},$$

(2.34)

TABLE 2.3
Coefficients Describing the Optimum Ratio $(w_0/\sigma)_{opt}$

Coefficient, a_i	Value
a_0	2.05613e+00
a_1	−1.33146e+00
a_2	−1.92403e−01
a_3	−2.51125e−02
a_4	−2.02818e−03
a_5	−8.87644e−05
a_6	−1.60597e−06

where σ_I^2 is the normalized variance called the scintillation index, which is a significant measure for atmospheric turbulence (as shown in Chapter 3). Following the same approach as in the previous section, the unconditional BER in the presence of atmospheric turbulence and random jitter is averaged with respect to the PDF of Eq. (2.34) and gives

$$\overline{BER}_a(Q_r) = \int_0^\infty p_{total}(I) BER\left(Q_r \frac{I}{\bar{I}}\right) dI,$$

$$= \frac{\beta}{4} \exp\left[\frac{\sigma_I^2}{2} \beta(\beta+1)\right] \int_0^\infty dx x^{\beta-1} \operatorname{erfc}\left(\frac{\ln x + \sigma_I^2\left(\beta + \frac{1}{2}\right)}{\sqrt{2}\sigma_I}\right) \operatorname{erfc}\left(\frac{Q_r x}{\sqrt{2}}\right),$$

$$(2.35)$$

where Q_r is the required SNR parameter for the desired average BER. The power penalty for the average BER, due to the variation of the received optical signal, is similarly defined as

$$L_a = \left(\frac{Q\big|_{BER(Q)=a_{BER}}}{Q_r\big|_{\overline{BER}_a(Q_r)=a_{BER}}}\right). \qquad (2.36)$$

Figure 2.9 shows the BER characteristics in the presence of both jitter and atmospheric turbulence, when $\sigma_I^2 = 0.05$ and the random jitter $\sigma = 3.8$ μrad rms.

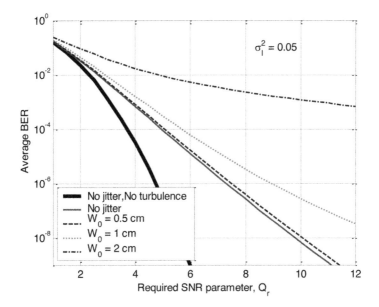

FIGURE 2.9 Average BER performances in the presence of jitter and atmospheric turbulence. W_0 denotes the optical beam radius at the transmitter.

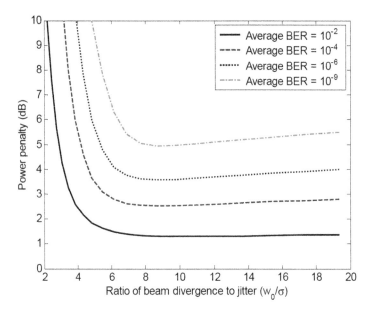

FIGURE 2.10 Power penalty L_a as a function of the ratio w_0/s in the presence of both jitter and atmospheric turbulence for different average BER, with scintillation index $\sigma_I^2 = 0.05$ and the random jitter $\sigma = 3.8$ µrad rms.

The BER performance degrades at the larger transmitter beam radius. Figure 2.10 shows power penalty for the average BER from Eq. (2.36), which corresponds to average power loss due to variation in the optical signals. One can find the bias power penalty (caused by lognormal distribution) by comparing with Figure 2.5, from which it is clear that more optical power is needed to achieve the required BER in the presence of atmospheric turbulence than what is needed when only random pointing jitter is present.

2.4.2 LINK EQUATION FOR OPTICAL COMMUNICATION CHANNEL

The average on-axis received optical power in the presence of jitter and atmospheric turbulence is

$$P_r = P_t \tau_t G_t \tau_a L_r G_r \tau_r \tau_j \frac{w_0^{\,2}}{w_e^{\,2}} L_a, \tag{2.37}$$

where w_e is the effective beam divergence angle in the presence of atmospheric turbulence, and τ_a is the atmospheric transmission due to air mass, which estimates received optical power for the average BER in the presence of jitter and atmospheric turbulence. Additional loss parameters of τ_j, w_0^2/w_e^2, and L_a should be incorporated into the conventional link equation. Figure 2.11 shows the relative power penalty of the optical uplink as a function of w_0/σ for different average BER with scintillation

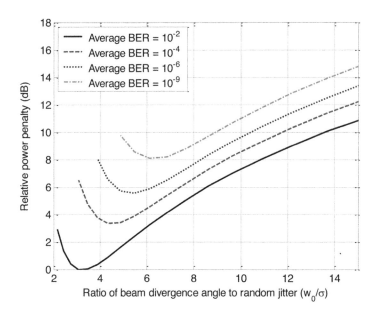

FIGURE 2.11 Relative power penalty for the optical uplink as a function of the ratio w_0/σ for different values of average BER and zenith angle $\zeta = 0°$.

index $\sigma_I^2 \sim 0.1$ at the zenith angle $\zeta = 0°$ and the random jitter $\sigma = 3.8$ μrad rms. The atmospheric coherence length r_0 is 5.4 cm at the zenith angle $\zeta = 0°$, which is calculated by using the C_n^2 structure parameter based on the H-V model between the altitudes 122 m to 590 km at a wavelength $\lambda = 0.819$ μm.[28] The average BER is degraded at small values of w_0/σ and large values of σ_I^2. For the downlink path, Figure 2.12 shows the relative power penalty at $\lambda = 0.847$ mm and $\sigma_I^2 = 0.025$. Power penalty is limited by the average BER degradation, due to both random jitter and the dependence of the received optical power on the divergence angle of the laser beam.

This phenomenon is more complicated than that observed in the presence of only jitter because the effective beam divergence angle depends on jitter and atmospheric turbulence, and the atmospheric turbulence varies with time. However, as can be seen from Figure 2.10, one can omit the pointing jitter effect for the value of w_0/σ above about 7. Note that we can thus get a key measure to predict the most efficient beam divergence angle for different average BER in the presence of atmospheric turbulence.

2.4.3 LINK EQUATION FOR OPTICAL TRACKING CHANNEL

The average received optical power at the tracking sensor is given by

$$P_{rs} = P_t \tau_t G_t \tau_a L_r G_r \tau_{rs} \tau_j \frac{w_0^2}{w_e^2}. \tag{2.38}$$

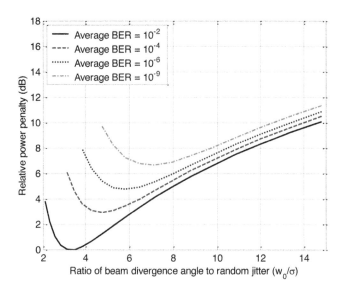

FIGURE 2.12 Relative power penalty for the optical downlink as a function of the ratio w_0/σ for different values of the average BER.

The PDF of the received signal is given by Eq. (2.34) in the presence of jitter and atmospheric turbulence. The allowable fade probability P_F at the fade level F_T from the average level is defined by

$$P_F = \int_0^{F_T \bar{I}} p_{total}(I)\,dI. \qquad (2.39)$$

The allowable surge probability P_S at the surge level S_T (from the average level) is similarly defined by

$$P_S = \int_{S_T \bar{I}}^{\infty} p_{total}(I)\,dI. \qquad (2.40)$$

Dynamic range for the tracking sensor should also cover D_{total} from Eq. (2.29). Figures 2.13 and 2.14 show the cumulative fade and surge probabilities calculated from Eqs. (2.39) and (2.40), as a function of the fade and surge levels. The fade margin should be allocated in order to keep the desired link quality that is specified by the allowable cumulative fade probability P_F. The surge margin should not be considered to saturate optical tracking sensors in the optical tracking channel.

2.4.4 MULTIBEAM GAIN

Let us consider N parallel beams, each of which is separated by a distance larger than the atmospheric coherence length r_0. The reduction in fluctuation in the optical signal is explained in Subsection 2.4.4. As the multibeam effect is independent of the random

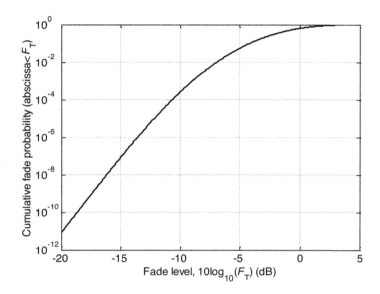

FIGURE 2.13   Cumulative fade probability P_F as a function of the fade level F_T. The scintillation index $\sigma_I^2 = 0.35$ at the zenith angle $\zeta = 60°$, the transmitter beam radius 1.2 cm, and the random jitter $\sigma = 3.8$ μrad rms are used.

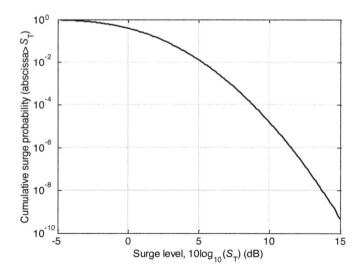

FIGURE 2.14   Cumulative surge probability P_S as a function of the surge level S_T. The scintillation index $\sigma_I^2 = 0.35$ at the zenith angle $\zeta = 60°$, the transmitter beam radius 1.2 cm, and the random jitter $\sigma = 3.8$ μrad rms are used.

pointing jitter effect and cannot be given by an analytical expression, the total PDF for multibeam situation is given by the convolution of the PDF of the random pointing jitter,

$$p_{multi,total}(I) = \mathfrak{S}^{-1}\left[\Psi_m(a) \cdot \Psi_j(a)\right], \tag{2.41}$$

$$\Psi_m(a) = \int_{-\infty}^{\infty} p_{multi}(x)\exp(jax)\,dx, \tag{2.42}$$

$$\Psi_j(a) = \int_{-\infty}^{\infty} p(x)\exp(jax)\,dx. \tag{2.43}$$

When the transmitted power for individual beams in a multibeam system is $1/N$ of the total power, the multibeam gain for average BER is defined as

$$G_{multi,BER} = -10\log\left(\frac{Q_r\big|_{\overline{BER}_m(Q_r)=a_{BER}}}{Q_r\big|_{\overline{BER}_a(Q_r)=a_{BER}}}\right), \tag{2.44}$$

where

$$\overline{BER}_m(Q_r) = \int_0^{\infty} p_{multi,total}(I)\,BER\left(Q_r\frac{I}{\bar{I}}\right)dI. \tag{2.45}$$

The fade and surge multibeam gains are also defined by

$$G_{multi,fade} = -10\log\left(F_{T,multi}/F_T\right), \tag{2.46}$$

$$G_{multi,surge} = 10\log\left(S_{T,multi}/S_T\right), \tag{2.47}$$

where

$$P_F = \int_0^{F_{T,multi}\bar{I}} p_{multi,total}(I)\,dI, \tag{2.48}$$

$$P_S = \int_{S_{T,multi}\bar{I}}^{\infty} p_{multi,total}(I)\,dI, \tag{2.49}$$

where

F_T and S_T represent fade and surge levels from the mean levels for the single beam, $F_{T,multi}$ and $S_{T,multi}$ represent those for the multibeam, and P_F and P_S are the allowable fade and surge probabilities of the optical link.

The fade level for the multiple beams is given by adding the fade level for a single beam and the multibeam gain. The multibeam gain is also used for improving the surge level of the optical signal, which is a useful notion in the link design for a multibeam laser transmission system.

2.4.5 APERTURE AVERAGING GAIN

The aperture averaging gain for the average BER is similarly defined as

$$G_{A,BER} = -10\log\left(\frac{Q_r\big|_{\overline{BER}_A(Q_r)=a_{BER}}}{Q_r\big|_{\overline{BER}_a(Q_r)=a_{BER}}}\right), \tag{2.50}$$

where

$$\overline{BER}_A(Q_r) = \int_0^\infty p_{A,total}(I)\,BER\left(Q_r\frac{I}{\overline{I}}\right)dI, \tag{2.51}$$

$$p_{A,total}(I) = \int_0^1 p_A\left(\frac{I}{x}\right)p_j(x)\frac{dx}{x}, \tag{2.52}$$

$$p_A(I) = \frac{1}{\sqrt{2\pi A\sigma_I}I}\exp\left(-\frac{\left\{\ln\left(I/\overline{I}\right)+\frac{1}{2}A\sigma_I^2\right\}^2}{2A\sigma_I^2}\right). \tag{2.53}$$

The fade and surge aperture averaging gains are also defined by

$$G_{A,fade} = -10\log(F_{T,A}/F_T), \tag{2.54}$$

$$G_{A,surge} = 10\log(S_{T,A}/S_T), \tag{2.55}$$

where

$$P_F = \int_0^{F_{T,A}\overline{I}} p_{A,total}(I)\,dI, \tag{2.56}$$

$$P_S = \int_{S_{T,A}\overline{I}}^\infty p_{A,total}(I)\,dI. \tag{2.57}$$

$F_{T,A}$ and $S_{T,A}$ represent fade and surge levels from the mean levels owing to the aperture averaging.

2.4.6 EXAMPLE OF A GROUND-TO-SPACE OPTIMUM LINK DESIGN

Table 2.4 shows an example of the link budget design when the OICETS is used.[29, 30] A four-laser-beams transmission system is employed to reduce the uplink optical scintillation. The coherence length r_0 is 5.4 cm at the zenith angle $\zeta = 0$, which is

TABLE 2.4

Link Budget Design for a Four-Laser-Beams Transmission System between the Ground and the OICETS Satellite

ITEM	UNIT	UPLINK		DOWNLINK	
		TRACKING	COMM.	TRACKING	COMM.
TX POWER, P_t	W	0.01	0.01	0.10	0.10
	dBm	10.0	10.0	20.0	20.0
TX BEAM WAIST ($1/e^2$)	cm	1.2	1.2	6.0	6.0
RATIO (w_0/σ)	-	5.20	5.20	5.18	5.18
POINTING JITTER, σ	μrad(rms)	3.80	3.80	0.87	0.87
TX BEAM DIVERGENCE, $2w_0$	μrad	39.5	39.5	9.0	9.0
TX OPTICS LOSS, τ_t	dB	-2.0	-2.0	-2.7	-2.7
WAVELENGTH, λ	m	8.19E-07	8.19E-07	8.47E-07	8.47E-07
AVERAGE POINTING LOSS, τ_j	dB	-0.2	-0.2	-0.2	-0.2
TX GAIN, G_t	dB	96.3	96.3	110.0	110.0
DISTANCE, z	m	1.00E+06	1.00E+06	1.00E+06	1.00E+06
SPACE LOSS, L_r	dB	-263.7	-263.7	-263.4	-263.4
ATMOSPHERIC TRANSMISSION, τ_a	dB	-7.8	-7.8	-7.0	-7.0
RX ANTENNA DIAMETER	cm	26.0	26.0	10.0	10.0
RX GAIN, G_r	dB	120.0	120.0	111.4	111.4
RX OPTICS LOSS, τ_r or τ_{rs}	dB	-7.6	-4.9	-6.0	-3.0
TRACKING SENSOR POWER, P_{rs}	dBm	-55.0	-	-38.0	-
RANGE VARIATION, $(z_{max}/z_{min})^2$	dB	3.0	-	3.0	-
FADE LEVEL, $F_T(P_F=10^{-2})$	dB	-6.9	-	-9.0	-
SURGE LEVEL, $S_T(P_S=10^{-2})$	dB	5.3	-	6.4	-
DYNAMIC RANGE, D_{TOTAL}	dB	15.2	-	18.4	-
RX POWER, P_r	dBm	-	-52.3	-	-34.9
POWER PENALTY FOR AVERAGE BER, L_a	dB	-	-9.1	-	-11.1
AVERAGE ATMOSPHERIC LOSS, w_0^2/w_e^2	dB	0.0	0.0	-0.1	-0.1
DATA RATE	bps	-	2.048E+06	-	4.937E+07
SENSITIVITY (@BER=10^{-6})	photons/bit	-	200	-	200
	dBm	-	-70.0	-	-56.4
AVERAGE MARGIN FOR BER	dB	-	8.6	-	10.2
MULTIBEAM GAIN FOR FADE, $G_{multi,fade}$	dB	3.6	-	-	-
APERTURE AVERAGING GAIN FOR FADE, $G_{A,fade}$	dB	-	-	5.8	-
MULTIBEAM'S FADE LEVEL, $F_{T,multi}(P_F=10^{-2})$	dB	-3.3	-	-3.2	-
MULTIBEAM GAIN FOR SURGE, $G_{multi,surge}$	dB	-2.3	-	-	-
APERTURE AVERAGING GAIN FOR SURGE, $G_{A,surge}$	dB	-	-	-4.0	-
MULTIBEAM'S SURGE LEVEL, $S_{T,multi}(P_S=10^{-2})$	dB	3.0	-	2.4	-
MULTIBEAM'S DYNAMIC RANGE, D_{TOTAL}	dB	9.3	-	8.6	-
MULTIBEAM GAIN FOR BER, $G_{multi,BER}$	dB	-	4.3	-	-
APERTURE AVERAGING GAIN FOR BER, $G_{A,BER}$	dB	-	-	-	6.5
AVERAGE MARGIN FOR BER	dB	-	12.9	-	16.8

calculated by using the C_n^2 structure parameter based on the H-V model between the altitudes 122 m to 590 km, at wavelengths 0.819 and 0.847 μm. The fade and surge probabilities of 10^{-2} are assumed in this case, and the 60-degree zenith angle is considered. The beam divergence angle of the optical ground station was selected to be optimum at the average BER of 10^{-6}. For the uplink path, fade and surge multibeam gains are estimated to be 3.6 dB and −2.3 dB. For the multibeam system, average BER performance is improved by 4.3 dB. Fade and surge levels will meet the required dynamic range, 10 dB for the OICETS, of the optical receiver. For the downlink path, BER performance will be improved by 6.5 dB via aperture averaging.

Table 2.5 shows the real link budget analysis measured on September 19, 2006. The average BER was about 10^{-5} for 10 sec at a link distance of 938 km, without the

TABLE 2.5

Link Budget Analysis for Optical Links between the Ground Station and the OICETS Satellite Measured on September 19, 2006

	Unit	Uplink			Downlink	
		Beacon	Tracking	Comm.	Tracking	Comm.
TX power	W	3.76	0.26	0.26	0.10	0.10
	dBm	35.7	24.2	24.2	20.0	20.0
Beam diameter at telescope	cm	1.7	72.1	72.1	12.0	12.0
Pointing jitter	μrad (rms)	23.1	23.1	23.1	0.12	0.12
TX beam divergence	μrad	9000	204.2	204.2	9.0	9.0
TX optics loss	dB	−0.05	−16.0	−16.0	−2.7	−2.7
Wavelength	m	8.08E-07	8.15E-07	8.15E-07	8.47E-07	8.47E-07
Average pointing loss	dB	0.0	−0.2	−0.2	−2.3	−2.3
TX gain	dB	53.0	85.8	85.8	116.0	116.0
Distance	m	1.50E+06	9.38E+05	9.38E+05	9.38E+05	9.38E+05
Space loss	dB	−267.3	−263.2	−263.2	−262.9	−262.9
Atmospheric transmission	dB	−3.2	−3.2	−3.2	−3.2	−3.2
RX antenna diameter	cm	26.0	26.0	26.0	31.8	31.8
RX gain	dB	120.1	120.0	120.0	121.4	121.4
RX optics loss	dB	−2.6	−7.6	−7.9	−15.4	−15.4
Tracking sensor power	dBm	−64.4	−60.2	-	−30.5	-
Fade level ($P_F=10^{-2}$)	dB	−8.0	−3.3	-	−3.5	-
Surge level ($P_S=10^{-2}$)	dB	8.5	2.7	-	2.2	-
Dynamic range	dB	16.5	6.0	-	5.7	-
RX power	dBm	-	-	−60.5	-	−29.0
Power penalty for average BER	dB	-	-	−11.7	-	−13.5
Average beam spreading loss	dB	0.0	0.0	0.0	−1.5	−1.5
Data rate	bps	-	-	2.048E+06	-	4.937E+07
Sensitivity (@BER of 10^{-6})	photons/bit	-	-	200	-	2200
	dBm	-	-	−70.0	-	−45.9
Average margin for BER	dB	-	-	**−2.2**	-	**1.9**

use of error-correcting code. The link margin of −2.2 dB was very consistent with a BER of 10^{-6} because the link margin for a BER of 10^{-5} is −1.3 dB, which is 0.9 dB smaller than that for 10^{-6}. For the downlink, the positive margin was estimated to be +1.9 dB, but the average BER was only about 10^{-4} without error-correcting code at the same propagation distance. The fluctuation in the received signal due to atmospheric turbulence apparently degraded the receiver's performance. Using forward error coding would produce even better BERs. These results demonstrate that optical communication links can be established between a ground station and a LEO satellite, even with atmospheric turbulence.

In the presence of atmospheric turbulence, the greater margin should be allowed for BER degradation because the strength of atmospheric turbulence will change with time, and with the variation in the zenith angle of the optical path during the communication. Further consideration will be needed for estimating the optical scintillation under the strong turbulence condition, since lognormal distribution changes into the gamma-gamma distribution.[28]

REFERENCES

1. A. Biswas, K. E. Wilson, S. Piazzolla, J. P. Wu, and W. H. Farr, "Deep-space optical communications link availability and data volume," Proc. SPIE. **5338**, pp. 175–183 (2004).
2. J. Sudey and J. R. Sculman, "In orbit measurements of Landsat-4 thematic mapper dynamic disturbances," 35th Congress of the International Astronautical Federation, Lausanne, Switzerland, IAF-84-117 (1984).
3. K. J. Held and J. D. Barry, "Precision pointing and tracking between satellite-borne optical systems," Opt. Eng., **27**(4), 325–333 (1988).
4. M. Wittig, K. van Holts, D. E. L. Tunbridge, and H. C. Vermeulen, "In-orbit measurement of microaccelerations of ESA's communication satellite OLYMPUS," in Free-Space Laser Communication Technologies II, D. L. Begley and B. D. Seery, eds., Proc. SPIE 1218, pp. 205–214 (1990).
5. K. I. Schultz, D. G. Kocher, J. A. Daley, J. R. Theriault, J. Spinks, and S. Fisher, "Satellite vibration measurements with an autodyne CO_2 laser radar," Appl. Opt. **33**(12), 2349–2355 (1994).
6. K. I. Schultz and S. Fisher, "Ground-based laser radar measurements of satellite vibrations," Appl. Opt. **31**(36), 7690–7694 (1992).
7. M. Toyoshima and K. Araki, "In-orbit measurements of short term attitude and vibrational environment on the Engineering Test Satellite VI using laser communication equipment," Opt. Eng. **40**(5), 827–832 (2001).
8. M. Toyoshima, Y. Takayama, and H. Kunimori, "In-orbit measurements of spacecraft microvibrations for satellite laser communication links," Opt. Eng. **49**(8), 083604-1–10 (2010).
9. R. M. Gagliardi and S. Karp, *Optical Communications*, Wiley, New York (1976).
10. V. A. Skormin, M. A. Tascillo, and T. E. Busch, "Demonstration of a jitter rejection technique for free-space laser communication," IEEE Trans. Commun. **33**(2), 568–575 (1997).
11. H. T. Yura, "Optimum truncation of a Gaussian beam in the presence of random jitter," J. Opt. Soc. Am. A **12**(2), 375–379 (1995).
12. C. C. Chen and C. S. Gardner, "Impact of random pointing and tracking errors on the design of coherent and incoherent optical intersatellite communication links," IEEE Trans. Commun. **37**(3), 252–260 (1989).

13. K. Araki and T. Aruga, "Space optical communication technologies—An overview," Rev. Laser Eng. **24**(12), 1264–1271 (1996).

14. G. P. Agrawal, *Fiber-Optic Communication Systems*, 2nd ed., Wiley, New York, (1997), Chap. 4.5.

15. K. Kiasaleh, "On the probability density function of signal intensity in free-space optical communications systems impaired by pointing jitter and turbulence," Opt. Eng. **33**(11), 3748–3757 (1994).

16. M. Toyoshima and K. Araki, "Far-field pattern measurement of an onboard laser transmitter by use of a space-to-ground optical link," Appl. Opt. **37**(10), 1720–1730 (1998).

17. M. Toyoshima and K. Araki, "Effects of time-averaging on optical scintillation in a ground-to-satellite atmospheric propagation," Appl. Opt. **39**(12), 1911–1919 (2000).

18. R. D. Nelson, T. H. Ebben, and R. G. Marshalek, "Experimental verification of the pointing error distribution of an optical intersatellite link," in *Selected Papers on Free-Space Laser Communications*, D. L. Begley and B. J. Thompson, eds., SPIE Milestone Series 30, SPIE Press, Bellingham, Washington (1991), pp. 218–228.

19. R. R. Hayes, "Fading statistics for intersatellite optical communication," Appl. Opt. **36**(30), 8063–8086 (1997).

20. D. L. Fried, "Statistics of laser beam fade induced by pointing jitter," Appl. Opt. **12**(2), 422–423 (1973).

21. M. Toyoshima, "Optical link equations and fade statistics in the presence of random pointing-jitter for optical space communications," in Research Signpost publishers review book, Recent Res. Dev. Opt. Eng. **5**, 18–33 (2003).

22. P. J. Titterton, "Power reduction and fluctuations caused by narrow laser beam motion in the far field," Appl. Opt. **12**(2), 423–425 (1973).

23. H. T. Yura, "Optimum truncation of a Gaussian beam in the presence of random jitter," J. Opt. Soc. Am. A **12**(2), 375–379 (1995).

24. M. Toyoshima, T. Jono, K. Nakagawa, and A. Yamamoto, "Optimum divergence angle of a Gaussian beam wave in the presence of random jitter in free-space laser communication systems," J. Opt. Soc. Am. A **19**(3), 567–571 (2002).

25. H. T. Yura and W. G. McKinley, "Optical scintillation statistics for IR ground-to-space laser communication systems," Appl. Opt. **22**(21), 3353–3358 (1983).

26. L. C. Andrews, R. L. Phillips, and P. T. Yu, "Optical scintillations and fade statistics for a satellite-communication system," Appl. Opt. **34**(33), 7742–7751 (1994).

27. L. C. Andrews, R. L. Phillips, and P. T. Yu, "Optical scintillations and fade statistics for a satellite-communication system: errata," Appl. Opt. **36**(24), 6068 (1997).

28. L. C. Andrews, R. L. Phillips, and Cynthia Y. Hopen, *Laser Beam Scintillation with Applications*, SPIE Press, Bellingham, Washington (2001).

29. Y. Suzuki, K. Nakagawa, T. Jono, and A. Yamamoto, "Current status of OICETS laser communication transceiver development," in *Free-Space Laser Communication Technologies IX*, G. S. Mecherle, ed., Proc. SPIE 2990, pp. 31–37 (1997).

30. M. Toyoshima, T. Takahashi, K. Suzuki, S. Kimura, K. Takizawa, T. Kuri, W. Klaus, et al., "Ground-to-satellite laser communication experiments," IEEE Aero. El. Sys. Mag. **23**(8), 10–18 (2008).

3 Pointing, Acquisition, and Tracking

Robert G. Marshalek

CONTENTS

3.1 INTRODUCTION AND CHAPTER OVERVIEW

Laser communications has long been viewed as an attractive means for linking wideband remote sensing and other scientific mission assets, with spatially dispersed field users of the data collected by these missions. System advantages of laser communications have been well-documented over the years, and include wide bandwidth, small antenna and swept volume, low weight and power, secure communications, jam resistance, and low recurring cost. The associated directionality of the transmit beam and narrow receive field of view (FOV) imply precise pointing of the optical line of sight (LOS), and several approaches have been developed that provide the desired pointing, spatial acquisition, fine tracking, and reliable communications functions. Major challenges in this regard include significantly reducing the combination of initial pointing knowledge, local platform base motion, and point-ahead angle that each greatly exceed the narrow optical beamwidth used to close the high-data-rate communications links. Once this is accomplished, the laser terminal must also precisely point the transmit beam and point the receive FOV despite local disturbances that are often more than an order of magnitude larger than either parameter.

This chapter describes the typical elements addressed to develop a low-risk pointing, acquisition, and tracking (PAT) subsystem that effectively establishes and maintains a precision link between two moving near-Earth platforms. Specific topics addressed are:

- *System description*: Describes a typical mission construct that implements optical terminals and the requisite high-precision PAT.
- *PAT overview*: Describes basic elements, functional flow, and interfaces.

- *General PAT requirements*: Describes key items evaluated to develop low-risk PAT concepts. Key design drivers are also identified based on past experience with prior design evolution and performance validation efforts.
- *PAT concept development*: Describes several point-of-departure concepts often evaluated and traded to best meet typical requirements.
- *PAT hardware development and integration*: Describes proven hardware elements that are readily available to support successful low-risk flight implementations today. Integration of these elements, appropriately modified for effective insertion into the overall PAT design construct, is currently providing low-risk demonstrations that validate the performance and maturity level of several technologies important to operational, high availability optical links.
- *PAT performance attributes*: Summarizes typical laser terminal pointing error allocations for major PAT functions, corresponding link budget performances, weight/power contributors, reliability, and relative cost drivers.

3.2 SYSTEM DESCRIPTION AND PAT OVERVIEW

Data relay concepts have been around as long as communications systems. Figure 3.1 shows a representative near-Earth mission scenario that integrates space-to-space, space-to-high-altitude aircraft, and other links to efficiently connect worldwide scientific, military, and intelligence community users within a common data routing thread. A significant portion of this evolving communications architecture assumes the use of laser communications links between constituent platforms owing to the wide bandwidth, low probability of intercept, and anti-jam features mentioned above that result from the high optical carrier frequency.

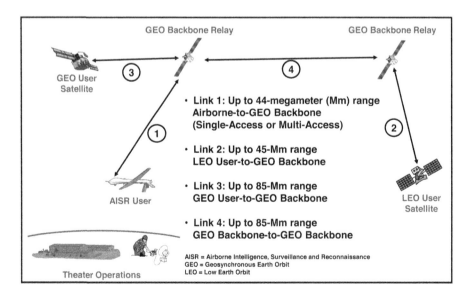

FIGURE 3.1   Several laser link types integrate to effectively route mission data.

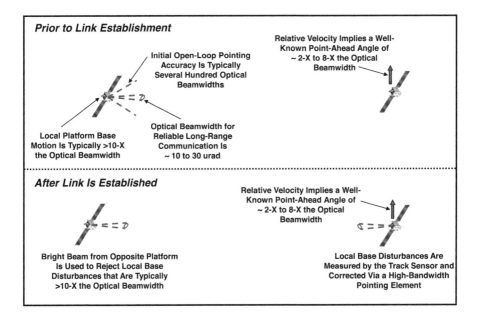

FIGURE 3.2 PAT system overcomes several obstacles to establish and maintain a link between two moving platforms.

To realize the advantages of optical communications, an efficient means of pointing narrow optical beams must be realized. Figure 3.2 illustrates the general PAT problem encountered to establish and maintain an optical link between any two moving platforms within the architectural infrastructure described above. Two major PAT obstacles are overcome. The first relates to the combination of initial pointing knowledge, local platform base motion, and point-ahead angle that each greatly exceed the eventual optical beamwidth used to close the high-data-rate communications links. Beacon light is initially sent between the two terminals to convey their relative angular positions, thus greatly reducing the pointing error. The second obstacle is met once the link is established by completing the functional steps described below, when the terminal must provide precise pointing of the transmit beam and receive FOV, despite local disturbances that are often more than an order of magnitude larger than either parameter. This is accomplished by using the relayed narrow light beam as a stable reference, and efficiently measuring and attenuating the disturbances via an optical feedback control loop.

Implementing an optical link involves an orderly process that reduces the initial pointing error and enables efficient use of the narrow beam. Figure 3.3 shows the top-level functional steps that establish and maintain the link between two platforms. Key elements include initial pointing information on each platform, completion of a beacon search routine, successful linking of the beacon beams between platforms, handoff to the narrow communication beam on each platform, communication data transfer, and periodic pointing error calibration to correct for slow transmit-receive alignment drifts. Each functional step is further described below.

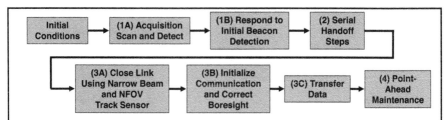

- (1A & 1B) Acquisition: Reduces pointing error from mrad to sub-urad levels
- (2) Handoff: Transfers line-of-sight (LOS) control from wide beam to narrow beam and from WFOV to NFOV optical sensors, to enable accurate pointing with the narrow transmit communications beam
- (3A, 3B, & 3C) Track: Uses high-bandwidth loop involving NFOV sensor and FSM to measure local base disturbances and point receive LOS back at the reference beam from the opposite platform
- (3A, 3B & 3C) Point Ahead: Offsets the transmit LOS to accommodate relative velocity between linked platforms
- (3A, 3B & 3C) Gimbal Offload: Supports wider field of regard by keeping the receive LOS close to the center of the FSM angular range of travel
- (4) Link Maintenance: Periodically corrects alignment drifts between transmit and receive LOS to improve the long-term link efficiency

FIGURE 3.3   Functional flow identifies critical PAT processes and interfaces.

Initial conditions: Each platform first defines the angular location of the opposite platform as determined by the following elements at the time determined to begin the link establishment process:

- Position knowledge typically provided by the host platform via the Global Positioning System (GPS) inputs for each terminal, on the order of ±100 to 250 m (3 sigma), imposing a significant impact only at a close range.
- Orbital trajectory based on ephemeris data, also used to determine the relative velocity between the two platforms (v) and the corresponding point-ahead angle given as $2v/c$, where c is the speed of light.
- Establishment of local pointing coordinate axes relative to a fixed reference, i.e., local attitude knowledge, typically up to several milliradians per axis (3 sigma) at the attach point.
- Local terminal errors, including mechanical and optical alignments, such as telescope boresight alignment, gimbal mounting accuracy, and associated thermal drifts, typically below 1 mrad.

Thus, the third item usually dominates the combined uncertainty cone so derived for the opposite terminal, and the combined errors are typically on the order of several milliradians (i.e., tenths of a degree). By comparison, the optical beamwidth used to close the communications link, approximated by the ratio of the wavelength to the aperture diameter, is circa 10 μrad (i.e., several tenths of a millidegree). Furthermore, since available laser power is smaller than the value required to flood this several-hundred optical beamwidth area and provide a detectable signal at maximum link range, a search routine is needed to begin the link-establishment process.

Acquisition scan and detection: In this step, each terminal uses a light beam to convey its angular position to the opposite terminal similar to a lighthouse at the seashore, or the rotating light atop that police car in your rearview mirror. The beacon beam is often broadened relative to the narrow communications beam and methodically scanned over the uncertainty area, eventually illuminating the opposite aperture. The beacon is sensed by a position-sensitive detector located within the opposite terminal. The most common approach has each terminal staring over the initial error cone using a wide field of view (WFOV) sensor that covers all possible angles of arrival, and uses the received beacon light to calculate the angle of arrival within the FOV. Since the transmit and receive functions are independent and performed simultaneously, many system designs implement a two-way acquisition sequence, whereby each terminal performs the scan and stare functions simultaneously to provide the shortest elapsed time before a beacon is detected by one of the terminals. To also support this desire, an outward raster or spiral scan is commonly implemented that begins at the center of the initial uncertainty cone, as this is the most likely location of the opposite terminal. Finally, some approaches have considered using the narrow communications beam for acquisition to eliminate extra optics and switching between the two divergence modes, but this also implies that the initial scan process is more dependent on the platform base disturbances, and the impact they place upon implementing a scan pattern that is free of intensity dropouts.

Respond to initial beacon detection: Once the beacon is detected by one side of the link, that terminal stops its scan and uses the WFOV sensor to determine the angle of arrival within the initial error cone, despite no longer sensing a receive signal based on the continuing scan by the opposite terminal. A beam-steering element, such as a fast-steering mirror (FSM), then points a steady beacon beam along the calculated angular LOS, often offset by the fixed point-ahead angle, to deliver the highest possible power density back to the opposite terminal. The opposite terminal receives the return beacon, stops its scan, and returns a steady beacon beam back to the partner. This completes the first portion of spatial acquisition process, but the power density relayed between the terminals is still well below the threshold for high-data-rate communications.

Serial handoff steps: The terminals use the steady beacon to cooperatively increase the optical signal delivered between them, and this enables an increase in the WFOV sensor readout rate and corresponding track loop bandwidth on each end that better attenuates local base disturbances to improve the pointing accuracy. As such, each terminal places a larger and larger signal onto the counterpart, first enabling a switch to the narrow communications beam, and then eventually exceeding the threshold for handing LOS control off to the narrow field of view (NFOV) track sensor. The improved pointing accuracy allows each terminal to proceed closer and closer to the peak intensity available from the narrow beam, eventually exceeding the power threshold for completing the handoff to the fine track function on each end, and enabling subsequent establishment of the communications link. An accurate point-ahead correction, as described below, is a critical element in this portion of the process.

Close link using the narrow beam and a NFOV track sensor: A steady narrow beam is now relayed between the two terminals, providing a receive power density at each end that enables NFOV track sensor sampling rates typically ranging from 10 kHz to 50 kHz. The high sampling rate allows effective rejection of platform jitter up through the 250 to 500 Hz range via commands to an FSM located in the common transmit-receive optical path that accurately points the FSM LOS directly back at the incoming reference beam. The 40-X to 100-X total oversampling ratio is consistent with standard controls theory, and represents the combination of two factors:

- A 4-X ratio between the LOS control loop bandwidth and the disturbance rejection bandwidth
- An optical feedback sensor readout rate in the range of 10-X to 25-X higher than the LOS control loop bandwidth for minimal phase margin and gain margin degradation

Residual radial pointing jitters on the order of 10% of the communications beamwidth (3-sigma) provide a receive power vs. time profile that supports reliable communications with an adequate average bit error ratio (BER). This level of performance is also consistent with an acceptable tracking loss in the communications mode (~ 1–2 dB) that couples with a suitably low number of random fades, or short-term BER increases, in a typical data relay mission.

Initialize communication, correct boresight, and transfer data: Communications initialization is completed by relaying a predetermined bit pattern between platforms to synchronize the clocks prior to transferring the high-rate communications data. After establishment of frame synchronization, and periodically throughout a communication event, a cooperative end-to-end power-on-target algorithm corrects the alignment between the laser terminal transmit and receive paths to optimize the link efficiency. This normally consists of a predetermined pointing pattern, with far-field receive power measurements vs. the pointing position provided by the link partner used via a "greatest-of" algorithm to correct the boresight. The end-to-end link maintenance process is completed without degrading the communications BER performance below the specified requirement.

Point-ahead/look-behind maintenance: One additional function is required to establish an effective relay between the two linked terminals that corrects for the relative velocity between the platforms and the finite velocity of light. Based on the above fine track description, the system commands the FSM element to point back at the incoming LOS. However, relative motion, that is almost always present between the two platforms, must be corrected to properly point the outgoing beam. Figure 3.4 illustrates this effect. Light that is sent from the opposite terminal, when it is located in position A, will be received by the local terminal when in position B, and the opposite terminal will receive the return beam when in position C. The transmit LOS must therefore be offset by the angle between positions A and C to properly illuminate the far-field target. This description assumes that the transmit LOS is pointed ahead to correct for this effect. As an aside, some designs perform

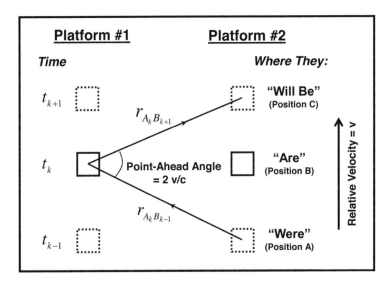

FIGURE 3.4   Point-ahead corrects for the relative LOS motion.

the fixed offset correction within the receive path in a look-behind implementation. Either approach accomplishes the same function. In either case, the process relies on accurate alignment between the transmit and receive LOS vectors vs. time. To provide the most efficient relay between two terminals, the transmit-to-receive alignment is frequently corrected via the link maintenance power-on-target algorithm previously described, and/or via an internal retroreflection device that couples light from the local terminal transmit path back into the local terminal receive path.

3.3   GENERAL PAT REQUIREMENTS

All PAT systems must meet a set of combined performance requirements in a low-risk, cost-efficient manner. Figure 3.5 summarizes the key requirements that typically drive the optical PAT design, together with proven techniques that have been evaluated analytically and demonstrated in risk-reduction hardware via several prior development efforts. In summary, the basic PAT requirements are:

- Acquire the link rapidly with a high probability of success: Circa 1 to 2 minutes with at least 95% success is typical
- Acquire close to the sun to increase acquisition opportunities: Circa a degree or two from the receive optical boresight is typical
- Track the opposite platform with sub-microradian residual LOS jitter to reduce the random link BER degradation (i.e., radial LOS jitter <10% of the transmitted beamwidth is typical)
- Maintain track lock despite a direct solar background to eliminate prolonged outage periods

Function	Driving Requirements	Baseline Design Options
Acquisition	• Wide scan area and FOV • Acquisition time and success probability • Optical background • Acquisition sensor dynamic range • Agile LOS pointing to establish and maintain links with multiple assets (e.g., GEO-GEO backbone link vs. GEO-User links)	• Moderate rate scan of broadened beacon using a spoiled communications beam vs. rapid scan of narrow communications beam • Wide-FOV, sensitive focal plane array sensor vs. quadrant sensor • Interactive scan, detect, and handoff process involving both terminals to be linked • Coarse LOS beam steering over large Az and El field of regard (FOR) to support any relative platform motion during acquisition timeline
Fine Track and Communications	• Beam divergence required to support data rate and range • Residual jitter and bias errors to provide low fade rate • FOV and FOR to link with all mission platforms • Optical background • Track sensor dynamic range	• Quadrant sensor with intensity modulated track beacon vs. nutate receive beam at fiber tip to eliminate need for separate intensity modulation • FOV to achieve high track and comm. SNRs amidst very large optical background: Singlemode vs. multimode receivers • Coarse LOS beam steering over large Az and El FOR to support any relative platform motion during communication engagement: Gimbaled payload, gimbaled flat mirror, coelostat gimbal, or suitable combination
Pointing	• Relative platform velocity • Transmit offset pointing accuracy	• Precise point-ahead mechanism vs. look-behind implementation • Periodic transmit-receive alignment correction
Platform Interfaces	• Attitude knowledge and update rate • Ephemeris accuracy and update rate • Base disturbance Power Spectral Density (PSD) • Host platform pointing stability • Size/Weight/Power accommodations • Thermal environment and control provisions • Radiation shielding for laser terminal elements located below the S/C deck	• Scalable acquisition designs accommodating several-mrad-class search and stare areas • Inertial measurement unit and feedforward LOS control correction prior to establishment of feedback control track loops, as required • Vibration isolation provided by mechanical interface structure • Gimbal FOR accommodates host platform pointing stability • Narrowband commands and telemetry via 1553 bus or equivalent
Other	• Contamination effects vs. mounting location • Conformal window vs. turret for airborne hosts • Boundary layer for airborne hosts • Optical turbulence and scatter for airborne hosts	• Aperture cover during post launch outgassing period and possibly whenever idle • Select mounting location away from thrusters and other contaminants • Adequate end-of-life design margins to support acquisition and track/comm functions amidst lifetime performance degradation effects

FIGURE 3.5   Key PAT requirement drivers are evaluated for the different link scenarios.

- Maintain sub-microradian alignment drift between the transmit and receive optical paths
- Provide the required elevation and azimuth field of regard (FOR)
- Perform all the above despite host platform base disturbances that greatly exceed the transmit beamwidth

Each application balances the design by evaluating and optimizing the items shown in the table to determine the lowest-risk approach that meets performance requirements with the lowest total cost.

As described above, one of the most important PAT interface requirements is the host platform base disturbance profile commonly represented by a power spectral density (PSD) plot of Energy/Hz vs. Frequency. Figure 3.6 shows representative PSD plots for a near-Earth host vehicle, indicating how the PAT design uses several proven techniques to measure and attenuate the base motion, allowing successful on-orbit operations. The right-hand and left-hand portions of Figure 3.6 indicate how these design approaches are used to attenuate typical rotational and translational base disturbances, respectively, taken from different platforms for illustrative purposes. High-frequency base disturbances are the most difficult to measure and attenuate, and the design risk is commonly reduced by attenuating these drivers using a passive isolation system. This reduces the burden placed on the optical bench LOS control mechanisms and loops described below. Although different combinations are commonly used, the low- and mid-frequency PSD content is generally attenuated using a local, hard-mounted inertial measurement unit (IMU) early in the acquisition process, supplemented by higher bandwidth optical feedback control loops as the sequence proceeds closer to the fine track mode.

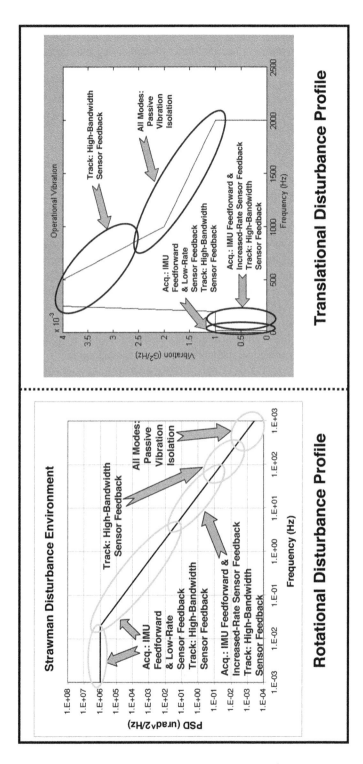

FIGURE 3.6   Standard techniques measure and attenuate host platform disturbances.

3.4 PAT CONCEPT DEVELOPMENT

Effective approaches for pointing narrow optical beams integrate the use of several mechanisms, optical sensors, and mechanical sensors. Figure 3.7 shows a representative laser communications terminal block diagram, and identifies critical PAT elements that combine to meet specific mission requirements. The terminal uses an externally mounted payload module that contains the elements shown in the figure, including (1) a telescope aperture to provide the narrow transmit optical beam and the receiver capture area, (2) an optical bench assembly that contains the transmit and receive routing optics, PAT mechanisms, and PAT sensors, (3) a coarse pointing gimbal assembly that provides a wide FOR to accommodate orbital trajectories between linked platforms, and (4) interface electronics needed within close proximity of the external elements. An internally mounted electronics module interfaces with the external elements, and contains the bulk of the control electronic elements needed to operate the laser terminal PAT sequence. Many laser terminal designs incorporate slight changes to the basic design shown here, but the same PAT elements are generally used. Finally, nonmechanical beam-steering devices have also been investigated for providing coarse LOS pointing. While this approach has the potential to further reduce laser terminal size and weight, its on-orbit use requires development in several areas. These include general technology readiness, supportable aperture, subaperture control required to meet link-driven aperture needs, compact packaging schemes, and such performance issues as optical throughput vs. the steering angle.

Several interrelated control functions are needed to optimally use the above elements in an end-to-end optical link. Figure 3.8 summarizes the general LOS control loops that support these key PAT functions. A point-ahead mirror (PAM) loop, a gimbal loop, and an FSM loop are fed by inputs from a variety of sensors to maintain

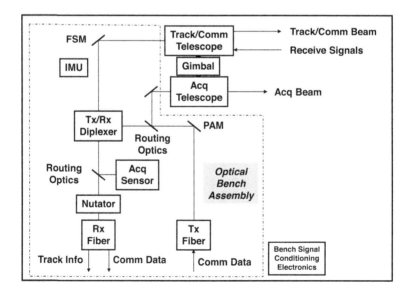

FIGURE 3.7 A generic block diagram defines specific PAT elements.

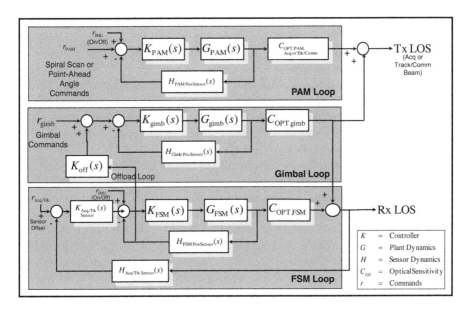

FIGURE 3.8 Interconnected control loops maintain the transmit and receive LOS pointing vectors.

the transmit and receive LOS vectors through different portions of the PAT sequence. The PAM loop first implements the beacon scan function early in the acquisition process, and this allows each terminal in the link to perform transmit and receive functions simultaneously and reduce the mean time required to successfully acquire the link with a corresponding high probability of success. After initial beacon detection and boresight correction, the PAM loop provides the optimum point-ahead angle between platforms moving with a nonzero relative velocity.

The gimbal loop adjusts the coarse terminal optical boresight to accommodate any change in the relative angle between two communicating platforms throughout an engagement. In some instances, this requires a gimbal range of travel that approaches a hemisphere; for example, from a low Earth orbit (LEO) platform to a geostationary Earth orbit (GEO) platform. The gimbal is driven by a low bandwidth control loop that uses the host spacecraft inertial navigation system and the gimbal angle sensor data to remove very-low-frequency base disturbances from the optical LOS. The gimbal loop also serves to offload the internal high-bandwidth FSM via feedback signals from position sensors located on the FSM, keeping the internal LOS close to the center of the limited FSM angular range.

Prior to acquisition and closing of the optical feedback control loops, the PAM and FSM loops sometimes use inputs from an IMU to stabilize the terminal LOS. This is often found beneficial to obtain an acquisition scan pattern without far-field intensity dropouts, as well as to stabilize the receive LOS for more efficient weak beacon signal detection. Once a beacon is detected by the WFOV acquisition sensor, centroiding algorithms determine the angle of arrival, and the FSM loop is closed about this data to point the common optical path directly back at the incoming signal. The FSM-WFOV sensor loop continues until the LOS accuracy is reduced to enable handoff to the fine track mode. The FSM commands are then switched to the NFOV track sensor (nutator in this example; alternatives are further described below) with a closed loop bandwidth near 1 kHz, to provide very effective LOS jitter rejection of a very large portion of the host disturbances. Even with these features, base disturbances near several hundred hertz remain as residual jitter on the optical LOS, and these residual errors are only a few percent of the optical beamwidth with common host platform and vibration isolator designs.

As mentioned above, several interrelated LOS control loops combine to meet all acquisition, tracking, and communications requirements. Figure 3.9 summarizes the design concept development process used to evaluate the basic PAT requirements summarized in the previous section, and define a balanced approach that effectively meets direct and allocated performance metrics. As shown in the figure, each PAT functional mode is evaluated to determine the best use of several LOS pointing approaches that provide a low-risk integrated concept. Standard techniques, such as vibration isolation, feedforward LOS control, and feedback LOS control, are traded to define an integrated design that efficiently supports all requirements identified earlier. The process also balances the performance challenges, and corresponding development risk, across such elements as the acquisition and track sensors, pointing and steering elements, and LOS control algorithms. Figure 3.10 illustrates how these

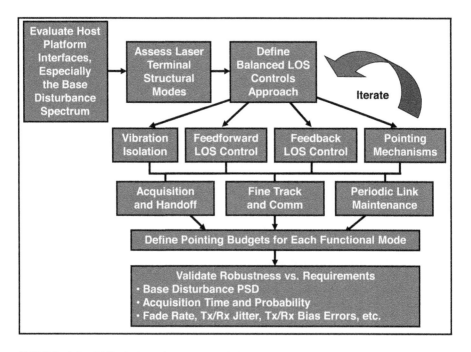

FIGURE 3.9 PAT design development balances key features to best meet specific requirements.

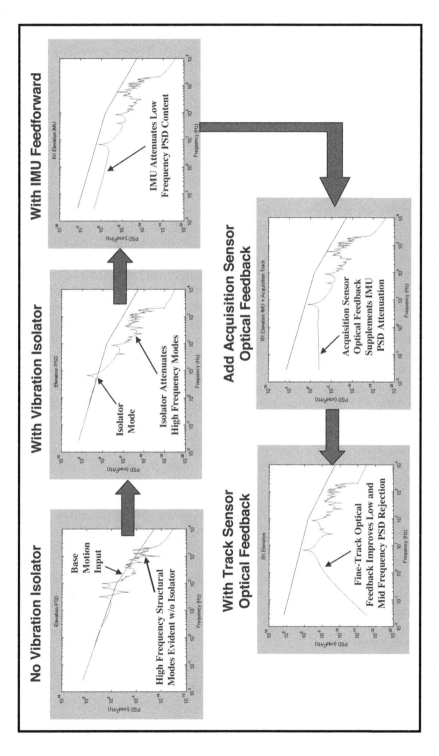

FIGURE 3.10 LOS control approaches measure and attenuate base disturbances to enable an effective PAT sequence.

techniques combine to effectively attenuate low-, mid- and high-frequency base disturbances throughout the end-to-end PAT sequence. This figure also indicates the trade-off between low-frequency disturbances (<20 Hz) that commonly dominate the acquisition process and high-frequency disturbances (>100 Hz) that typically dominate the fine track accuracy and corresponding communications BER performance vs. time (i.e., the random fade statistics and tracking loss in the communications mode briefly addressed earlier).

An example is the best way to clearly show how the key contributors are analyzed to define an effective PAT design. Figure 3.11 provides an example simulation indicating a specific application of the general flow diagram shown in Figure 3.9. The host platform PSD profile is applied to the structural finite element model of the optomechanical laser terminal assembly to determine the impact of the structural modes on the base motion content vs. frequency. A detailed LOS control approach is defined for the acquisition, handoff, and track modes to provide a seamless interface between these PAT functions. Since the acquisition and track modes almost always include an optical sensor to measure and attenuate the local base motion, some sensor noise is added to the LOS residual jitter from the closed-loop controls system. As such, the root sum square (RSS) of the residual errors from the feedback system and the filtered sensor noise spectrum provides the total LOS jitter on the incoming and outgoing optical beams. The designs are iterated in a terminal-to-terminal performance analysis to optimize the parameter selections that meet the acquisition time, acquisition probability, handoff probability, and long-term track jitter requirements.

Any interdisciplinary system design balances several critical trade-off evaluations. Figure 3.12 summarizes the series of interrelated design trades that are always completed to specifically define the laser terminal items that integrate into the PAT LOS controls approach developed using the above process. Iterative performance evaluations and design modifications are routinely completed to achieve the desired balanced approach. Critical PAT-specific design trades focus on the following items:

- *Optomechanical assembly*: Compares the advantages and disadvantages of a Gimbaled Telescope and Optical Bench, a Gimbaled Telescope with Coudè Path to Fixed Optical Bench, and a Gimbaled Flat Mirror to Fixed Telescope and Optical Bench
- *Telescope*: Compares the advantages of obscured, off-axis and refractive designs
- *Materials selection*: Evaluates the combined PAT effects of material choices for the telescope, gimbal, optical bench substrate, and other structural elements
- *Acquisition beacon and sensor*: Compares the advantages and disadvantages of a Scanned Broadened Beam, a Scanned Narrow Communication Beam, and a Separate Pulsed Beacon Source combined with receiver implementations that utilize a Focal Plane Array (FPA) or a Quadrant Sensor

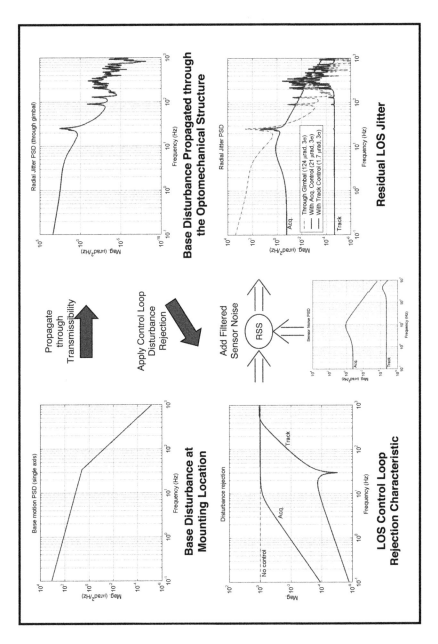

FIGURE 3.11 Simulations evaluate base motion rejection and closed-loop pointing performance.

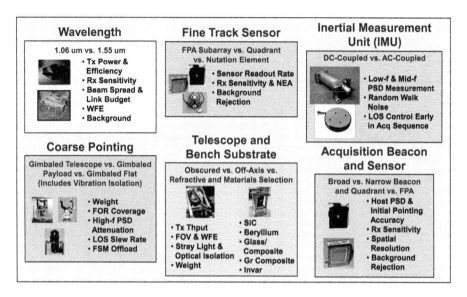

FIGURE 3.12   Key laser terminal trades optimize the design.

- *Fine track sensor*: Compares the advantages and disadvantages of a FPA Subarray, a Quadrant Sensor, and a Nutation Element
- IMU *Sensor*: Compares the advantages and disadvantages of a DC-Coupled (e.g., Fiber-Optic Gyro) vs. an AC-Coupled (e.g., a Magnetohydrodynamic or MHD device), clearly influenced by the energy content vs. frequency contained in the host platform PSD

The general nature of orbiting near-Earth platforms demands coarse LOS pointing over a large FOR. Figure 3.13 indicates several of the key considerations for the optomechanical assembly that provides this function, by comparing the advantages and disadvantages of the three most common approaches listed above. Major items are the FOR, vibration isolation from the host spacecraft PSD, across-gimbal interfaces, optical interfaces between the telescope and optical bench, and induced torque back to the host platform. The relative importance of these interrelated items is very often application specific, and the design balancing process is adjusted accordingly. No single approach always emerges as the clear-cut winner.

Proper materials selection is necessary for many key elements of a balanced, low-risk PAT design. Figure 3.14 indicates several of the key considerations for materials selection within key laser terminal assemblies. Major items such as weight, thermal properties, stiffness, strength, manufacturing, and cost are assessed for each assembly shown, and then the integrated PAT performance is critically evaluated. Iterative design modifications are completed at the assembly

Pointing Approach	Pro	Con
Gimbaled Telescope and Optical Bench 	• Near-hemispherical angular coverage • Stabilization between laser terminal and platform base motion • Smaller laser terminal envelope • Fewer optical components • Short telescope-to-optical bench separation eases CTE mismatch effects, alignment, and accommodates acquisition FOV • Straightforward telescope baffling	• Larger gimbaled mass increases payload mass and power (bearing size, motor size, etc.) • Data, electrical, power, and some thermal transfer across gimbal, requiring many long cables • Larger ratio of swept volume to stationary volume • Higher pointing-induced momentum and torque
Gimbaled Telescope with Coude' Optical Path to Fixed Optical Bench 	• Near-hemispherical angular coverage • Lighter gimbaled mass • No data transfer across gimbal • Limited thermal, electrical, and power transfer across gimbal • Straightforward telescope baffling	• Coude' fold mirrors decrease throughput and increase complexity • Polarization control for high throughput has proven difficult • Separated telescope and optical bench introduces CTE mismatch difficulties and complicates the acquisition FOV • Initial assembly and alignment are more difficult • Base motion directly imparted to optical bench assembly, typically demanding a separate vibration isolation system • Moderate pointing-induced momentum and torque

Gimbaled Flat Mirror to Fixed Telescope and Optical Bench 	• Lightest gimbaled mass
	• No data transfer across gimbal
	• Limited thermal, electrical, and power transfer across gimbal
	• Short telescope-to-optical bench separation eases CTE mismatch effects, alignment, and accommodates acquisition FOV
		• Less-than-hemispherical angular coverage
		• Favors aperture below 8-inch diameter
		  - Additional large optic and associated thermal control
		  - Requires large envelope to separate gimbaled flat and telescope
		• More-difficult telescope baffling, usually adds structure and weight
		• Base motion directly imparted to optical bench assembly, typically demanding a separate vibration isolation system
		• Moderate pointing-induced momentum and torque

FIGURE 3.13 Optomechanical design trades evaluate FOR, across-gimbal interfaces, mass, and induced torque.

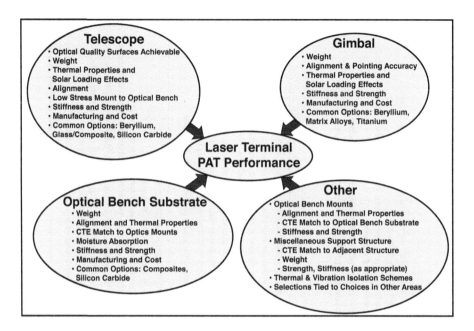

FIGURE 3.14 Materials trades evaluate weight, stiffness, thermal properties, alignment, and associated impacts to integrated PAT performance.

and terminal levels to define the final concept. A key element of the optomechanical evaluations considers the physical properties of candidate materials for the telescope, gimbal, optical bench, and support structure. Figure 3.15 summarizes several key parameter values for common materials considered in the design trades, including weight, stiffness, and thermal performance metrics. The best

Characteristic	Aluminum	Beryllium	Composite M55J-954	Silicon	Silicon Carbide	ULE	Zerodur
E (GPa) Modulus of Elasticity	68	303	50 (z axis) 110 (x,y axes)	131	331	67	90
ρ (g/cc) Density	2.7	1.85	1.62	2.33	2.7	2.2	2.5
E/ρ Specific Stiffness	25	164	31 (z axis) 68 (x,y axes)	56	123	30.5	36
κ (W/mK) Thermal Conductivity	237	210	20 (longitudinal) 0.4 (transverse)	156	165	1.3	1.6
α (ppm/K) Thermal Expansion	11	11.3	+/- 0.1	2.6	2.0	0.03	0.02
κ/α Thermal Stability (Steady State)	21.5	18.6	40-200	60	82.5	43	80

FIGURE 3.15 Mechanical and thermal properties influence optimum design choices.

performers in each category (row) are shaded. It is clear from this comparison why many space payloads use beryllium even with its well-documented manufacturing difficulties. The low thermal expansion offered by Ultra Low Expansion (ULE™) and Zerodur glasses offer clear advantages for a telescope assembly, but their selection implies mass and stiffness penalties compared to some alternatives. Silicon carbide has long been considered as a primary choice for telescope assemblies, but it still suffers from lack of operational time on-orbit, although that perception is easing over time.

Specific acquisition design trades compare approaches for reducing initial milliradian-class LOS errors to levels acceptable for effective handoff to the fine track elements. Figure 3.16 indicates the advantages and disadvantages of several common acquisition design options, including key considerations for the interrelated transmit and receive elements. Major items are component count and corresponding complexity, the far-field uncertainty cone coverage amidst local base motion, detection sensitivity and background discrimination over the large acquisition FOV, sensor angular resolution, beacon modulation requirements, and transmit/receive optical interfaces. As discussed above, application-specific conditions commonly determine the best solution.

Fine track design trades similarly compare approaches for maintaining the residual LOS error at a small fraction of the transmit beamwidth for effective communications. Figure 3.17 indicates the advantages and disadvantages of several common fine track design options, again including key considerations for the interrelated transmit and receive elements. Major items are the sensor readout rate, detection sensitivity and background discrimination, beacon modulation requirements, LOS transfer error, and corresponding alignment calibration.

An additional significant element of the PAT design optimization process and subsequent performance validation via demonstrations uses well-established modeling and simulation tools. Key examples include Simulink/MATLAB for LOS controls, CODE V and ZEMAX for optical performance, NASTRAN for finite-element models, and various thermal analysis programs (SINDA, etc.). These constituent analysis elements are often integrated into an end-to-end performance model, commonly supplemented by inputs from detailed excel spreadsheets, which allows for efficient completion of design trades and to support model validation of performance predictions with measured performance data. The measured data are commonly used to upgrade the performance model, as appropriate, and reduce development risk for operational systems. This process develops a well-founded design with low development risk consistent with (1) successful flight implementation, (2) robust operations amidst a range of host platform disturbances, and (3) flexible LOS controls design commensurate with evolving requirements in all areas to maintain robust spatial acquisition and a low probability of pointing-induced communication link outages.

Several additional design trades are clearly part of defining an integrated laser terminal concept, and these are beyond the scope of this brief overview (but are nonetheless important). Obvious examples are the optical trades that define

Acquisition Approach	Transmit Configuration	Pro	Con
Raster- or Spiral-Scanned Broad Beacon with Imaging Array Detector	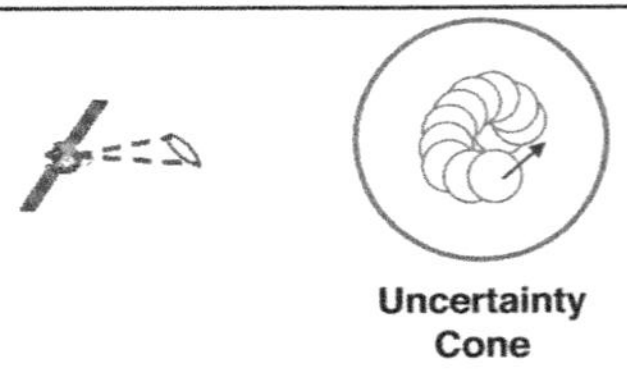 Uncertainty Cone	• Scanned broadened beam greatly increases probability of painting initial uncertainty cone without intensity dropouts • Acquisition shares comm laser • High detector spatial resolution pinpoints beacon angle of arrival once detected, and eases wide FOV and f/# • Narrow pixel IFOV reduces detected background from extended sources, providing excellent sensitivity of unmodulated beacon signals	• Inertial measurement unit commonly needed to measure and suppress low frequency base disturbances during scan period • Defocusing optics or rapidly scanned narrow beam needed to share comm laser for acquisition • More-complex sensor processing and lower full-frame readout rate than quadrant option • Higher sensor non-recurring cost and flight development
Raster- or Spiral-Scanned Narrow Beacon with Quadrant Detector	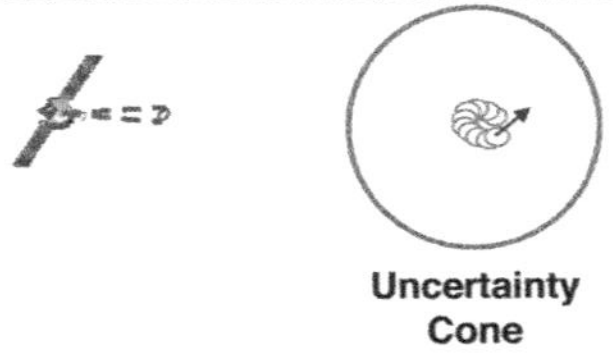 Uncertainty Cone	• Acquisition shares comm laser • Eliminates defocusing optics • Rapid scan reduces need for inertial sensor • Rapidly scanned beam is detected as a pulse train by the quadrant sensor, providing good background discrimination • Less expensive sensor	• More difficult to eliminate intensity dropouts in scan pattern • Rapid scan stresses point-ahead mirror bandwidth and thermal load • Spatial information limited to quadrant data, requiring an iterative receiver pointing error reduction scheme based on multiple pulses • Linear performance only near quad center • Wide FOV and f/# more difficult than array
Pulsed, High-Power Beacon with Quadrant Detector	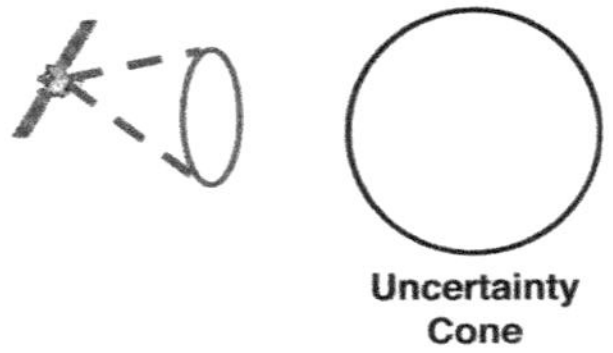 Uncertainty Cone	• High beacon power per pulse eases detection and/or scan velocity • Receive pulse train provides good background discrimination	• Separate acquisition laser source • Spatial information limited to quadrant data, requiring an iterative receiver pointing error reduction scheme based on multiple pulses • Wide FOV and f/# more difficult than array • Linear performance only near quad center • Longest range links require some scanning

FIGURE 3.16   Acquisition Tx scan and Rx sensor trades evaluate uncertainty coverage, background discrimination, and wide-FOV optical interfaces.

Tracking Approach	Receive Configuration	Pro	Con
Sense Portion of Rx Comm Beam with Acquisition Sensor Subarray	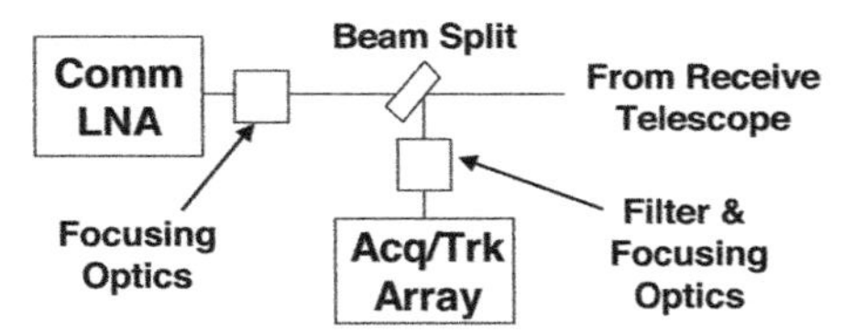	• Fine track shares acquisition sensor • Narrow pixel IFOV reduces detected background from extended sources, providing excellent sensitivity with unmodulated signals	• Power split to support fine track impacts comm link • Sub-pixel resolution limit often exceeds residual fine track dynamic error allocation • Limited subarray readout rate • Pixel saturation with large backgrounds • Non-zero transfer error from track sensor to comm receive fiber, requiring periodic alignment calibration
Sense Portion of Rx Comm Beam with Quadrant Sensor	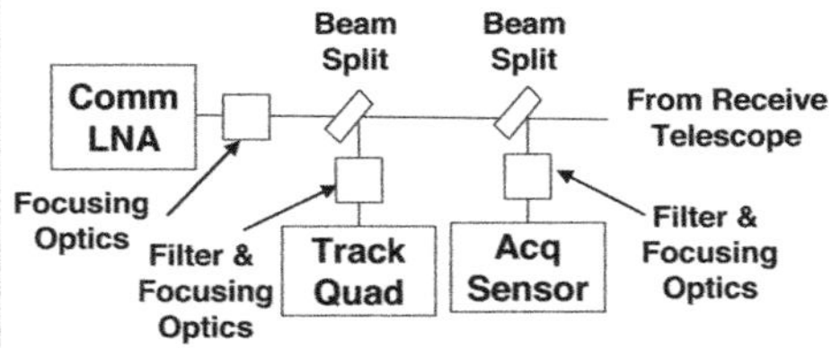	• Readout rate limited only by signal split allowed from comm channel • Simple sensor • Simpler track receiver pointing and handoff	• Intensity modulated beam required to suppress background • Power split to support fine track impacts comm link • Non-zero transfer error from track sensor to comm receive fiber, requiring periodic alignment calibration
Nutate Receive LOS on Fiber and Sense with Post-LNA PIN Photodiode	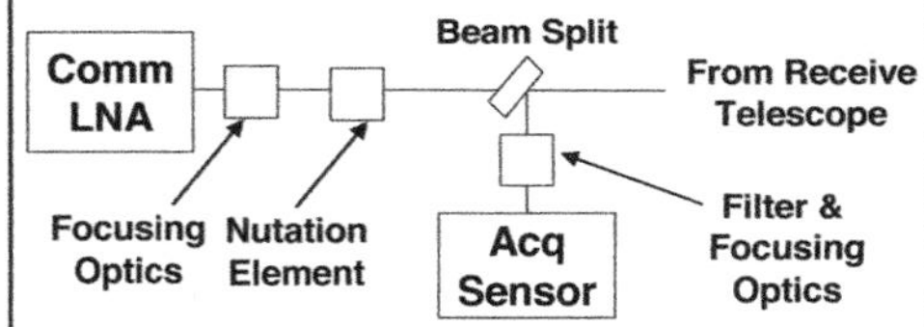	• Directly corrects receive fiber LOS error • No power split for track • Single-mode fiber spatially filters receive background • Receiver amplification boosts off-axis signals	• Adequate nutation frequency more difficult to achieve, but within the state-of-the-art • More-difficult track handoff, but proven reliable in laboratory and field demonstrations

FIGURE 3.17 Fine track sensor trades evaluate readout rate, background discrimination, and residual LOS pointing error.

the telescope type, transmit/receive diplexing method, optical isolation performance, and other similar elements.

3.5 PAT HARDWARE DEVELOPMENT AND INTEGRATION

A significant element of the interactive design process requires careful evaluation of the development risk associated with each and every PAT element. Figure 3.18 evaluates candidate design approaches for critical PAT functions, summarizes the current capabilities in each area, and provides a top-level assessment of ongoing efforts that are qualifying these items for flight implementation. At the mid-2018 publication time of this article, these efforts were driving toward a Technology Readiness Level (TRL) that demonstrates integrated terminal performance in a relevant environment (i.e., TRL = 6 using the NASA standard) for specific data-relay system implementations.

Following careful evaluations of these components and subassemblies, the integrated design trades described in the previous section evolve toward a laser terminal hardware concept. Figure 3.19 shows an example laser terminal originally developed for a LEO-to-LEO mission. The optical head uses a 12.7-cm (5-in) transmit/receive fixed telescope in series with a gimbaled flat mirror to provide the mission-derived azimuth and elevation FORs of ±180 deg and 30 deg total, respectively. The terminal supports up to 2500 Mbps per wavelength at a 6000-km range with a 10^{-8} BER, expandable up to 4-X higher rates using wavelength multiplexing through the same optical elements. The design uses a 0.5-W, single-channel, non-return-to-zero (NRZ), on-off keying (OOK) erbium-doped fiber amplifier (EDFA) transmitter, and a 1-W Indium Gallium Arsenide (InGaAs) acquisition beacon source. Other key PAT elements include an InGaAs octal cell acquisition sensor (a close cousin to a quadrant sensor) with a 0.4-deg FOV, an InGaAs octal cell track sensor with a 300-μrad FOV, a kHz-class, common-path FSM, and an accurate point-ahead mechanism. The optical head shown in the upper right-hand portion of Figure 3.19 is approximately 25 cm (W) × 66 cm (H) × 30 cm (D). The host spacecraft attach point is located approximately 28 cm from the top of the optical module, so the remaining hardware extends about 38 cm below the spacecraft deck. The internal control electronics module shown in the lower right-hand portion of Figure 3.19 is estimated at approximately 30 cm (W) × 25 cm (H) × 30 cm (D) in its final packaging configuration.

Based on the developments that led to the prototype hardware shown in Figure 3.19, as well as other very successful space demonstration efforts, such as the European Semiconductor Intersatellite Link Experiment (SILEX) and Alphasat programs, and the US Geosynchronous Lightweight Technology Experiment (GeoLITE) and Lunar Atmosphere Dust and Environment Explorer (LADEE) programs, laser communications has certainly reached a TRL that makes it a viable solution to high-data-rate space communications missions. Future developments for evolving communications architectures will build upon these successful activities to optimize the design for specific mission needs, including extensive use of proven components and subsystems, and performance testing to specific interfaces and environments.

Component	Function	Description	Features	Hardware	Status
FPA/ROIC	Acquisition Sensor Option	• InGaAs array sensor with capacitive transimpedance readout circuit • FLIR Santa Barbara, Sensors Unlimited, or Princeton Lightwave units are off-the-shelf • 640x512 or 320x256 formats typical • Compatible with subµrad-class-NEA operations to support acquisition and handoff pointing budgets	• Integration time, readout rate, sensitivity and sub-windowing functions support multiple acquisition stages • Small IFOV suppresses background, enabling detection of an unmodulated beacon		• Prototype hardware performance validated • Qualification and reliability development in process for flight applications
Electro-Optic Nutator Crystal or Nutation Mirror	Fine Track Sensor Option	• EO crystal or mechanical resonance mirror • Provides multi-kHz-class beam modulation at receive fiber tip • Synchronous demodulation measures and corrects fiber error using common-path FSM • Local position control system monitors small nutation angle (approx 10% of the beamwidth) • Compatible with subµrad-class-NEA operations to support residual LOS jitter budgets	• Multi-kHz nutation frequency supports LOS controls jitter rejection • Small nutation angle meets fade rate allocations • Fiber spatial filtering enables high track SNR even with a direct solar optical background	 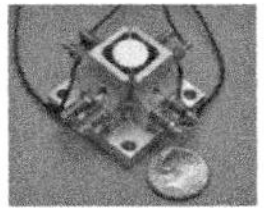	• Prototype hardware performance validated • Qualification and reliability development in process for flight applications
Quadrant Sensor	Acquisition or Fine Track Sensor Option	• InGaAs sensor available from several suppliers (EOS, UDT, etc.) • Compatible with subµrad-class-NEA operations to support residual LOS jitter budgets • Trades simplicity against better FPA angular resolution and sensitivity	• Greater than multi-kHz readout rate supports LOS controls jitter rejection • Intensity modulated beacon enables high track SNR even with a direct solar optical background		• Prototype hardware performance validated • Qualification and reliability development in process for flight applications

FIGURE 3.18　Proven hardware elements support integrated PAT concepts.

Component	Function	Description	Features	Hardware	Status
IMU (AC-Coupled)	Local Base Motion Sensor Option	• Two-axis angular rate sensor used in feedforward configuration with common-path FSM • Applied Technology Associates (ATA) magneto-hydrodynamic device (MHD) available off-the-shelf • 1 Hz – 800 Hz bandwidth typical • Some designs consider blending with S/C IRU output • Some designs implement an inertially stabilized platform configuration. An example is the magneto-hydrodynamic inertial reference unit (MIRU) flown on the NASA LADEE Moon-to-Earth demonstration.	• Off-the-shelf angular rate noise, bandwidth and latency support acquisition and handoff functions • Ultra-light and small package easily integrated onto optical bench assembly	 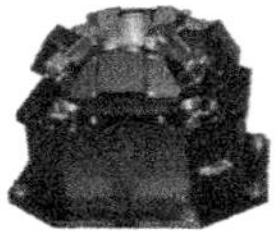	• Prototype hardware performance validated • Some units flown in successful demonstration systems • Qualification and reliability development in process for flight applications
IMU (DC-Coupled)	Local Base Motion Sensor Option	• Two-axis angular rate sensor used in feedforward configuration with common-path FSM • DC to 500 Hz bandwidth typical • Fiber-optic gyro available from several suppliers [KVH Industries, NG Navigation Systems (former Fibersense), etc.]	• Off-the-shelf angular rate noise, bandwidth and latency support acquisition and handoff functions • Small and lightweight package easily integrated onto optical bench assembly		• Prototype hardware performance validated • Qualification and reliability development in process for flight applications

FIGURE 3.18 (*Continued*)

Component	Function	Description	Features	Hardware	Status
FSM	LOS Jitter Correction	• Two-axis common path steering element critical to effective LOS jitter suppression • Works with track sensor to provide sub-µrad-class residual LOS jitter	• >1-kHz bandwidth • 2-deg steering range • DIT position sensors support gimbal off-load loop	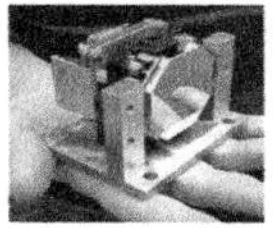	• Heritage FSM mechanisms are flight proven • Additional qualification and reliability development in process for specific mission applications
PAM	Transmit Point Ahead and Beam Scanning	• Provides sub-µrad pointing accuracy in point-ahead mode • Supports beam scan or dither for other operating modes • Same basic mechanism as FSM	• < 100 nrad resolution over typical required angular range • Several hundred Hz bandwidth provides effective transfer of receive LOS to transmit	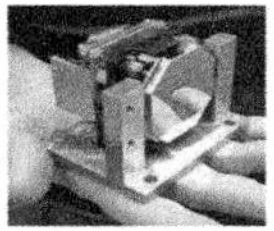	• Heritage PAM mechanisms are flight proven • Additional qualification and reliability development in process for specific mission applications
Mechanical Isolator	Passive Vibration Isolation	• Attenuates high frequency base motion PSD content, reducing burden on optical control loops • Uses passive mechanical struts or other interfaces	• ~10-Hz corner frequency • Acceptably low amplification of low frequency structural modes		• Basic concept derived from flight proven hardware
Two-Axis Gimbal	Coarse LOS Pointing	• Provides large angular dynamic range in azimuth and elevation axes • Gimbaled payload, gimbaled telescope, and gimbaled flat mirror(s) evaluated for each specific application	• Close to hemispherical FOR possible, as required • Multi deg/sec LOS slew rate • Up to 100 Hz class bandwidth possible, as desired (lower typical)		• Several heritage units are flight proven • Design modifications are common to optimally meet mission needs

FIGURE 3.18 (*Continued*)

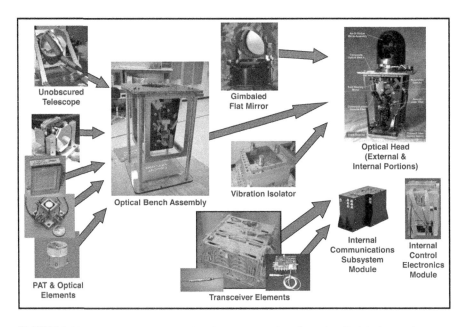

FIGURE 3.19   Integrated laser terminal hardware evolves from detailed design trades.

3.6   PAT PERFORMANCE ATTRIBUTES

This section provides top-level overviews of several performance attributes that quantify integrated laser terminal PAT performance, as well as allocations to the key components and subassemblies described earlier.

Pointing budget and timeline/probability allocations: These performance metrics define the orderly approach used to reduce the LOS pointing error and enable successful full-time communications operations with a low BER, including the corresponding durations and success probabilities for each PAT step. Figure 3.20 provides an example of the pointing error allocations for the key Acquisition and Fine Track modes, indicating how individual bias errors and jitter values are flowed down to PAT elements. It is often instructive to convert angular errors into actual physical distances by way of an example. Consider a 30-cm system with a transmit beamwidth near 10 µrad. The sum of the RSS static and RSS dynamic terminal error contributors then corresponds to 45 µrad, 2.5 µrad, and 0.75 µrad in the acquisition detection (just after beacon light has been detected), handoff, and fine track modes, respectively. For a typical 42,000-km GEO-GEO link, the corresponding physical distances are approximately 1900 m, 100 m, and 30 m for the same three operating modes. If we consider a shorter link distance, say, equivalent to the distance from New York to Los Angeles, or approximately 4000 km (2500 miles), the three distances scale to about 180 m, 10 m, and 3 m, respectively. In such a case, beacon detection corresponds to selection between two buildings in Lower Manhattan from Sunset Boulevard, handoff selects a floor of the building, and fine track picks out an office window into which to send the center of the laser beam.

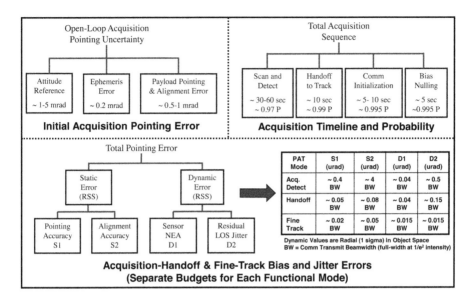

FIGURE 3.20 LOS analyses define detailed parameter allocations and component-level requirements.

Link budget performance: This important assessment characterizes the end-to-end factors that balance the relay of optical power between two separated laser terminals and thereby support successful completion of each PAT mode. Figure 3.21 summarizes typical link budget allocations for the Acquisition Detection, Acquisition Handoff, Fine Track, and Communications modes in a long-range GEO-to-GEO

Parameter	Acquisition Detection	Acquisition Handoff	Tracking	Communications
Average Laser Power (dBW)	+7.0 (5000 mW)	+7.0 (5000 mW)	+7.0 (5000 mW)	+7.0 (5000 mW)
Transmit Losses (dB)	-2.5	-3.0	-3.0	-3.0
Transmitter Gain $[\pi D/\lambda]^2$ (dBi)	+89.7 (1.5 cm)	+112.3 (20.3 cm)	+112.3 (20.3 cm)	+112.3 (20.3 cm)
Defocusing/Truncation Loss (dB)	-0.9 (200 urad)	-0.9 (14.3 urad)	-0.9 (14.3 urad)	-0.9 (14.3 urad)
Pointing and Tracking Loss (dB)	-4.5	-9.0	-3.0	-1.5
Beam Spread Loss $[1/(4\pi R^2)]$ (dB/m^2)	-169.6 (85 Mm)	-169.6 (85 Mm)	-169.6 (85 Mm)	-169.6 (85 Mm)
Receive Aperture Area $[\pi D^2/4]$ (dB-m^2)	-14.9 (20.3 cm)	-14.9 (20.3 cm)	-14.9 (20.3 cm)	-14.9 (20.3 cm)
Receive Optics Losses (dB)	-10.0	-10.0	-3.5	-3.5
Receive Coupling Loss (dB)	-0.0	-0.0	-1.8	-1.8
Received Signal (dBW)	-105.7	-88.1	-77.4	-75.9
Required Signal (dBW)	-110.0 (10 pW)	-93.0 (500 pW)	-83.0 (5 nW)	-82.2 (6.1 nW)
End-of-Life Link Margin (dB)	+4.3	+4.9	+5.6	+6.3

(85,000-km Separation, 1550-nm Wavelength, 8-inch Transmit/Receive Telescope, 2.5-Gbps Data Rate, DPSK Modulation, 10^{-9} BER, 60 photons/bit with 5-dB Coding)

FIGURE 3.21 Crosslink performance summary (2.5-Gbps data at 85-Mm range with 5-dB coding).

crosslink. Specific entries are the result of the design balancing process described earlier that trades many different implementations to meet integrated PAT performance needs. The calculation uses standard link equations, portions of which are provided in the table, and assumes required signal levels for each function that are derived for the specific terminal configuration and function. Key metrics include the probability of detection and probability of false alarm for acquisition, the sensor noise contribution to the residual LOS jitter for fine track, and the BER for communications. A few words regarding the pointing and tracking loss assumed for each operating mode are instructive:

- The −4.5 dB loss for Acquisition Detection assumes beacon reception at the 1/e point of the scanned beam (rounded up to the nearest half dB)
- The −9.0 dB loss for Acquisition Handoff assumes the narrow transmit beam and increased mispointing at that stage of the process
- The −3.0 dB loss for Tracking enables the fine track loop to maintain lock even when the communication BER increases during a fade event
- The −1.5 dB loss for Communications is consistent with the combined transmit and receive coupling losses given the fractional static and dynamic (3-sigma) pointing errors described above

In this example, the acquisition links exhibit a lower residual link margin than the tracking and communications links, a result that is clearly dependent on the performance parameters and design choices assumed for the assessment. Alternative choices could alter this conclusion.

Reliability allocations: This performance assessment captures the contributors to the predicted laser terminal lifetime and the corresponding probability of mission success. Figure 3.22 summarizes typical top-level probability allocations for critical PAT elements that support a 10-year operational lifetime at a full-duty cycle. The figure also indicates redundant elements assumed for the preliminary assessments that use a standard exponential model with constant failure rates:

$$R(t) = \exp[(-(\lambda \times 10^{-9} \times T)],$$

where

λ = failure rate in FITs (failures in 10^9 hours)
T = operating time in hours = $8760 \times Z \times Y$
Z = operating duty cycle
Y = lifetime in years

The combined PAT reliability in this example is approximately 0.95 for 10 years or 585 equivalent FITs. These initial allocations are combined with similar estimates for the communications transceiver to determine the laser terminal-level reliability, and then iterated to achieve a terminal design that balances the reliability and redundancy with the corresponding weight, development risk, and cost.

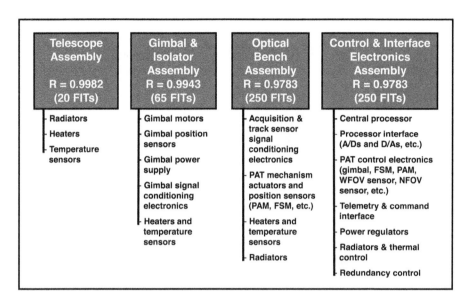

FIGURE 3.22 PAT reliability allocations are iterated to determine redundant elements (assumes a 10-year operation with a 100% duty cycle).

Weight and power: These performance metrics capture the relative contributors to the combined laser terminal accommodations that must be provided by the host platform. Figure 3.23 summarizes weight and power estimates by percentage for the elements that combine to provide the laser terminal PAT functions, but excluding the communications elements. Redundant elements identified in Figure 3.22 are

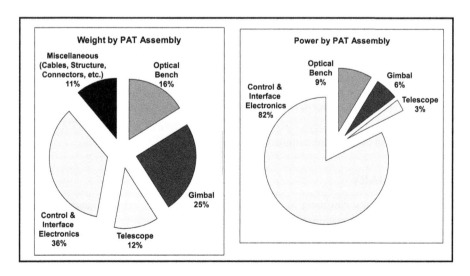

FIGURE 3.23 PAT weight and power contributors by key assemblies.

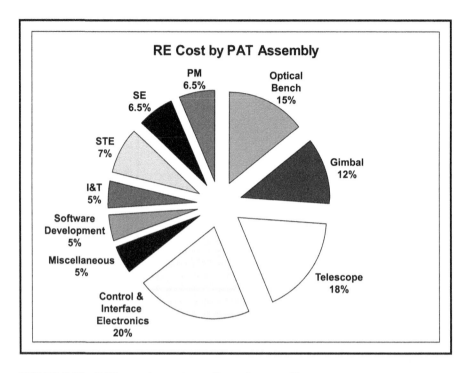

FIGURE 3.24   PAT recurring cost contributors by assembly.

included in the estimates. As discussed above, the weight and power of each element depend on the performance parameters and design choices assumed for the assessment. The values shown in this example represent actual results for prototype PAT hardware capable of supporting a multi-Gbps link over a 2xGEO range (i.e., 85,000 km).

Recurring cost: This assessment accounts for the relative cost contributors for developing the PAT portion of an operational laser terminal. Figure 3.24 summarizes recurring cost estimates by percentage for the PAT elements, again excluding the communications elements. As discussed above, percentages contributed from each element depend on the performance parameters and design choices assumed for the assessment. These values also represent actual results for prototype PAT hardware capable of supporting a multi-Gbps link over a 2xGEO range.

3.7   SUMMARY

Today's technology can meet or exceed all the typical PAT requirements imposed by near-Earth laser communications systems. What's missing is a collection of long-term performance data to support precise reliability assessments, well-defined redundancy requirements, and associated weight implications. The earliest terminals will, therefore, likely incorporate extra redundant elements to ensure long-life mission requirements are not compromised, and the future offers the promise for even more efficient designs once the lifetime drivers are clearly defined from this performance data over the mission life. This normal evolution will surely reduce recurring engineering cost as well. Even with this, proven, low-risk PAT system designs exist today that are robust and adaptable to meet defined mission applications and support reliable high-bandwidth flight optical links. The community brings more than 40 years of development experience in precision pointing and tracking systems for laser communications links, ensuring mission success. These developments have also defined several lessons learned in the areas of design for manufacturability and assembly (DFMA) and design-to-cost (DTC) objectives paramount to future system cost effectiveness.

ACKNOWLEDGMENTS

The author wishes to thank many former and current colleagues at Ball Aerospace, including Bob Baisley, Dave Begley (since deceased), Jeff Chu, Homero Gutierrez, Steve Harford, Jino Jo, Bob Kaliski, Gerhard Koepf, Mike Lieber, Roy Nelson, and RJ Smith.

ACRONYMS

Acq	Acquisition
A/D	Analog-to-Digital Converter
AISR	Airborne Intelligence, Surveillance, and Reconnaissance
ATA	Applied Technology Associates
Az	Azimuth
BER	Bit Error Ratio
BW	Beamwidth
Comm	Communications
CTE	Coefficient of Thermal Expansion
D/A	Digital-to-Analog Converter
DFMA	Design for Manufacturing and Assembly
DIT	Differential Impedance Transducer
DPSK	Differential Phase Shift Keying
DTC	Design to Cost
El	Elevation
EO	Electro Optic
EOS	Electro-Optical Systems, Inc.
FIT	Failure in 10^9 hours
FOR	Field of Regard
FOV	Field of View
FPA	Focal Plane Array
FSM	Fast-Steering Mirror
GEO	Geostationary Earth Orbit
GeoLITE	Geosynchronous Lightweight Technology Experiment
GPS	Global Positioning System
IFOV	Instantaneous Field of View

IMU	Inertial Measurement Unit
IRU	Inertial Reference Unit
I&T	Integration & Test
LADEE	Lunar Atmosphere Dust and Environment Explorer
LEO	Low Earth Orbit
LNA	Low Noise Amplifier
LOS	Line of Sight
MHD	Magnetohydrodynamic Device
MIRU	Magnetohydrodynamic Inertial Reference Unit
Mm	Megameter
NEA	Noise Equivalent Angle
NFOV	Narrow Field of View
PAM	Point-Ahead Mirror
PAT	Pointing, Acquisition and Tracking
PM	Program Management
PSD	Power Spectral Density
ROIC	Read Out Integrated Circuit
RSS	Root Sum Square
Rx	Receive
S/C	Spacecraft
SE	Systems Engineering
SILEX	Semiconductor Intersatellite Link Experiment
SNR	Signal-to-Noise Ratio
STE	Special Test Equipment
TRL	Technology Readiness Level
Tx	Transmit
UDT	UDT Sensors, Inc.
ULE™	Ultralow Expansion (registered trademark of Corning Incorporated)
WFE	Wavefront Error
WFOV	Wide Field of View

REFERENCES

Barry, J.D. and Mecherle, G.S. "Beam pointing as a significant design parameter for satellite-borne, free-space optical communications systems," Optical Engineering, Vol. 24, No. 6, November/December 1985, pp. 1049–1054.

Bielas, Michael S., "Stochastic and dynamic modeling of fiber gyros," SPIE Proceedings, Vol. 2292, Fiber Optic and Laser Sensors XII, San Diego, CA, 1994, pp. 240–254.

Boroson, Don M., Robinson, Bryan S., Murphy, Daniel V., Burianek, Dennis A., Khatri, Farzana, Kovalik, Joseph M., Sodnik, Zoran, and Cornwell, Donald M., "Overview and results of the lunar laser communication demonstration, SPIE Proceedings, Vol. 8971, Free-Space Laser Communication and Atmospheric Propagation XXVI, San Francisco, CA, 2014, pp. S-1–S-11.

Borrello, Michael A., Santina, Mohammed S., and Weight, Thomas H., "Jitter stabilization experiment for a precision pointing optical system," SPIE Proceedings, Vol. 641, Acquisition, Tracking, and Pointing, Orlando, FL, 1986, pp. 94–101.

Casey, W.L. and Phinney, D.D, "Representative pointed optics and associated gimbal characteristics," SPIE Proceedings, Vol. 887, Acquisition, Tracking, and Pointing II, Los Angeles, CA, 1988, pp. 116–123.

Chan, Vincent W.S., "Optical space communications," IEEE Journal on Selected Topics in Quantum Electronics, Vol. 6, No. 6, November/December 2000, pp. 959–975.

Chen, Chien-Chung and Gardner, Chester S., "Impact of random pointing and tracking errors on the design of coherent and incoherent optical intersatellite communication links," IEEE Transactions on Communications, Vol. COM-37, No. 3, March 1989, pp. 252–260.

Clarke, Ernest S. and Brixey, Harley D., "Acquisition and tracking system for a ground-based laser communications receiver terminal," SPIE Proceedings, Vol. 295, Control and Communication Technology in Laser Systems, San Diego, CA, 1981, pp. 162–169.

Deadrick, Robert B., "Design and performance of a satellite laser communications pointing system," American Astronautical Society Proceedings, Vol. 57, Advances in Astronautical Sciences, Keystone, CO, 1985, pp. 155–166.

Duncan, T.M. and Ebben, T.H., "Measurement of pointing error distributions in tracking loops of optical intersatellite links," SPIE Proceedings, Vol. 756, Optical Technologies for Space Communication Systems, Los Angeles, CA, 1987, pp. 54–61.

Eckelkamp-Baker, Dan, Sebesta, Henry R., and Burkhard, Kevin, "Magnetohydrodynamic inertial reference system, SPIE Proceedings, Vol. 4025, Acquisition, Tracking, and Pointing XIV, Orlando, FL, 2000, pp. 99–110.

Franklin, Gene F., Powell, J. David, and Workman, Michael L., Digital Control of Dynamic Systems, 3rd ed., Prentice Hall, Inc., Englewood Cliffs, NJ, 1997.

Hayes, Robert R., "Fading statistics for intersatellite optical communication," Applied Optics, Vol. 36, No. 30, 20 October 1997, pp. 8063–8068.

Held, K.J. and Barry, J.D., "Precision optical pointing and tracking from spacecraft with vibrational noise," SPIE Proceedings, Vol. 616, Optical Technologies for Communication Satellite Applications, Los Angeles, CA, 1986, pp. 160–173.

Hughes, Robert O., "Pointing performance of space-based crosslink laser communication systems," AIAA Space 2001 Conference and Exposition, Albuquerque, NM, 28–30 August 2001, Paper 2001-4710.

Katzman, Morris, Editor, *Laser Satellite Communications*, Prentice Hall, Inc., Englewood Cliffs, NJ, 1987.

Kazovsky, Leonid G., "Theory of tracking accuracy of laser systems," Optical Engineering, Vol. 22, No. 3, May/June 1983, pp. 339–347.

Kern, R.H. and Kugel, U., "Pointing, acquisition and tracking (PAT) subsystems and components for optical space communication systems," SPIE Proceedings, Vol. 1131, Optical Space Communication, Paris, France, 1989, pp. 97–107.

Killian, Kevin, Burmenko, Mark, and Hollinger, Walter, "High performance fiber optic gyroscope with noise reduction," SPIE Proceedings, Vol. 2292, Fiber Optic and Laser Sensors XII, San Diego, CA, 1994, pp. 255–263.

Klein, Bernard J. and Degnan, John J., "Optical Antenna Gain. 1: Transmitting Antennas," Applied Optics, Vol. 13, No. 9, September 1974, pp. 2134–2140.

Knibbe, Todd E., Stevens, Mark L., Kaufmann, John E., Boroson, Don M., and Swanson, Eric A., "An integrated heterodyne receiver and spatial tracker for binary FSK communication," *SPIE Proceedings, Vol. 2123, Free-Space Laser Communication Technologies VI*, Los Angeles, CA, 1994, pp. 188–199.

Knibbe, Todd E. and Swanson, Eric A., "Spatial tracking using an electro-optic nutator and a single-mode fiber," *SPIE Proceedings, Vol. 1635, Free-Space Laser Communication Technologies IV*, Los Angeles, CA, 1992, pp. 309–317.

Lambert, Stephen G. and Casey, William L., *Laser Communications in Space*, Artech House, Boston, MA, 1995.

Lopez, J.M and Yong, K., "Acquisition, tracking and fine pointing control of space-based laser communication systems," *SPIE Proceedings, Vol. 295, Control and Communication Technology in Laser Systems*, San Diego, CA, 1981, pp. 100–114.

Mankins, John C., "Technology Readiness Levels," A White Paper, 6 April 1995. Available on the Internet at http://fellowships.teiemt.gr/wp-content/uploads/2016/01/trl.pdf.

Nelson, R.D., Ebben, T.H., and Marshalek, R.G., "Experimental verification of the pointing error distribution of an optical intersatellite link," *SPIE Proceedings, Vol. 885, Free-Space Laser Communication Technologies*, Los Angeles, CA, 1988, pp. 132–142.

Ogata, Katsuhiko, *Modern Control Engineering*, 4th ed., Prentice Hall, Inc., Englewood Cliffs, NJ, 2001.

Perez, E., Bailly, M., and Pairot, J.M., "Pointing, acquisition and tracking system for SILEX inter satellite link," SPIE Proceedings, Vol. 1111, Acquisition, Tracking, and Pointing III, Orlando, FL, 1989, pp. 277–288.

Popescu, A.F., Huber, P., and Reiland, W., "Experimental investigation of the influence of tracking errors on the performance of free-space laser links," SPIE Proceedings, Vol. 885, Free-Space Laser Communication Technologies, Los Angeles, CA, 1988, pp. 93–98.

Russell, Donald, Ansari, Homayoon, and Chen, Chien-C, "LaserCom pointing acquisition and tracking control using a CCD-based tracker," SPIE Proceedings, Vol. 2123, Free-Space Laser Communication Technologies VI, Los Angeles, CA, 1994, pp. 294–303.

Southwood, Dana M., "CCD based optical tracking loop design trades," SPIE Proceedings, Vol. 1635, Free-Space Laser Communication Technologies IV, Los Angeles, CA, 1992, pp. 286–299.

Swanson, Eric A. and Chan, Vincent W.S., "Heterodyne spatial tracking system for optical space communication," IEEE Transactions on Communications, Vol. COM-34, No. 2, February 1986, pp. 118–125.

Swanson, Eric A. and Roberge, James K., "Design considerations and experimental results for direct-detection spatial tracking systems," Optical Engineering, Vol. 28, No. 6, June 1989, pp. 659–666.

Teplyakov, I.M, "Acquisition and tracking of laser beams in space applications," Acta Astronautica, Vol. 7, 1980, pp. 341–355.

Trondle, D., Pimentel, P. Martin, Rochow, C., Zech, H., Muehlnikel, G., Heine, F., and Meyer, R., et al., "Alphasat—Sentinel-1A optical inter-satellite links: Run-up for the European Data Relay Satellite System," SPIE Proceedings, Vol. 9739, Free-Space Laser Communication and Atmospheric Propagation XXVIII, San Francisco, CA, 2016, pp. 1–6.

Ulich, Bobby L., "Overview of acquisition, tracking and pointing system technologies," SPIE Proceedings, Vol. 887, Acquisition, Tracking, and Pointing II, Los Angeles, CA, 1988, pp. 40–63.

Van de Vegte, John, *Feedback Control Systems*, 2nd ed., Prentice Hall, Inc., Englewood Cliffs, NJ, 1990.

Wittig, M., van Holtz, L., Tunbridge, D.E.L., and Vermeulen, H.C., "In-orbit measurements of microaccelerations of ESA's communication satellite OLYMPUS," SPIE Proceedings, Vol. 1218, Free-Space Laser Communication Technologies II, Los Angeles, CA, 1990, pp. 205–214.

Young, Philip W., Germann, Lawrence M., and Nelson, Roy, "Pointing, acquisition and tracking subsystem for space-based laser communications," SPIE Proceedings, Vol. 616, Optical Technologies for Communication Satellite Applications, Los Angeles, CA, 1986, pp. 118–128.

4 Laser Transmitters: Coherent and Direct Detection

Klaus Pribil and Hamid Hemmati

CONTENTS

4.1   LASER TRANSMITTER: COHERENT DETECTION

4.1.1   INTRODUCTION

The move from microwave to optical frequencies is a logical step in the evolution of communication systems to cope with the ever-increasing amount of data to be exchanged. Even at optical frequencies, we see the need to move to terabit-per-second (Tbps) capacities on a single channel. This dynamic also sets the direction for the development of free-space communication systems. Coherent communication, now commonly used in terrestrial fiber communications, provides a valuable basis for low-Earth-orbit (LEO) free-space data communication systems. As coherent communication systems are the most sensitive receivers, they allow to select the best trade-off within the triad of transmitter output power, telescope aperture and data rate in free-space communication system design. Different systems and scenarios are being explored today, which shall take advantage of optical free-space communication to increase data transmission performance and to enable new system features. Although not all of them use coherent data transmission techniques[1] and not all are used exclusively in LEO, they all together establish a portfolio of building blocks that can be used for the transmitters of coherent systems in high-altitude platforms (HAPs) to satellites in geosynchronous (GEO) orbit. In particular a noncoherent differential phase-shift keying (DPSK) and a coherent binary phase-shift keying (BPSK) transmitter can share many of their building blocks.[2]

Among the LEO applications envisaged are primarily data downlinks from LEOs to ground stations,[3] LEOs to airborne platforms,[4] as well as the respective uplinks, LEO to LEO links, e.g., in LEO networks, and links between airborne platforms. Scenarios that are similar and can also be considered are data relay systems with the

---

[1] In its strict definition with a local oscillator laser in the receiver.

[2] For clarification it is assumed that any coding of the data stream is done before the transmitter.

[3] The term ground station shall stand for any stationary, terrestrial mobile, maritime, and any similar counter terminal.

[4] The term airborne platform shall comprise aircraft, UAVs, balloons, and similar which are operated at different altitudes.

relay node in GEO, and links between GEO to ground stations, between GEO and LEOs and others of the platforms above, as well as links between GEOs of different distance. There is no fundamental difference between these latter applications and LEO systems. The main distinction is the magnitude of the free-space loss in the link budget, which has to be scaled appropriately for satellites beyond LEO.[5]

Demonstrations with LEO spacecraft have been performed [1, 2] with the European Data Relay System (EDRS) and operational data relays for Earth observation LEOs are in use [3]. Transmission experiments from small LEOs to Earth have been conducted [4] while developing the next generation of transceiver [5, 6]. Flight experiments with airborne platforms have also been performed, e.g. [7] with balloons.

Data rates of currently operational space systems start as low as 1 Mbps [4] and as high as to 2.8 Gbps in EDRS [2] with 5.7 Gbps in a previous demonstration mission between LEOs [2]. For the planned LEO satellite networks data rates four to five times above that may be expected. Building blocks available in terrestrial communications already allow to serve data rates in the multi-100 Gbps range. Links between airborne platforms at high altitudes and to a ground station have already been demonstrated with a data rate of 100 Gbps [8]. For in-orbit applications, however, some further development will be necessary to close this gap to make such data rates available on satellites. Wavelength division multiplex (WDM) is an option to enhance the link capacity. In a WDM system, several laser oscillators are operated at different frequencies in parallel, subsequent modulators modulate the input data streams on their continuous-wave carriers, which are then combined and amplified by the transmitter power amplifier, see Figure 4.1.

Terminals on moving platforms are practically concerned with size, weight, and power (SWaP) limitations. The transmitter output power for current flight transceivers is typically chosen at below 10 W. In contrast, for ground station uplinks different considerations can be applied as they are not restricted in size, weight, and power and can use different concepts and output power levels. It shall be noted that the above values shall give a coarse orientation only. Due to the continuously increased demand for higher data-rate delivery from smaller satellites, miniaturized flight optical terminals at greater output power levels are becoming necessary - note also [9–11]. Currently two primary wavelengths are in use for free-space optical links: 1550 nm, which is used by terrestrial telecommunication (C-band erbium-doped fiber amplifier) technology; and 1064 nm, based on Nd:YAG solid state laser technology as used by EDRS satellites. Besides those two, there are implementations, e.g., at ~900 nm wavelength and research and development is ongoing at other wavelengths, which will extend the technology portfolio for the systems and the transmitter design.

To summarize the above considerations, a transmitter's system design has, among others, to consider the following aspects that have direct impact on the transmitter: output power, wavelength and modulation scheme, link dynamics, SWaP and

---

[5] The same considerations apply to systems with satellites in MEO whereas deep space communication is different.

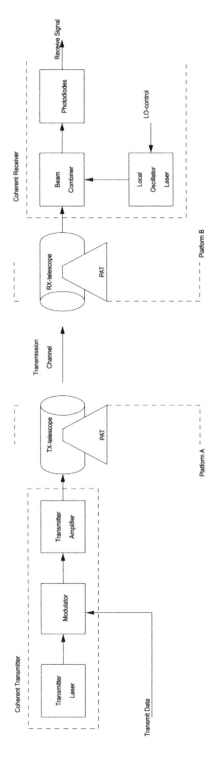

FIGURE 4.1 Transmitter with wavelength division multiplex (WDM) architecture comprising of several oscillators which generate CW carriers of different frequencies, each one followed by a modulator which modulate the input data streams on the CW carriers. Output light from the modulators is combined and amplified by the power amplifier. Note that all frequencies have to fit into the transmission bandwidth of the amplifier.

interface constraints, and mission specific criteria like reliability, availability, and radiation hardness for applications in space.

4.1.2 COHERENT TRANSMISSION AND COHERENT TRANSMITTERS

Coherent data transmission is an alternative transmission principle to the non-coherent scheme. It has several advantages, primarily better sensitivity and less vulnerability to stray light or background light. In a coherent receiver, incoming light is mixed with light that is generated inside the receiver.[6] While in a non-coherent direct detection transmission system the transmitted light is modeled in its corpuscular nature, coherent data transmission uses the representation as electromagnetic wave. Coherent systems can use modulation formats that impact on amplitude, frequency, phase, and polarization information as well as combinations thereof. Using these degrees of freedom available to a coherent system for the encoding of information allows to maximize spectral and power efficiency and outperform direct detection and DPSK [12].

In a non-coherent system, the received light (which in general consists of signal as well as of background and system noise power) is directly put onto the photodetector. Individual photons generate electrons in the photodetector, producing the signal and related noise currents. The signal current is proportional to the number of signal photons received.

In a coherent system, the received light wave is superimposed to a second light wave, which is generated in a local light source inside the receiver, the local oscillator (LO) laser. These two light waves are then put onto the photodetector. The nonlinearity of the photodetector generates an output current, which comprises a term with the mixing product of received and locally generated light waves with the transmitted information. Due to this mixing process, several requirements for the transmitter also apply to the LO laser. The main requirements for the transmitter of a coherent optical system are therefore:[7]

- *Single frequency operation:* As the transmitter works as an oscillator, all energy has to be concentrated in an output signal at one single frequency with a sufficiently narrow linewidth. For the laser, this means that only one mode must exist in which all the energy is contained. A change in the oscillation mode would cause a phase and frequency change of the transmitted light signal, and a loss of signal in the receiver. Single frequency mode has to be ensured over the entire frequency band.
- *Optical output power:* Transmitter output power must be sufficiently high as requested by the link budget under all operational cases considered to provide enough signal for the receiver and include margin to compensate for various losses, e.g., component aging. The majority of the applications

---

[6] Note that based on this definition, DPSK is a noncoherent transmission scheme; some authors consider it as a self-standing scheme between direct detection and coherent with a dedicated LO laser.
[7] Note that these requirements are typical for a LEO to LEO ISLs and may vary for other applications such as UAV links and space to ground links.

envisaged at present require an output power in the range from 100 mW up
to 10 W for the transmitter onboard the moving platform.

- *Modulation bandwidth:* According to the data rate and modulation scheme
  of the data signal, sufficient transmission bandwidth has to be provided.
  In practically all high data rate implementations, a dedicated modulator
  is used for data signal modulation. In addition, also the bandwidth of the
  power amplifier, and of any intermediate amplifier, has to be sufficiently
  large.
- *Low phase noise:* Laser oscillator phase noise has to be low enough to allow
  for high-sensitivity operation of the receiver. The performance of coherent
  transmission systems is influenced by the linewidth of transmitter laser and
  local oscillator laser. Oscillator phase noise will create a noise floor which
  will limit the sensitivity of the system. As a rule of thumb, laser linewidth
  should be less than 10^{-4} of the signal bandwidth, cf. [13] or [14] for details.
  Similar considerations apply to other noise sources as in the amplifiers.
- *Polarization:* Changes in polarization result in loss of received signal
  power. Therefore, polarization of the emission has to be stable, linear, well
  defined, and adjustable if necessary.
- *Stable operation:* All parameters must remain within their specified tolerance bands over the lifetime of the equipment. They must withstand the
  environmental conditions during operation as well as during ascent and
  descent or launch.

For operation on board a satellite—and to a lesser extent for others, e.g. airborne
platforms—additional requirements have to be considered or further accentuated
and special design solutions may be necessary to cope with them. In contrast,
requirements are usually less demanding for ground-based terminals. Among the
requirements to be considered are

- Size, mass (weight), and volume (SWaP)
- High overall efficiency
- Stable operation over the entire lifetime
- Design to withstand the loads impacted during launch
- Resistance to space environment
- High reliability and availability as needed
- Options for the implementation of sufficient redundancy
- Reasonable effort for thermal control
- Quick startup time for operation

In a coherent system, the LO laser in the receiver has to fulfill similar requirements
as the transmitter laser, except that the optical output power level can usually be
smaller. It is therefore advisable to balance requirements of the transmitter and
the LO laser. For example, the LO laser may be chosen with a wider tuning range,
which covers the frequency uncertainties of both lasers, compensating for the entire
Doppler shift between satellites that move relative to each other. For fast frequency

acquisition and good tracking performance, both lasers or just the LO laser need sufficiently high tuning agility. As transmitter and LO laser have to operate in close cooperation, it may be an advantageous solution to use similar technologies for both building blocks.

A block diagram comprising the main building blocks of a typical coherent transmission system is shown in Figure 4.2. The figure shows the transmitter unit in one optical terminal and the receiver unit in the remote terminal. The transmitter oscillator laser generates a continuous wave (CW) laser light signal at the (transmit-) operation wavelength of the system, which is guided into the data modulator. The (preprocessed) electrical data signal is also fed into the modulator that modulates the optical CW signal with the data signal according to the modulation scheme chosen.

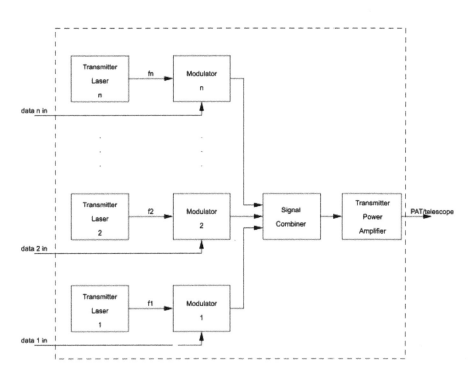

FIGURE 4.2 Block diagram of the key building blocks of a typical coherent transmission system. The coherent transmitter is hosted on platform A and comprises the transmitter laser oscillator, the modulator and the transmitter power amplifier. The receiver with the LO laser is hosted on platform B. The received signal has to be processed further according to the receiver concept chosen to obtain the received data signal and the control signals for the LO laser. While the transmitter part will be similar for various architectures and modulation concepts, the architecture of the receiver may vary considerably. Also, the layout and the arrangement of the telescopes, the pointing, acquisition, and tracking (PAT) system, and transmitter and receiver blocks on the respective platform depicted in the figure are to illustrate the concept in principle. Note that the graphic comprises no control components to supervise the operation of the transmitter.

Typical modulation schemes affect the CW light in amplitude or phase or both,[8] for digital data transmission between distinct amplitude and phase states.

A widely used modulation format is DPSK, which uses a delay line in the receiver to compare phase changes in the signal. As a DPSK system has no LO laser, it is considered a noncoherent system which is less sensitive than a homodyne BPSK system, cf. e.g. [12]. Although the receiver is different, the transmitter of a DPSK system is rather similar to the one of a coherent system.

To bring the level of the modulated signal to the transmitter output power required by the link budget, the output signal from the modulator is amplified by an optical power amplifier. Output signal of the power amplifier is routed by the optical system and the PAT system of the terminal to the transmit telescope, which emits it into the free-space channel toward the receive telescope of the counter terminal toward the receiver.

A laser emits electromagnetic waves in the 300 THz domain.[9] Receiver electronics work in a band up to several gigahertz. In order to create an electrical beat signal at the photodiode(s) of the receiver that can be processed by the subsequent electronics, transmitter, and LO laser frequencies have to be set in an initial frequency acquisition phase as close enough to each other so that an electrical beat signal is produced by the photodiodes. This beat signal must fall into the lock-in range of the frequency or phase control loop to lock onto the transmitter. To obtain such a beat signal, the wavelengths of both lasers have to be in adjacent or overlapping frequency bands and at least the LO laser has to be tunable. For fast frequency acquisition, absolute wavelengths of both lasers should be known within small frequency bands. In addition to the width of the tuning range, the tuning process must be dynamic enough to allow the frequency control loop to quickly achieve lock and maintain proper lock throughout the available time for communication. For a phase sensitive homodyne transmission scheme as, e.g., BPSK,[10] laser phase noise requirements are more stringent than for a heterodyne scheme. The utilization of digital coherent receiver with a digital signal processor (DSP) with tailored algorithms will allow to relax some of the above requirements for the receiver and its LO laser.

In systems where the linked optical transceivers move at a substantial speed relative to each other, Doppler frequency shift needs to be considered. The exact amount of frequency shift that must compensated between a terminal, onboard a LEO, and a ground station or a slowly moving airborne platform terminal depends on the orbit. A typical value is 14 to 20 GHz. In most systems, such frequency and phase compensation are achieved by tuning solely the LO laser. A cooperative frequency tuning strategy involving transmitter and LO laser frequency has been verified on TerraSAR and NFire and is used in EDRS.

An example of a tuning scheme for transmitter and LO laser tuning is depicted in Figure 4.3. The frequency acquisition process starts with the LO being tuned with open frequency control loop toward the transmitter frequency until an electrical beat

---

[8] Amplitude shift keying (ASK), phase shift keying (PSK), and particular binary phase shift keying (BPSK) or more advanced QPSK and QAM schemes.

[9] 300 THz corresponds to a laser wavelength of 1 μm typical for the Nd:YAG lasers discussed here.

[10] Unless explicitly mentioned, analog BPSK shall be meant.

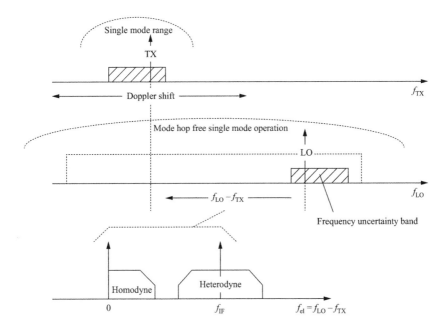

FIGURE 4.3 Example of a tuning strategy for transmitter and LO laser: At the beginning of the frequency acquisition process, both lasers are located inside their frequency uncertainty bands. During frequency acquisition, local oscillator frequency f_{LO} is tuned toward the transmitter frequency f_{TX} such that an electrical signal either around f_{IF} for heterodyne systems or around zero for homodyne systems occurs. If the satellites move relative to each other, an additional Doppler offset is introduced which also has to be compensated by the LO laser. Mode-hop-free stable single-mode operation is a must for this scheme.

signal at $f_{el} = f_{LO} - f_{TX}$ is obtained. When the beat signal approaches the locking range of the heterodyne frequency control loop at f_{IF} or of the homodyne loop, the respective tuning control loop can be closed and set into frequency tracking mode. From this point on, communication can start. The figure also shows the locations of the bands in which the laser wavelengths should be located when the frequency acquisition process is initiated. The absolute laser frequencies should be repeatable within these uncertainty bands in order to allow for fast frequency acquisition.

Frequency or phase lock has to be maintained during the entire transmission period. Transmitter and LO laser must operate as stable single-frequency sources and provide single-frequency emissions under all operational and environmental conditions. In particular only one single longitudinal mode must exist in the lasers and no hops of modes must occur. Figure 4.3 also shows the bands in which both the lasers must operate in single mode. The width of the bands is based on the design of the lasers. All of the parameters have to be kept within tight tolerances. It can be challenging to achieve this over the lifetime of the equipment and under the temperature conditions to which the equipment may be exposed particular onboard of satellites and on some airborne platforms. Hereby the lasers must not only withstand the nominal operation temperature range, which can be kept rather small, but also a large nonoperational temperature range and thermal gradients, without damage.

Aging of components and materials and degradation and failure mechanism have to be considered, in particular in-orbit applications. Reliability, availability, lifetime, and redundancy play a major role in transmitter design. In addition, in LEO radiation hardness of components and materials has to be considered. In particular laser diodes, other semiconductors like photodiodes, modulators and amplifier fibers have to be chosen to withstand the space environment.

Different forms of laser oscillators, modulators, and laser amplifiers, based on various technologies, exist which may not have the same technology readiness levels (TRLs) at different wavelengths. Careful trade-offs on transmitter level, which have to be raised up to system level, are necessary to select the most appropriate wavelength for the transmitter for a particular application. Considerations among them include selection of laser wavelengths that meet laser safety requirements for the link's operations environment, performance characteristics, overall efficiency, SWaP, reliability, survival of launch and space environment, commercial availability, TRL and cost. Preference should be given to laser wavelengths with high TRL. At present, the wavelength band of 1064 nm is mainly exploited for coherent BPSK, while 1550-nm band is used for DPSK and other noncoherent formats. Due to the commercial availability of 1550-nm coherent transceivers for the fiberoptics industry, these devices are finding increased applicability for free-space optical communications applications.

4.1.3 Systems and Applications

A brief overview of transmitters for free-space data transmission systems is provided by Table 4.1. The table lists the type of application, the data rate and the modulation format used and the optical output power of the transmitter. To extend the view, also links to GEO and noncoherent links are included for reference.

4.1.4 Architecture

The transmitter is often the key contributor to the transceiver's overall power budget and transmitter efficiency can significantly affect the design of an optical terminal. The optical output power level that the transmitter has to provide is the product of the receiver input power required for the specific data rate at a certain signal quality, the channel loss, and the gains and losses of the telescopes and the optical systems of the two terminals.[11] Receiver input power scales with the data or symbol rate multiplied by the receiver sensitivity in photons per single bit or symbol. In a free-space system[12] the transmitter has to provide the transmit power for the entire signal path so that much higher output power levels than in terrestrial systems may be necessary.

---

[11] The terminals can have can be identical or different layouts.
[12] In contrast to a fiber-based system in which intermediate amplifiers can be used.

TABLE 4.1

Overview of Transmitters for Free-Space Data Transmission Systems

System type	Coherent						Noncoherent			
Reference	[2]	[2]	[15]	[8]	[8, 15]		[10]	[16]		
Name	TerraSar-NFire	EDRS	EDRS	Face-book	Face-book	LEO-net			LCRD	
Application	LEO-LEO	LEO-GEO	Earth-GEO	HAP-HAP	HAP-Earth	LEO-LEO	Earth-GEO	GEO-Earth	GEO-LEO	
Link range	6000	45,000	39,000	250	50	6000	39,000	39,000	n/a	km
Principle	BPSK	BPSK	BPSK	16QAM+	16QAM+	n/a	DPSK	DPSK	DPSK	
Wavelength	1064	1064	1064	1550	1550	1064	1550	1550	1550	nm
Data rate (total)	5.6	2.8	2.8	100	100	10+	100	40	1.24	Gbps
P_{out}	1	5	50	1	0.1	4	$40 \times 4 \times 10$	16×0.2	0.5	W
Operation	Tested	Active	Active	Tested	Tested	Assumed	Study	Study	Planned	
Remark	1st gen.	2nd gen.								

Note: In addition to coherent low-Earth systems and links between and to HAP systems which link to GEO and DPSK systems are included for reference. Also included are assumptions for the coherent and a noncoherent transmitters of a LEO (-net) satellite network.

There exist some basic layouts for the realization of the transmitter of a coherent system. All of them have their specific advantages and the choice of a particular concept is often the result of a careful trade-off on system level, including requirements imposed by the transceiver, the receiver and the transmission format chosen, the host platform, and the availability, maturity, and reliability of key components. Most realizations of today favor the layout depicted in Figure 4.2, which consists of:

- A low-power oscillator laser that is operated in CW-mode with typical output powers between 1 and 100 mW. Preferred components as of today are based on laser diodes or solid-state lasers.
- A modulator device, which modulates the (pre-processed) electrical data signal onto the CW signal coming from the laser. Depending on the modulation scheme chosen, this modulator block can be fairly complex.
- A power amplifier that boosts the modulated signal up to the required output power level, e.g., up to several watts. Most systems at present use a fiber amplifier pumped by laser diodes. Another technology that can be considered is semiconductor optical amplifier (SOA). Several amplifier stages may be necessary to achieve the overall gain.[13]

---

[13] The individual stages can be of the same or different technology and different terms can be coined to name and describe them.

Often, optical output power, bandwidth, frequency, and noise behavior are key criteria for selecting a transmitter architecture. In particular, for use onboard the satellite and on certain airborne platforms, transmitter's overall efficiency is a key determinant in a transmitter's architecture. Wall-plug efficiency is the ratio of the optical output power produced by the transmitter and the required input electrical power consumption. For satellites and HAPs, dissipation of the thermal load generated by a high-power laser transmitter could become a major issue. That is another reason to pay close attention to the overall efficiency of the transmitter module. Additional accommodation efforts are involved if a thermal link to the platform or a special-purpose radiator is required.

For transmitter efficiency calculation, the contributions from all building blocks of the transmitter have to be taken into account. Besides the oscillator and the power amplifier, this includes the efficiency of the power supplies for various modules of the transmitter, thermal management devices such as thermoelectric coolers (TECs), the modulator driver, and a variety of control and power electronics. An overview of the individual contributors of the transmitter laser oscillator is shown in Figure 4.21.

4.1.5 Modulation Formats

The coherent modulation format that is used for optical data transmission in the EDRS and that was used in its precursor between the TerraSar and the NFire LEOs [17] is binary phase shift keying (BPSK). EDRS terminals have a homodyne receiver with an LO laser, which is locked by an optical phase locked loop (OPLL).[14] The CW light signal from the transmitter laser is modulated in its phase by an optical modulator with the digital (pre-processed) data signal whereby two-phase states, separated by 180° of the transmitter light, are created. Signal from the modulator is amplified by a fiber power amplifier and coupled into free-space by the telescope. In the receiver, the local oscillator laser creates a reference phase to recover the data signal. The configuration corresponds to the block diagram depicted in Figure 4.1. Homodyne BPSK provides the best signal-to-noise ratio of all PSK schemes and is the transmission format that yields the highest receiver sensitivity [12, 13]. To apply modulation formats with efficient bandwidth utilization like quadrature phase shift keying (QPSK) or 16QAM and beyond, different and perhaps more complex modulators have to be applied along with sufficiently linear power amplifiers.

Another modulation format which is used in terrestrial as well as in optical free-space communication systems is DPSK. DPSK is, e.g., used in with NASA's Laser Communications and Relay Demonstration (LCRD) [16, 18].[15] In a system with differentially encoded BPSK, the phase change of 180° occurs at a one (1) but in the already preprocessed data signal while it remains unchanged for a zero (0). In contrast to the BPSK system, the receiver does not require a local oscillator. But, DPSK modulated systems are less sensitive than BPSK modulated versions. In this case, the received optical signal from the telescope is amplified, split, and one part is fed into an optical delay line and from there to the photodiodes. Although DPSK is not

---

[14] Also referred to as analog homodyne technique [18].
[15] Note that the LCRD terminal also supports pulse position modulation (PPM).

a coherent system, the link budget is not identical, and phase noise requirements are less stringent as for the analog OPLL, most of the transmitter building blocks can be considered for utilization in a coherent transmitter too. More advanced data transmission concepts have been discussed where M-ary pulse-position modulation (PPM) is combined with frequency-shift keying (FSK), polarization, and or phase modulation [19].

A modulation format that uses a combination of pulse-width and DPSK to allow for scaling the peak output power of the (average power limited) fiber power amplifier against data rate was developed by [20].[16] Data rate scaling is achieved by burst-mode operation of the transmitter. While the transmitter laser is operated in CW mode, a specific modulator[17] provides signal modulation as well as cutting and forming of the DPSK bursts. PPM-QPSK concepts to scale data rate are discussed in [21].[18] In particular, for open systems scalability may become requirement.

While the above concepts and systems are tailored to be used onboard the satellite, for which usually high development efforts have to be made, coherent communication has also been tested between high altitude airborne platforms (HAP) and from an HAP to a ground station using commercial off-the-shelf (COTS) equipment from terrestrial telecommunications. Commercial 1550 nm hardware was used for the transmission of a total data rate of 100 Gbps using two carriers, 16QAM and dual polarization [8].

4.1.6 LASER-DIODE-BASED SOURCES

Good candidates for the source in a coherent data transmission system are diode lasers and solid-state lasers (including fiber-based devices).

Laser diodes are widely used in terrestrial telecom systems, in particular as signal source at the wavelengths 1550 nm (C-band) but also at 1310 nm and as pump source, e.g., at 808 and 980 nm. Laser diodes are used in various other applications at wavelengths from IR to UV with research ongoing.

Solid-state lasers for communication are primarily used at the main wavelength of the neodymium-doped yttrium aluminum garnet (Nd:YAG) at 1064 nm, which generates the highest power among the available options. Doped fiber lasers and amplifiers can also generate multi-watt to tens of watt of power form a single-mode fiber.

Besides monolithic designs, also hybrid laser designs are available in which primarily the laser diode (as the active element) is embedded in a passive resonator cavity with higher Q. The spectral performance of a diode laser oscillator can be improved by this external high-Q cavity, giving a source with better spectral performance.

Frequency accuracy, stability, and tuning agility are essential requirements for laser diodes as well as for solid-state lasers, in particular with respect to fast frequency acquisition or in multi-carrier systems.

---

[16] Data rate variation is 1:40.

[17] A consolidated burst-mode modulator as it is the combination of three individual modulators.

[18] In [18] data rate variation is from 2.5 Gbps down to 312 Mbps.

Both types of lasers can be used as transmitter sources in heterodyne and homodyne systems provided that output signal has sufficient spectral purity whereby homogenous systems have higher sideband noise requirements. Many DPSK systems today use distributed feedback (DFB) laser diodes at 1550 nm whereas, e.g., EDRS which uses BPSK is the domain of solid-state lasers at 1064 nm [2]. The following sections give an overview of some laser types that have been realized for coherent free-space communication.

4.1.7   LASER DIODE OSCILLATORS

4.1.7.1   DFB Laser Diode Oscillators

The DFB laser diode is a device in which the resonator cavity is implemented by a periodically structured element. The structure builds an interference grating and thus provides the feedback for the oscillator. Unlike a classical laser design, there is no distinction between the laser cavity and the mirrors, the mirrors are "distributed" along the cavity. The grating itself provides all the back reflection and ensures that energy is confined in only one very narrow laser mode. Both facets are anti-reflection (AR) coated but provide no reflection, which makes the laser quite stable. Laser wavelength changes with device temperature and with current, both need to be controlled. Several different designs of DFB lasers exist which have different properties, e.g., for precise tuning over the entire band. DFB lasers are widely used in terrestrial telecommunications, in C-band at around 1550 nm for long-distance communication applications and at 1310 nm for shorter ranges. For free-space communication, mostly 1550 nm is used. Linewidths are typically in the 100 to 200 kHz range.

The typical layout for the coherent transmitter is to put the DFB laser diode into a separate package, add an optical isolator, attach a fiber pigtail, and equip the building block with a temperature sensor, a photodiode for output power monitoring, and a TEC. Redundancy can be implemented, if necessary, on laser diode package level.

Depending on the target application, additional qualification programs have to be run, e.g., radiation tests for space applications. The development and testing of a space-qualified DFB laser diode for a 1550-nm transmitter is described in [22]. These DFB laser diodes are packaged in a 14-pin butterfly that is fully hermetically sealed and have been exposed to the space flight typical mechanical, radiation, and electrooptical tests. An overview on the performance of the (pre-qualification) devices is provided in Table 4.2.

4.1.7.2   External-Cavity Laser Diode Oscillators

The external cavities diode laser (ECDL) is an oscillator which use an external resonator to define the laser wavelength. Various designs of the external resonator exist. ECDLs are available at 1550 nm as well as 1064 nm. The advantage of the ECDL is a narrower full width at half maximum (FWHM) linewidth down to kilohertz. Some ECDLs can also be built compact and are available in 14-pin butterfly packages. However, the typical mode-hop free tuning range of current ECDLs is too small to compensate the Doppler shift in satellite-communication applications and the effort to increase the mode-hop-free tuning range is too complex for space hardware.

TABLE 4.2

Performance of DFB Laser Diodes for Space Flight Applications

Parameter	Value	Unit
Output power ex fiber	80	mW at 350 mA
Output power ex fiber	100	mW at 500 mA
Side mode suppression	55	dB
RIN	150	dBc/Hz
Linewidth	650	kHz
Isolation	32	dB
Power consumption	1.3	W at 25°C
	4.1	W at 65°C

4.1.7.3 Monolithic DBR Laser Diode Oscillators with Micro-Heaters

Distributed Bragg reflector (DBR) oscillators for space coherent lasercom have been under development since 2011 in Germany. DBR lasers do have characteristic mope-hops as a drawback, but DBR lasers have two significant advantages over DFB lasers: the spectral stability of DBR lasers improves with higher facet reflectivity and DBR lasers are less sensitive to optical feedback, enhancing FWHM and intrinsic linewidths and noise performance. DBR lasers use the standard design of high-power InGaAs/AlGaAs/GaAs quantum-well (QW) lasers with a high front facet reflectivity. The wavelength stabilization is achieved by a Bragg grating implemented in a part of the cavity. Wavelength tuning is achieved by variation of refractive index due to selective heating.

For coherent communication applications, the monolithic DBRs comprise three sections: gain, phase, and Bragg. The current injected into the gain section provides the gain to generate optical power. DBR Bragg and phase sections have micro-heaters to allow selective heating and thus a variation of the wavelength of the reflectivity peak and to allow mode-hop free frequency tuning. A monolithic DBR laser is shown in Figure 4.4. The laser can be tuned over a frequency range of more than 20 GHz at a speed of 800 MHz/s without mode-hops, see Figure 4.5, and has a typical FWHM linewidth of 150 kHz, as shown in Figure 4.6. They can deliver between 50 and 100 mW of optical output power at 970 nm. The monolithic design also yields an absolute frequency setting accuracy and repeatability of 500 MHz by setting the base/substrate temperature. Current produced monolithic DBR models with micro-heater control are limited to the wavelength range from 940 to 1070 nm.

4.1.8 Solid State Lasers

A proven alternative source to the laser diodes is the solid-state laser which is pumped by light from laser diodes -. Such lasers are, e.g., used in the coherent BPSK transceivers of the EDRS. This laser diode-pumped solid-state laser (DPSSL) can be considered as a device that absorbs incoherent laser light from laser diodes and emits

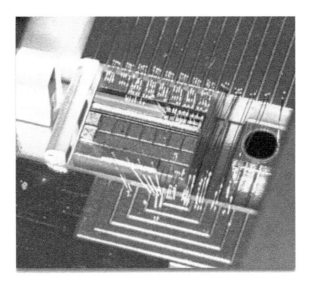

FIGURE 4.4 Monolithic DBR laser with a wavelength of 974 nm, a mode-hop-free tuning range of 24 GHz, an intrinsic linewidth down to 2 kHz and a FHWM linewidth of 150 kHz. (Courtesy of Delos Space.)

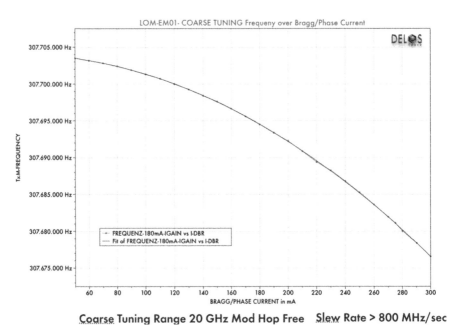

FIGURE 4.5 Mode-hop-free tuning performance of a monolithic DBR. (Courtesy of Delos Space.)

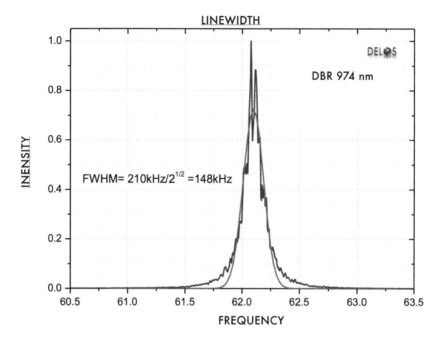

FIGURE 4.6   Typical FWHM linewidth of a monolithic DBR. (Courtesy of Delos Space.)

coherent light of high spectral purity. The output light from a (high-power) pump laser diode with its inherently poor spectral and spatial quality is converted into a light with significantly better properties. Combination of the pump light from several pump laser diodes can be easily facilitated. Although this conversion process costs some efficiency, it yields a bright, highly coherent light with excellent spectral and spatial beam properties as its output. An increase in output power can be achieved by using more or more powerful pump sources whereby thermal limitations at high optical output powers have to be considered. By combining the output of several laser diodes to pump a passive crystal the solid-state laser provides inherent redundancy. The device can be built compact and rugged. The DPSSL has demonstrated its suitability for coherent and in particular homodyne optical communication in the EDRS terminals, which are operating since several years on a routine basis in-orbit. The following sections will focus on this laser type and its designs.

There exists a range of host materials [23] with which solid-state lasers can be built. More than 30 potential candidate garnets are known to deliver output light in the 1-μm band with good conversion efficiency when they are pumped with laser diodes.[19] For a material trade-off, the doping level of the Nd ions, symmetry of the crystal, fluorescence lifetime, effective emission and absorption cross sections, linewidth of spontaneous emissions, gain coefficients, quantum efficiency, index of

---

[19] The most widely used pump wavelength for Nd:YAG is 808.4 nm, but the crystal can also be pumped at smaller absorption bands at other wavelengths like 804.4, 812.9, and 817.3 nm.

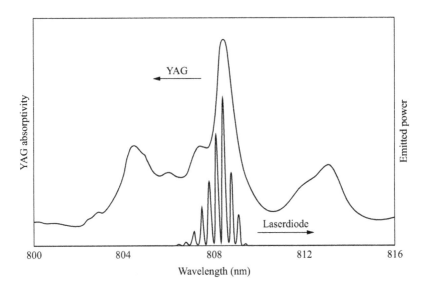

FIGURE 4.7 Emission spectrum and temperature drift direction of a laser diode array and absorption band of a Nd:YAG rod of 1% doping (arbitrary units).

refraction and its temperature variation, thermal conductivity, and the Verdet constant should be taken into account. In addition, crystal quality, availability, and compatibility between different manufacturers, as well as accessibility to production lots are the issues to be considered for a use in space.[20]

Nd:YAG[21] has both good optical and thermal material properties. It offers a well-balanced mixture for many applications, including communication systems. It is, therefore, widely used in the industry and its production is well established. A disadvantage of Nd:YAG is its relatively narrow absorption band for the pump light at 808 nm. This makes the overlap between laser diode pump source and crystal and so the efficiency of the laser sensitive to changes in wavelength. In Figure 4.7, the 1- to 2-nm absorption band of Nd:YAG and the typical emission spectrum of a laser diode pump array are shown. With increasing temperature, the laser diode emission drifts upward at 0.3 nm/K, while the absorption band of the solid-state crystal moves in the opposite direction. This will need a careful preselection of the laser diodes of the pump, good thermal design and as necessary active control during operation.[22] Nd:YAG is rather resistant to space radiation. Cr co-dopants can be added to further increase radiation insensitivity [24].

The solid-state laser basically consists of three elements: the laser crystal, the resonator, and the pump source(s). The resonator design determines frequency stability,

---

[20] It has to be noted that materials of transmitter oscillator and local oscillator should preferably be matched that they offer identical and close as possible performance.

[21] Nd:Y$_3$Al$_5$O$_{12}$

[22] Other Nd-doped host materials like yttrium vanadate Nd: YVO$_4$ (Nd:YYO) offer broader absorption bands, which relaxes temperature control requirements but lack other advantages of Nd:YAG for communication applications.

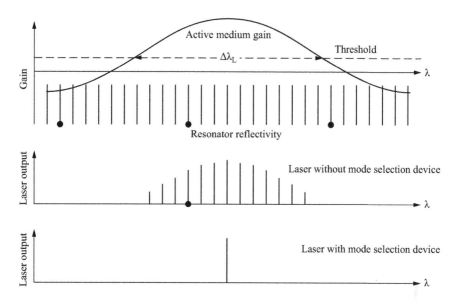

FIGURE 4.8 Spectral output of a laser without and with mode selection components. Resonator and laser modes with a dot represent modes of a microchip laser.

mode-hop-free tuning range, maximum optical output power, and efficiency. Furthermore, it determines complexity, ruggedness, and the reliability of the device. The range in which laser oscillation is possible is limited by the properties of the active gain medium to a wavelength band $\Delta\lambda_L$ above the lasing threshold. In the example shown in Figure 4.8 oscillation is possible whenever the loss inside the resonator is compensated by the gain of the active medium. This condition can be fulfilled for several resonator modes as depicted in the figure. Therefore, in this example, the laser could emit at several longitudinal modes. To achieve the desired single frequency operation additional, wavelength selection elements have to be introduced into the resonator.

Single-longitudinal-mode operation is essential to concentrate all power at one frequency and narrowband emission is vital for coherent data transmission systems [13]. There are several means to accomplish single longitudinal mode operation, including short resonators (which have fewer modes inside the gain band), filters inside the resonator, one-way ring resonators, and low-power operation. However, achieving reliable single-mode operation over a large tuning bandwidth still is a challenge. Appropriate choice of material, temperature control, and other tuning/ stabilization mechanisms are necessary to build a high-quality oscillator, which can be operated over a large portion of the gain bandwidth of the Nd:YAG host crystal of approximately 120 GHz and produce a clean laser line with only a few kilohertz width and jitter.

For a coherent communication system, operation in the fundamental transverse mode is also essential. The laser must concentrate all its transmitter power in one single beam of Gaussian shape, which is in most designs coupled into a single-mode polarization maintaining fiber that guides the light further through the system toward the telescope. Such single transverse operation can be achieved by imaging the pump light into the volume of the fundamental mode of the active medium.

There are several arrangements of the laser crystal, the resonator, and the pump source: Resonators can be built as a monolithic block or with discrete elements in linear or ring configuration. Pumping can be achieved via the end surfaces, or from the side of the active medium, or for a ring resonator at an appropriate point along it. The most relevant solid-state laser concepts are

- *Non-monolithic linear/twisted mode laser:* It is a basic linear laser design that is simple and straightforward but very sensitive. Optical output power is rather limited due to spatial hole burning effects. The latter can be prevented by the twisted mode configuration that allows for higher output power and is less sensitive to pump power variations.
- *Microchip laser:* This is a design that is derived from the linear laser. With its very short linear resonator it provides a large separation of the laser modes. As a consequence, only one resonator mode overlaps with the gain bandwidth of the active material, as shown in Figure 4.8, and the device is inherently operating in single mode. The resonator can be realized in a monolithic or semi-monolithic configuration with plane mirror surfaces. Spatial single-mode operation can be achieved by gain guiding. The laser can provide output powers of several 100 mW and a fairly high-tuning bandwidth beyond 100 GHz.
- *Non-monolithic ring laser:* This is design is more complex in its optical layout, more sensitive to misalignments and sensitive to linewidth broadening due to acoustic coupling. It has some advantages when several detached pump sources have to be accommodated.
- *Monolithic ring laser:* The design is less sensitive to optical feedback and does not suffer from spatial hole burning and offers several facets at which the pump source can be coupled. Ring lasers need a selection of only one unidirectional mode. Compared to other concepts, tuning range is more limited. Monolithic Nd:YAG ring lasers are proven technology, and a preferred source for coherent space communication systems.

4.1.9 LINEAR LASERS

When a linear laser resonator is operated in continuous mode, a standing wave pattern occurs and causes gain saturation for the required laser cavity mode and hinders single-mode operation. This can be overcome by using the twisted mode cavity (TMC) design in which the light passing through the active medium is circularly polarized. Polarization rotation is chosen such that no spatial intensity modulation occurs. This is achieved by two λ_4-wave plates at the input and the output of the active medium that transform linearly polarized light into circularly polarized light and back. Such lasers have been realized for coherent transmitters but could not offer the performance that the ring lasers can.

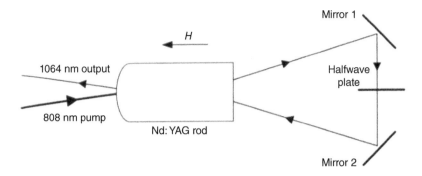

FIGURE 4.9 Concept of a non-monolithic ring laser with end-pumping. The Faraday rotator is built by the laser crystal and an external magnet which is represented in the graphic by the H vector.

4.1.10 NON-MONOLITHIC RING LASER

In a non-monolithic ring laser, light travels along a closed path that consists of several spatial segments. Such ring structure is another way to avoid spatial hole burning. A design, as shown in Figure 4.9, can be used to overcome output power limitations and to build stable high-power transmitter lasers. The pump beam generated by a laser diode at 808 nm is focused into the Nd:YAG rod. The resonator is formed by one end-facet of the Nd:YAG rod and by two folding mirrors 1 and 2. In the configuration shown, 1064 nm output is coupled out at the laser rod. A Faraday rotator and a $\lambda/2$-wave plate ensures stable unidirectional operation. No further components for wavelength selection are necessary in this design.

A prototype that was directly pumped by two laser diodes that were attached at one side of the mechanical flange of the crystal delivered 540 mW optical output power at 1064 nm in mode with a frequency stability of better than 100 MHz/h, cf. [25].

The laser was extensively characterized for its single-mode performance during tuning and change in output power by several consecutive tests between which the laser was switched off. The result depicted in Figure 4.10 shows the areas in which the laser did not work reliably in single mode so that further improvements of the design by additional mode selection elements would be necessary to use the device in free-space communication.

4.1.10.1 Monolithic Nonplanar Ring Laser

The nonplanar monolithic ring oscillator (NPRO) laser, in combination with the laser diode (as a pump source), allows to realize very compact lasers in which most functional elements (Faraday rotator, $\lambda/2$-wave plate and polarizer) are already part of the Nd:YAG resonator crystal. The concept of the resonator[23] from [26] is depicted

---

[23] The design is also known as monolithic isolated single-mode end-pumped ring (MISER).

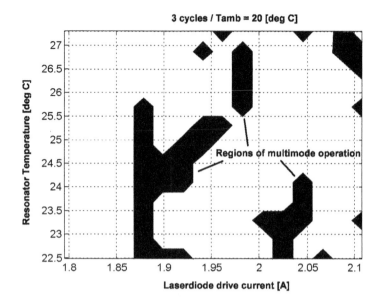

FIGURE 4.10   Results of the evaluation of the single-mode performance of a non-monolithic ring laser.

in Figure 4.11 and a photo of the laser crystal is shown in Figure 4.12. A magnetic field is applied to the crystal to ensure unidirectional operation. Due to the nonplanar beam propagation, together with the magnetic field induced polarization rotation based on the Faraday effect, a slightly higher loss is induced for one of the beam directions than for the other, finally resulting in suppression of the higher loss mode such that the lower loss mode direction is selected and builds up as the only (unidirectional) laser output. The laser is very compact and offers single-mode operation, combined with very good frequency and amplitude stability.

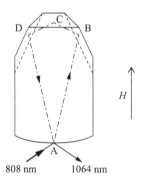

FIGURE 4.11   Principle of the monolithic ring laser: pump light at 1064 nm enters at A, laser light at 1064 nm is output at A. Polarization selection takes place at A, magnetic rotation takes place at AB and DA.

FIGURE 4.12   Monolithic nonplanar ring laser MISER crystal in a test fixture. (Courtesy of Innolight.)

Space designs for transmitter and local oscillator lasers, based on the laser concept, have been developed, and this type of laser has gained sound in-orbit heritage in the EDRS. A photo of a space qualified laser head is shown in Figure 4.13: Pump light at 808 nm from a redundant pump module is delivered to the laser by a multimode fiber (MMF), collimated and focused into the Nd:YAG crystal. The amount of pump power to be delivered depends on the optical output power required by the transmitter signal budget. Output light at 1064 nm is coupled into a polarization maintaining single-mode fiber (PMF) and guided further to the respective transmitter or receiver units. Frequency adjustment of the laser is achieved by two mechanisms: a low-bandwidth thermal control loop provides tuning over a range of several gigahertz and a piezo element bonded to the crystal provides fast tuning over a smaller range.

FIGURE 4.13   Photo of a space laser head; pump light is input via MMF to the right; laser light is output via the polarization maintaining fiber (PMF) in front. (Courtesy of Tesat.)

TABLE 4.3

Performance of a Monolithic Nonplanar Ring Laser
(MISER) as Transmitter and LO Laser in Space Applications

Parameter	Value	Unit
Wavelength	1064	nm
Frequency tuning range	30	GHz
Linewidth	10	kHz
Output power	100	mW max
Output	PPF	
Envelope	$25 \times 25 \times 20$	mm^3

The parameters of a laser qualified for space application are summarized in Table 4.3. The redundant pump sources are contained in an additional hermetic package of $40 \times 45 \times 20$ mm^3, which delivered the pump light by a 100-μm MMF. The mechanical, thermal, and radiation design matches the requirements of different orbits in which the terminals are operated. Tests have been developed to ensure the performance of the laser head and its pump source over the projected lifetime.

4.1.11 MICROCHIP LASER

Diode-pumped miniature solid-state lasers are also attractive devices for coherent communication applications due to their inherent single-mode operation over a broad frequency range. The basic design principle is to build a (linear) laser with a very short resonator so that the longitudinal modes are widely separated, and that only one mode can fulfill the criterion for oscillation, as depicted in Figure 4.8.

This can be achieved by the monolithic microchip laser design that consists of a short resonator with the Nd:YAG crystal and with two mirrors directly coated onto it, as shown in Figure 4.14. Light from the pump laser diode is focused into the

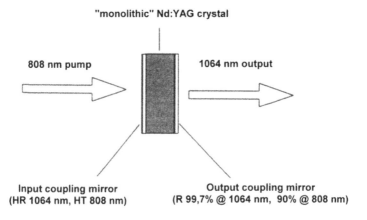

FIGURE 4.14   Monolithic microchip laser.

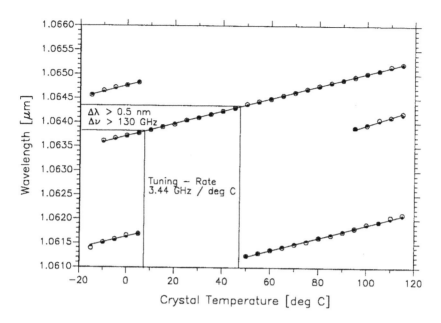

FIGURE 4.15   Free spectral range of a 300-μm microchip laser at 23 mW optical output power.

resonator through one mirror, which is highly transparent at the pump wavelength of 808 nm and highly reflective at the laser wavelength of 1064 nm. The other side of the resonator is coated with a mirror, which allows a small portion of the laser light to exit the cavity. This mirror is also reflective at the pump wavelength and reflects pump energy back into the cavity. Tuning of the monolithic microchip laser can be achieved by a variation of the length of the cavity, e.g., by changing the temperature of the crystal. Temperature tuning allows to change the wavelength over a wide band while single-mode operation is maintained. The thermal tuning performance of a microchip laser with a 300-μm resonator is shown in Figure 4.15 [27]. At a maximum output power of 23 mW, a free-spectral tuning range of 130 GHz can be obtained for a temperature change of 38 K, which leads to a thermal tuning coefficient of 3.44 GHz/K. If optical output power is reduced, the free-spectral range can be extended further, as shown in Figure 4.16.

Both the transmitter and the local oscillator have to provide reliable and stable single-mode operation at the transmitter wavelength whereby in addition, in particular, the LO laser has to be tunable over a broad band in order to allow and maintain frequency or phase locking of the receiver of the coherent link. Thermal tuning offers a large tuning coefficient but is a rather slow mechanism to change the laser frequency. Preferably for a local oscillator using the microchip laser, an additional fast tuning mechanism has to be implemented. Several means to achieve that are possible like the modulation of the pump light or the application of mechanical stress to the cavity, e.g., by a piezo element. Another alternative is to change to a semi-monolithic design and put one of the mirrors onto an actuator, which allows one to modulate the mechanical length of the cavity, as depicted in Figure 4.17.

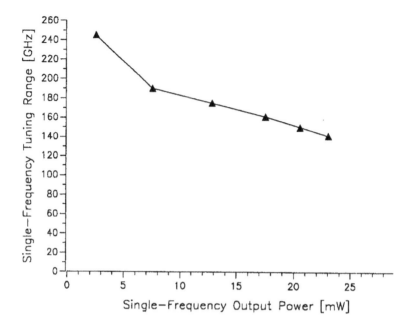

FIGURE 4.16   Extension of free spectral range by reduction of output power.

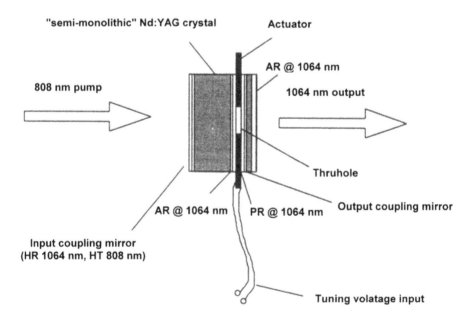

FIGURE 4.17   Semi-monolithic microchip laser with tuning element.

TABLE 4.4
Performance of a Microchip Laser Prototype

Parameter	Value	Unit
Output power	30	mW
Output power	5	mW
Wavelength	1064	nm
Linewidth	<10	kHz
Frequency drift	<450	kHz/h
Low tuning band	±40	GHz
Fast tuning band	±200	MHz
Fast tuning bandwidth	60	kHz
Power dissipation	1.5	W
Size	12	cm³
Mass (including fiber)	50	g

The actuator can be of piezo or micromechanical elements [28]. In the design shown, the output mirror is bonded onto a piezo foil that changes the length of the cavity according to the tuning voltage applied to it. The performance of a semi-monolithic microchip laser prototype that has been integrated into a coherent communication system [29] is summarized in Table 4.4. A photo of the laser with its proximity electronics, an external isolator, and a PPF pigtail is provided in Figure 4.18.

The emission spectrum of the laser is shown in Figures 4.19 and 4.20, respectively,[24] and reveals that the relaxation oscillations are comparable with the MISER lasers [30]. With its broad free-spectral range, the microchip laser is a good candidate for a low-power transmitter oscillator in particular for systems in which a wide tuning range is required. With the addition of a fast tuning element, it can also be used as the local oscillator.

FIGURE 4.18 Photo of microchip laser. (Courtesy of Airbus.)

---

[24] Both spectra were recorded as the beat signal with a MISER laser.

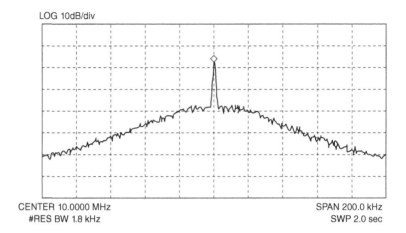

FIGURE 4.19   Emission spectrum of the microchip laser close to carrier.

4.1.12 LASER EFFICIENCY

Overall efficiency of a solid-state laser, defined as the ratio between optical output power and electrical input power, is determined by several contributors. It takes into consideration of generation of the pump light by the laser diodes and the losses of the pump driver electronics, pump to absorption band overlap, pump light transfer into the laser gain medium, pump light absorption, energy transfer into the upper stage, and conversion of the upper state energy to the laser output power. Furthermore, coupling efficiency into an optical fiber, isolators, and polarization controllers may need to be considered. In addition, various control elements, e.g., thermistors and coolers for thermal control, power monitoring photodiodes, and control electronics, must be taken into account. If the thermal control of the transmitter is split into internal and

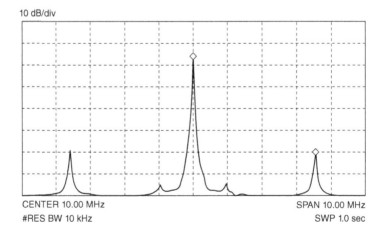

FIGURE 4.20   Microchip laser emission spectrum with relaxation oscillations at 3.6 MHz and −43 dBm.

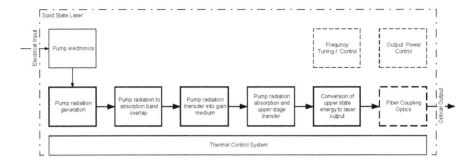

FIGURE 4.21 Contributors to the overall efficiency of a solid-state laser from the electrical input power via the pump to the optical output power. Building blocks for fiber coupling and control depend on the particular application.

external parts, whereby the latter can be provided by the optical terminal or the platform, also the external parts have to be considered in the model. A block diagram of various contributors from the electrical input through the diode pump to the output is depicted in Figure 4.21, cf. [31].

The optical-to-optical conversion efficiency of a Nd:YAG laser depends on several factors but can be theoretically as high as 76%. Laser diodes of the pump typically have an electrical-to-optical conversion efficiency beyond 45%, so that the efficiency for practical systems will be 15%. For a detailed efficiency budget, see [32]. The value applies to devices without fiber coupling, without control electronics, and without thermal control. Complex pump modules will further decrease this figure.

4.1.13 MODULATORS

The modulator packs the information to be transmitted onto the CW carrier. From a system design point of view, the issues that have to be considered are: with respect to performance: which modulation scheme shall be used, with respect to feasibility; by which type of device can this be done; and at which TRL are such components available for space, terrestrial, or ground systems.

There are two main possibilities to apply the modulation in the transmitter: by the direct modulation of the light source and by the external modulation of its CW output by using an additional modulator. Not all modulation formats are equally well suited for a particular oscillator or for an oscillator-modulator combination. Low-power laser diodes can be directly amplitude or frequency modulated (up into the gigabit regime); however, their spectral behavior is influenced by the modulation process. Frequency chirping and similar effects have to be carefully investigated and may prohibit the use of such direct modulation. For coherent high data rate transmission systems, a dedicated modulator is used to modulate the CW signal from the laser diode. As solid-state lasers do not offer higher modulation bandwidths, they do in general require an external modulator.

Basically, modulation can be applied to intensity, phase, frequency, and polarization of the monochromatic laser light. The most common modulation formats for coherent free-space communication are amplitude and phase modulation and combinations thereof. The main parameters that characterize an optical modulator are as follows:

- Bandwidth for modulation
- Optical insertion loss
- Modulation characteristics, e.g., linearity
- Drive power/voltage
- Optical power handling capability

4.1.13.1 Integrated Optical Modulators

Integrated optical modulators (IOMs) are standard devices in telecommunications industry (since over three decades) and are available for a broad range of wavelengths. Early devices could only handle a few milliwatts of optical power while higher power levels caused a drastic increase in attenuation. Newer processes like proton exchange with subsequent annealing could eliminate this disadvantage and provide long-term stable devices with low loss and increased damage threshold [33] under radiation conditions.[25] At present, waveguides that can handle high optical power level in excess of 1 W at 1550 nm and 0.5 W at 1064 nm are available [34].

Amplitude and phase modulators as well as more complex modulators for QPSK and QAM can be produced for various wavelengths whereby the basic building block is a phase modulator section and $LiNbO_3$ is the most widespread technology for its realization. The typical performance of modulators for 1064 and 1550 nm using $LiNbO_3$ and other materials is listed in Table 4.5. Polarization maintaining fiber pigtails are often attached to both sides of the waveguide chip to allow for a simple integration into the transmitter, but free-space coupling is also possible, cf., e.g., Figure 4.31.

TABLE 4.5
Typical Performance Values of Several Different Integrated Optical Phase and Mach-Zehnder (MZ) Modulators

Reference	[36]	[35]	[35]	[35]	[35]	[37]	[38]	
Type	Phase	Phase	Phase	MZ	MZ	MZ	MZ	
Wavelength	1064	1064	1550	1064	1550	1550		nm
Material	$LiNbO_3$	$LiNbO_3$	$LiNbO_3$	$LiNbO_3$	$LiNbO_3$	$LiNbO_3$ on SiO_2	GaAs	
Bandwidth	3	20	33	18	30	210	45	GHz
Modulation depth	—			30	22		23	dB
Insertion loss	4.5 dB	3	2.5	3.5	3.5	0.5	6	dB
Throughput power	500	100	100	100	100			mW
V_π		5.5	6	6	5	1.4	4.6	V
Dimension		85×15	85×15	85×15	85×15	10×5	45×20	

Note: Dimensions are without connectors, fiber pigtails, and mounting flanges.

---

[25] For transmitters using high-gain fiber amplifiers, power handling capability of the modulator is not considered to be critical.

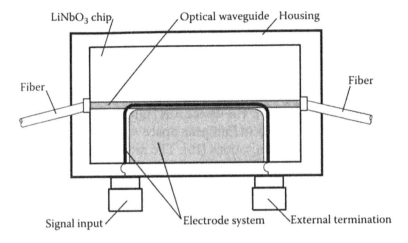

FIGURE 4.22   Principle of a LiNbO$_3$ phase modulator.

The concept of a LiNbO$_3$ phase modulator is shown in Figure 4.22. The device consists of an optical waveguide on the LiNbO$_3$ chip and an electrode system. The operating principle of the IOM is that the index of refraction is changed by the applied electrical field (electrooptical effect). The electrical modulation signal from the modulator driver is traveling for some distance alongside the optical waveguide and is then absorbed by a termination. The phase of the optical carrier is retarded along the waveguide according to the electrical field built up by the traveling wave electrode. For high modulation bandwidths, more advanced electrode structure with matched driver amplifiers are used and drivers and termination are mounted close to the electrodes.

With the phase modulation section, a Mach-Zehnder (MZ) modulator for efficient amplitude modulation can be realized as depicted in Figure 4.23. The device consists

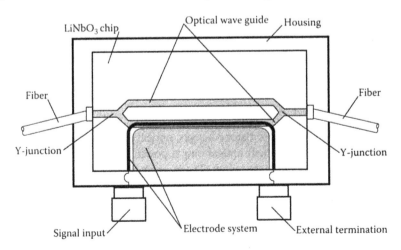

FIGURE 4.23   Principle of a Mach-Zehnder amplitude modulator.

of a Y-splitter which splits the incoming light equally into two branches. The part of the CW carrier passing through the phase modulation section suffers a change of its phase according to the modulation signal while the part in the other branch remains unaffected. Light from both branches is then combined in a second Y-adder. At the adder junction, light from both branches destructively and constructively combine according to its phase difference introduced, which results in an amplitude variation of the output light. To allow for symmetric amplitude excursion, the operating point of the MZ interferometer has to be set by an additional bias signal and controlled during operation to compensate the inherent drift of a MZ modulator due to environmental variations and aging. Different designs for such operation point stabilization, e.g., with and without pilot tone, are available, cf. [35].

Radiation tests that have been conducted, revealed that LiNbO$_3$ is resistant to the radiation dose of a typical LEO space mission. In combination with an optical fiber amplifiers, IOMs are for most cases the preferred choice for the transmitter chain of a coherent free-space communication system.

For complex modulation formats envisaged, which use combinations of phase and amplitude modulation, cf. e.g. [19], several of the above modulator blocks can be combined.[26] The block diagram of a general purpose IQ modulator with two MZ stages which can be used, among others, for QPSK modulator is depicted in Figure 4.24. The modulator of a transmitter that allows to scale different data rates or output power levels is described in [39][30] and its implementation is reported in [20].

While BPSK is the modulation format used in current coherent space systems, QPSK and quadrature amplitude modulation QAM are standard in terrestrial telecommunications industry; cf. for many [40]. The building blocks like MZ modulator with integrated drive amplifier modules, 16QAM modulators and others are

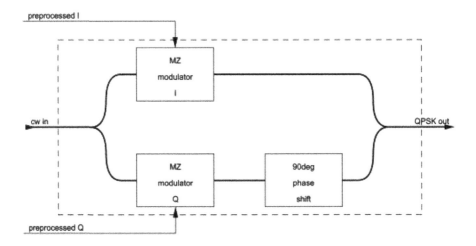

FIGURE 4.24   Block diagram of a QPSK modulator.

---

[26] The authors describe in general possible combinations of pulse-position modulation (PPM), phase modulation (QPAK and QAM), polarization modulation, and frequency shift keying.

available as COTS for 1550 nm up to many 10 Gbps, also using other technologies than $LiNbO_3$ like InP [41], GaAs, or Si. Such advanced modulation schemes were already used by terminals onboard of airborne platforms [8]. For space missions, IQ modulators as main building blocks are announced [38].

Also, research on the modulator itself continues, a significantly smaller $LiNbO_3$ with a length of only 10 mm, a data rate of 210 Gbps at a drive voltage of 1.4 V with potential for further upscaling has been developed by [37].

$LiNbO_3$ modulators that can withstand the radiation and reliability requirements of space missions are available in the main frequency bands of 1064 and 1550 nm, e.g. [35].

4.1.13.2 Other Modulator Types

For the modulation at high optical power levels, development and testing of a high-power bulk modulator for an optical power of 1 W at 1064 nm, and a bandwidth of 1 Gbps has been developed [42]. The modulator design used $LiNbO_3$ crystals forming an oversized waveguide through which the CW light from the input was led. The electrical modulation signal was provided by a strip line alongside the crystals and dissipated in a matched termination. Incoming light was polarization modulated in the crystal and converted into amplitude modulation by a subsequent polarizer. Phase modulators using this technology were also investigated.

Another type of modulator that may be considered as a part of a complex modulator is the electro-absorption modulator, which can be easily integrated together with a distributed feedback (DFB) lasers and is a standard building block for terrestrial telecommunications systems.

4.1.14 OPTICAL POWER AMPLIFIERS

For most transmitters with medium- and low-power oscillators, optical power amplifiers are required to achieve the optical output power level requested by the link budget. The power amplifier is another key device of an optical communications terminal, as optical output power is the foremost parameter for the transmission capacity of a free-space link.[27] The main characteristics of an optical power amplifier for a coherent free-space data transmission system are, therefore, as follows:

- Maximum optical output power that can be achieved and from which any further increase in input power will no more result in an increase in output power
- Amplifier gain and linearity as the ratio between optical output and optical input power
- Noise, primarily amplified spontaneous emission (ASE) from the amplifier, as a background noise source for the receiver
- Polarization stability

---

[27] Assuming that the modulation format has been chosen and the maximum telescope apertures are given by envelope, mass, and other boundary conditions.

- Efficiency, as the ratio between optical output power and electrical input power to operate the amplifier; like for the oscillator various contributors have to be taken into account
- Fundamental transverse mode output and a low beam quality factor M^2

Other criteria such as optical bandwidth, gain flatness over the frequency band, or channel crosstalk may also need consideration, particularly in dense wavelength division multiplexing (DWDM) systems whereas for a single channel system of several gigabits, these parameters are usually not critical.

The function principle of an optical amplifier is similar to the one of the laser: Injected photons in the input signal release more photons in the active material of the amplifier. Main difference between amplifier and laser is the absence of the resonator. Three different types of optical power amplifiers are subsequently discussed in more detail: solid-state optical bulk amplifier, optical fiber amplifier, and optical semiconductor amplifier. A survey of amplifiers used or proposed for space missions is comprised in Table 4.1. The table comprises both coherent systems and noncoherent DPSK data transmission systems as well as different application scenario, including links to GEOs to allow some more scaling.

4.1.14.1   Solid-State Optical Bulk Amplifier

The principle of optical amplification can be best seen at the simple solid-state bulk amplifier for 1064 nm, which is shown in Figure 4.25. In this amplifier, the 1064-nm input light passes through a Nd:YAG solid-state laser crystal, which is pumped at its absorption wavelength. Pump light at 808 nm is focused into the Nd:YAG rod so that a diameter somewhat larger than the signal beam is activated over the absorption length, which is typically around 3 mm in Nd:YAG doped with 1%. The signal beam at 1064 nm is coupled into the rod so that it is roughly collinear with the volume stimulated by the pump light beam. The facet at which the pump beam enters the crystal is highly reflective coated and the signal beam is reflected back and passes a second time through the activated region of the rod. Gain of such an amplifier is

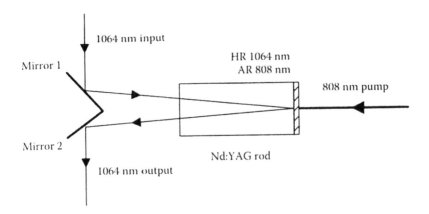

FIGURE 4.25   Simple double-pass solid-state bulk optical amplifier for 1064 nm.

small and lies between 2 and 3 for Nd:YAG and can be somewhat higher for other materials, e.g., for Nd:YVO4. For higher gains, multi-pass configurations have been developed like a six-pass amplifier comprising of three of the above double-pass stages with a gain of 14 dB and an optical output power of 1.5 W at 1064 nm in [25] or a 28-pass amplifier in [43].

4.1.14.2 Optical Fiber Amplifiers

An elegant way to achieve good overlap between the signal to be amplified and the volume stimulated by the pump light is to confine both in an optical fiber that is doped accordingly with the rare earth elements, e.g., for 1550 nm with erbium (Er) and for 1064 nm with neodymium and ytterbium (Yb). Fiber amplifiers with medium optical output power levels up to several 10 mW are widespread in terrestrial telecommunications industry. High-power amplifiers up to several watts are used for free-space applications and the technology for amplifiers with power levels above is available. In other types of applications, e.g., in advanced materials processing, single-mode CW laser systems in the 10-kW range and beyond are used, which indicates the potential of the technology. For coherent space communication systems, high-power devices with optical output powers of several watts have been developed and have gained several years in-orbit heritage, c.f. Table 4.1. The amplifier in general has to preserve the polarization state of the input signal, which is usually achieved by a polarization maintaining active fiber. For satellite onboard applications, the amplifier and, in particular, the fiber has to be insensitive to space radiation.

There are two main schemes to pump the doped fiber: direct pumping of the doped single-mode fiber core that can be accomplished by single mode (SM) pump modules and cladding pumping of the doped fiber with a multimode high-power pump module. Design and control of the pump modules during operation depend on the fiber material. If it has a sufficiently broad absorption band, e.g., as in erbium-doped fibers, no additional thermal control of the pump modules may be necessary.

The optical fiber amplifier in a core- or cladding-pumped design consists of a device, which couples the signal in or out of the doped SM core of the amplifying fiber and the pump light also into the SM core or into the multi mode (MM) cladding. The length of a typical amplifier fiber is several meters up to a few 10 m. The amplifier fiber is bent in order to facilitate good overlap between the pump light modes and the amplifying core. Typical designs use front-side pumping in which pump and signal light are coupled into the amplifier fiber at the low-power end, backward pumping in which the pump light enters at the high-power output end of the fiber or double-side pumping, which uses both methods. The principle of a double-side pumped amplifier is shown in Figure 4.26. Optical isolators at the input and output are used to avoid optical feedback. To maintain the polarization state of the input light signal throughout the amplifier, a polarization maintaining active fiber must be used. The usual designs of polarization maintaining fiber (PMF) can be used to induce stress in the core of the amplifier fiber like the elliptical, panda, and bow-tie shapes [44].

A 1064-nm optical fiber amplifier prototype for a coherent satellite transceiver has been presented in [45]. The modulated signal at 1064 nm enters the amplifier via a polarization maintaining fiber. Light from the fiber is collimated and combined

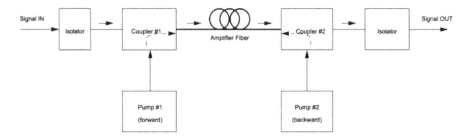

FIGURE 4.26 Principle of a double-side-pumped optical fiber amplifier with pumping from both sides.

with the light from the pump source at 977 nm. Combined input and pump light are focused into a doubled clad Yb-doped fiber. Along the amplifying fiber, the communication signal is amplified to the optical output power of the system. At the end of the amplifier fiber, the light is collimated again and coupled into a polarization maintaining output fiber. As the amplifier has a double-side pump design in which pump light is fed from both sides into the amplifying fiber, there is a second pump light coupler at the communication signal output. The optical path of the device consists of the Yb-doped amplifier fiber, optical isolators, and the pump light combiners. Operation has to be monitored and controlled by a controller. All components have been mounted in a stable housing, which also provides radiation protection. The interior of the optical fiber amplifier is depicted in Figure 4.27, main technical parameters of a prototype are summarized in Table 4.6.

A different design of an amplifier in which signal and pump light are reflected at the end of the fiber and pass through it for a second time is described by [20] and depicted in Figure 4.28. In contrast to the above designs, this amplifier uses no polarization maintaining amplifier fiber. The input signal at 1550 nm passes a polarizing beam splitter, which is the only polarizing determining element in the amplifier. It is then amplified by two amplifying stages and toward a narrowband Faraday mirror. This Faraday mirror reflects the light back into the active fiber, rotates its polarization direction and removes out-band ASE noise. After passing again through the active fiber, the output signal is extracted by the polarizing beam splitter.

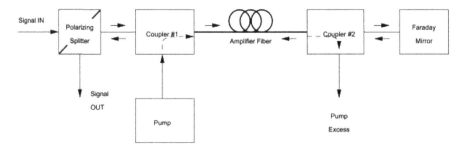

FIGURE 4.27 Principle of an amplifier in which signal and pump light are reflected at the end of the fiber and pass through it a second time.

TABLE 4.6
Performance Parameters of the 1064-nm Optical Fiber Amplifier

Parameter	Value	Unit
Signal wavelength	1064	nm
Input power	10	mW
Output power	1250	mW
Pump wavelength	977	nm
Pump power	5.5	W
Polarization extinction	20	dB min
Envelope	$200 \times 100 \times 40$	mm^3
Mass	1.7	kg
Operating temperature	+15 to + 50	°C

A small amplifier for 1550 nm with an envelope of $90 \times 60 \times 15$ mm^3, which is based on COTS equipment is described by [46]. The amplifier provides an output power of 200 mW at a power consumption of 8 W of the entire transmitter.

Er-doped optical amplifier fibers are very sensitive to space radiations, so their degradation can impact longer term space missions. A trade-off at system level instead only of component level is proposed [47].

Radiation tests on mid-power erbium-doped fiber amplifiers (EDFAs) have been performed [48] with a total ionizing dose (TID) of 0.1 Gy (10 krad).[28] During irradiation the amplifier was unpumped until the total dose was reached. Tests then

FIGURE 4.28   Photo of the interior of the amplifier. (Courtesy of Contraves Space.)

---

[28] Withdrawn IEEE standard 1156.4-1997 defines the TID in LEO between 4 and 40 krad(Si) per year for an effective shielding thicknesses of 10 mils Al.

revealed at the lower band end of 1540 nm, a gain drop of 0.45 dB and an increase in noise figure (NF) of 0.16 dB. Values were slightly higher at the upper end at 1565 nm with a gain reduction of 0.55 and 0.21 dB increase of the NF. Tests with a second amplifier that was pumped during irradiation revealed a similar deterioration at 1530 and 1550 nm but 1 dB at 1565 nm. It has to be noted that not only the radiation dose but because of annealing effects. Also, the dose rate plays a major role in the radiation degradation behavior of the fiber. For high-output power amplifiers up to 10-W, Er/Yb-doped fibers with a gain degradation of less than 1.5 dB at 1 kGy (100 krad) are available [49]. However, at these high-power levels also thermal effects in the fiber and the interrelation between radiation and thermal effects have to be considered and further maturation remains [50].

4.1.14.3 Amplifier Pump Modules

Amplifier pump light, of the selected wavelength, can be provided by many means, common are laser diodes or fiber lasers. Most pump source use single laser diodes or laser diode arrays. Designs that use an intermediate fiber laser in the pump path have also been proposed [51]. Depending on the power and reliability requirements, output from two laser diode can be polarization combined. If more output power is necessary, several laser diodes or single emitters of a high-power laser diode bar can be combined.

A single-mode pump module for core pumping of EDFAs is described by [22]. The model was radiation and environment tested for use in space flight applications. Each laser diode is packaged into a separate hermetically sealed standard small form-factor 8-pin Mini-DIL package. The module also comprises a monitor photodiode and a thermistor. Laser light is coupled into a polarization maintaining fiber (PMF) so that two modules can be orthogonally polarization combined. For wavelength stabilization of the pump module, a grating at the 976-nm EDFA pumping band is implemented. The EDFA pump module needs no TEC. In de-rated operation, the pump module delivers an optical output power of 235 mW out of the fiber using 480 mA drive current at a case temperature of 65°C and a maximum of 325 mW at 600 mA drive current at room temperature without derating.

For high-power EDFAs with cladding-pumping with several watts of optical output power, high-power pump modules for space applications have been developed [22] and tested for space qualification conditions. Laser diode chips are packaged into a hermetically sealed 14-pin butterfly package. The wavelength of the chips is selected to match the absorption of Yb and Er-Yb co-doped fibers at 915 nm. Pump light is delivered by an MMF. The pump module employs a photodiode power monitoring and a thermistor to measure the temperature. Also, for this module, a TEC is not necessary. The pump modules deliver around 10 W output power out of the MMF at a driving current around 12 A at 25°C, which drops to around 9 W at 50°C at which the chips reach 65°C.

The inside of a multi-emitter laser diode bar pump module is shown in Figure 4.29. The pump module for 977 nm is located in a separate housing detached from the amplifier and is optically coupled to it via an MMF. The pump module comprises of a set of laser diode arrays that are optically combined into the MMF. The number

FIGURE 4.29 Photo of the interior of the amplifier pump module for the amplifier characterized in Table 4.6. (Courtesy of Contraves Space.)

of laser diodes implemented is determined by the optical pump power required as well as by the reliability concept of the amplifier. In the prototype shown, 12 laser diode arrays are used. Each of them contains 24 or 25 emitters on a 10-mm wide bar. The output of a set of six arrays is combined and coupled into the fiber to provide pump light along the forward pumping direction. The other set delivers the backward pump light. The operation concept of the unit is such that three arrays of each source are operated concurrently at one-third of their nominal output power. This allows increased reliability and lifetime of the laser diodes and the pump module. The laser diode arrays, the optical elements for beam combination, and the fiber couplers are mounted on a common baseplate. Particular attention had to be paid to the thermo-mechanical design of the device. Operating several laser diodes in parallel at lower output power also simplifies the thermal design, as the dissipated energy from the arrays is spread over a larger surface of the baseplate. The main technical properties are summarized in Table 4.7.

TABLE 4.7

Performance Parameters of the 977-nm Pump Module for the 1064-nm Optical Fiber Amplifier of Table 4.6

Parameter	Value	Unit
Wavelength	977	nm
Spectral linewidth	3	nm
Optical output power	16	W nominal
Beam divergence	7	mrad
Pointing stability	±0.06	mrad
Envelope	$210 \times 130 \times 30$	mm^3
Mass	1.1	kg
Power consumption	88	W max
Temperature non-operational	−40 to +65	°C
I/F temperature operational	+25 ± 3	°C

The calculation of the overall efficiency of the fiber amplifier with power supply, drive electronics, thermal control system, and operation control can be obtained in analog application of the scheme depicted in Section 4.1.12.

4.1.14.4 Semiconductor Optical Amplifier

Semiconductor optical amplifier (SOA) are based on the same principle as laser diodes, they have an active region in which stimulated emission is achieved by inversion of the active material directly by electric energy, but SOAs have no resonant structure. They are available for different wavelengths, and 920, 1064, and 1550 nm are standard. The advantage of the SOA is its direct pumping, which results in better efficiency and its smaller size. It can be used as pre-, intermediate/ gain control and booster amplifier. Pending on the configuration used the SOA has a potentially higher noise figure and stronger nonlinear distortions compared to fiber amplifiers.

4.1.14.5 Linear SOA

The simplest form of such an amplifier is a linear laser diode structure with antireflective coatings on both sides. Standard commercial devices achieve a gain of more than 20 dB or an output power of around 13 dBm. The device is a single-pass amplifier that amplifies one polarization state only. The chip is housed in a standard 14-pin butterfly package with a PPF or single-mode fiber on input and output.

4.1.14.6 Ridge Waveguide Power Amplifier

A very attractive (low noise) device is a ridge-waveguide power amplifier (RWA): the RWA is following a standard layer design of high-power semiconductor laser with an etched-ridge-waveguide structure with index guiding for stable fundamental lateral mode operation. AR coatings are applied on both facets. The typical longitudinal structure length of 4 to 8 mm yields a large amplification by which optical output power levels between 0.5 and 1 W at 10 to 50 mW optical input are feasible. Output beam quality is excellent with a M^2 of 1.1. Currently produced RWAs are limited to a wavelength range between 940 and 1070 nm. Highest reliability is achieved at 970 nm.

4.1.14.7 Tapered Power Amplifier (TPA)

For higher output power levels, tapered structures are used. Such a device can produce an optical output power of up to 5 W and a nearly diffraction-limited beam profile. The schematic drawing of a tapered traveling wave semiconductor power amplifier as described in [52] is shown in Figure 4.30. The amplifier was combined with a 25-mW output microchip master oscillator laser to a transmitter that provided up to 1.0 W output power at 1064 nm with good spectral quality. Devices that achieved several watts optical output power have been developed [53].

4.1.14.8 Transmit Hybrid (TXM)

A high degree of miniaturization can be achieved by the integration of oscillator, modulator, RWA-pre-amplifier and TPA-power-amplifier in one hermetic hybrid

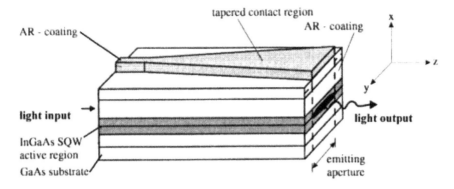

FIGURE 4.30 Schematic of a tapered semiconductor power amplifier.

as shown in Figure 4.31. A complete transmit chain with free beam output for an in-orbit application can so be realized with a weight of less than 250 g mass. Currently, TXMs are produced in the wavelength range between 974 and 1064 nm.

4.1.14.9 Optical Amplifier Summary

The majority of coherent communication system designs for near-Earth applications will require optical output power levels between 100 mW and 10 W, which can

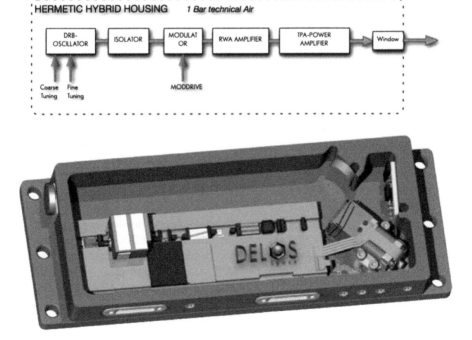

FIGURE 4.31 Block diagram and 3D-drawing of a hermetic TXM, comprising oscillator, modulator, RWA-pre-amplifier, and TPA-power-amplifier. (Courtesy of Delos Space.)

preferably use a layout with a low-power oscillator, integrated optical modulator, and a booster amplifier. For such configuration, based on the technology portfolio available today, the optical fiber amplifier will be the preferred solution. Output power and reliability requirements can be met by scaling the pump module configuration, the amplifying fiber itself has sufficient potential for that. Both wavelengths in focus 1064 and 1550 nm can use this device. However, size and efficiency of the fiber amplifier are disadvantages and attention has to be paid to radiation hardness.

For future systems with focus on SWaP as well as on cost, the semiconductor amplifier may become an interesting choice as it allows to build very compact transmitters, with inherently better efficiency. Reliability requirements can thereby be met easily on the unit level. A change in wavelength to use the most reliable technology may be envisaged in the course of this step.

In any case, for near future requirements of 100 Gbps and beyond in LEO satellite networks output power levels beyond 10 W can be expected.

4.1.15 TRANSMITTER CONTROL ELECTRONICS

Control electronics to operate the transmitter has to set and control the frequency of the laser and the output power of the power amplifier. In addition, it has to fulfill a variety of other command and control functions and to handle communication with the terminal control computer. Frequency tuning, stabilization, and Doppler tracking can be facilitated by changing the temperature of the DFB laser diode or the solid-state laser crystal, or by applying a mechanical force to the crystal. Cooling is achieved by a temperature control loop consisting of a temperature sensor and a TEC attached to the laser diodes or crystals and a controller in the laser electronics. Frequency/temperature set point can be input externally. Thermal tuning works over a broad frequency range but offers only low bandwidth. For fast frequency tuning of solid-state lasers, often a piezo element attached to the crystal is used. A voltage signal applied to the piezo element by the control electronics imposes some mechanical stress onto the crystal and changes the optical cavity length and so the frequency of the laser by stress-induced birefringence and a variation of the mechanical length. It can be necessary to implement relative frequency control within a WDM architecture or absolute frequency stabilization. The latter can be achieved by an absolute frequency reference, e.g., an atomic vapor cell. Laser control also has to provide sufficient current to drive the laser diodes, monitor their operation and ensure best overlap between laser diode emission and crystal or fiber absorption bands. Depending on the host crystal chosen, more or less complex thermal control functions may be necessary that may include a thermal interface with the terminal and/or the platform thermal control system, e.g., via fixed or variable conductance heat pipes. Transmitter output power has to be stabilized, for which several power detectors and several control loops to adjust the drive currents of the individual elements of the transmitter have to be implemented. Depending on the redundancy scheme implemented, the electronics also needs to control the operation of the individual laser diode or emitters and change configurations as necessary—either by commands from a terminal control computer or autonomously. Solid-state laser oscillators also show relaxation oscillations in their emission spectrum, which are not to be acceptable for certain applications. By using a high bandwidth feedback loop from the laser output to the pump laser diode, this effect can be substantially reduced. Depending on the application, additional features may have to be provided by the control electronics, may be necessary. Tasks of the transmitter control electronics also comprise control of the modulator (e.g., operation point stabilization) and its driver (e.g., signal pre-distortion).

4.2 LASER TRANSMITTERS FOR DIRECT DETECTION

4.2.1 INTRODUCTION

Highly efficient laser transmitters capable of multi-gigabits per second (Gbps) bandwidth modulation, at moderate average power levels and high spatial beam quality, are required for near-Earth laser telecommunications in the direct detection mode. Certain fiber amplifiers and semiconductor diode lasers are well suited for this application. Owing to the demands of fiberoptic telecommunication market, fiber lasers and amplifiers have been extensively developed at the 1.5 μm wavelength. Today, data rates exceeding 10 Gbps are possible via semiconductor lasers whose peak or average output power can be boosted from milliwatt to thousands of milliwatt power levels by utilizing optical amplifiers such as an erbium-doped fiber amplifier (EDFA) [54, 55]. Today's off-the-shelf laser transmitters meet the output power and beam quality requirements for near-Earth laser communication (lasercom) links. Except for radiation tolerance and operation in vacuum, Telecordia-qualified commercial off-the-shelf lasers meet nearly all of the demanding requirements for airborne and spaceborne operations. Desired improvement areas include higher overall efficiency (including laser drivers and thermal management), and improved lifetime span for missions lasting 10 to 15 years.

Due to device availability, adequate laser power, and ease of modulation, several laser wavelengths are prominent in lasercom. Among them, the 1550-nm wavelength is highly desirable for its commercial availability, as well as reduction in atmospheric attenuation and eye hazard. Optical amplifiers re-amplify a weak signal without optical-to-electrical signal conversion. The amplifying medium may be an impurity-doped bulk crystal or fiber, or a semiconductor diode. Whether in the visible or near infrared, in a master oscillator power amplifier (MOPA) configuration, high-power semiconductor lasers optically pump the amplifier medium. Amplifiers are capable of boosting a compatible and sufficiently stable master oscillator's output power by tens of decibels, while maintaining its modulation and beam quality characteristics at the multi-Gbps modulation rates. Selection of the transmitter's wavelength is driven by

- Divergence of the transmit beam (space loss effect)
- Desire to minimize background light or turbulence (for ground-based telescopes)
- Minimizing atmospheric attenuation (for ground-based telescopes)
- Required quality for optics
- Ground detector technology readiness/performance

- Transmitter efficiency (power consumption)
- Achievable transmitter peak and average power
- Flight qualification requirements (e.g., radiation hardness, reliability)
- Commercial availability (e.g., technology readiness level)
- Cost

In addition to output power and overall efficiency, some of the characteristic-defining parameters of optical amplifiers include

- *Gain:* The ratio in decibels, of input power to output power
- *Gain coefficient:* The ratio of the gain in decibels to the launched pump power in milliwatts
- *Gain ripple:* Shape of the gain spectrum
- *Noise figure:* Ratio of signal-to-noise ratio (SNR) at input to SNR at output
- *Bandwidth:* The range of wavelengths over which the amplifier operates

A discussion of laser transmitter options, trades among available options, characteristics of typical fiber amplifiers, modulation devices, and future directions will follow.

4.2.2 Laser Transmitter Options

Most lasercom links utilizing space-based transmitters require laser light sources with average output power levels below 10 W. Given this assumption, the pros and cons of viable laser transmitter options are traded in Table 4.8.

The merits and potential drawbacks of each of the viable laser transmitter options for flight are summarized below.

TABLE 4.8
Viable Flight Laser Types Comparing Some of Their Most Significant Merits and Drawbacks

Source	Pros	Cons
Diode lasers modulated directly or externally	Compact efficient	Difficult to achieve Gbps modulation at watt levels
Diode-pumped solid-state lasers	Multi-watt power possible	Inefficient
		Modulation rates are limited Heat dissipation a problem
Doped fiber amplifiers (direct-pumped or Raman)	Tens of watt power and tens of Gbps modulation rate in single spatial mode possible	Nonlinear effects may occur
SOAs	Compact and simple	Lower output power capability
		High coupling loss
		Higher noise figure

4.2.2.1   Semiconductor Laser Diode Transmitters

Laser diode transmitters for optical communications have matured significantly in a short time to multi-watt output levels, and in some cases with efficiencies exceeding 60%. Table 4.9 summarizes the current status of laser diode transmitters and compares them to the capability of MOPA systems. Laser diode transmitters are now delivering power more than 0.5 W in a single mode and several watts of output power in a multiple spatial mode. The output of diode lasers can be amplitude modulated (directly through rough current modulation) or externally via a waveguide or bulk crystal.

Laser diodes with an output in the range of a fraction of 1 W are not suitable for multi-gigabit communication rates from spacecraft, due in large part to space loss. Laser diodes with watt-level output have inadequate spatial mode quality and are difficult to modulate to the Gbps rates. Laser diodes with output power in the sub-watt regime may be large signal modulated at 1 GHz rate by direct modulation of the diode current [56]. However, a large-signal modulation of applied current to higher power lasers is very challenging, particularly as the packaging of higher output power lasers is typically not suited for higher speeds

TABLE 4.9
Merit Comparison of Commercially Available Laser Transmitters

Laser Type	Laser Power (W)	Single or Multi-Spatial Mode	Efficiency (up to, %)
370–535 nm diode	<0.2	SM	~40
	<2	MM	
600–795 nm diode	~1	SM	~60
	>100	MM	
800–895 nm diode	<10	SM	~60
	>100	MM	
850 nm VCSEL* 50 Gbps modulation	>0.01–0.1	SM	~60
940 nm VCSEL	50 (pulsed)	MM	~45
900–1195 nm diode	<10	SM	~60
	>200	MM	
1200–1300 nm diode	0.1	SM	~40
	<1.5	MM	
1310–1695 nm diode	~10	SM	~40
	>100	MM	
1700–1950 nm diode	<1	SM	~40
	<50	MM	
2000–2400 nm diode	~0.01	SM	~30

* VCSEL stands for vertical cavity surface emitting laser.

of modulation. Typically, external modulators require a high-quality input beam (not commonly available from multimode lasers). Most external modulators cannot tolerate several watts of input power and may introduce more than 3 dB of loss into the system.

4.2.2.2 Fiber Amplifier Transmitters

Rare-Earth-doped fiber lasers and amplifiers are a key subsystem in today's telecommunication systems. They are well suited for free-space laser communication links [57]. Depending on the desired characteristics, a large number of different rare-Earth ion dopants (Nd, Yb, Er, Pr, etc.) are available. The host fiber is glass based (SiO_3–GeO_2), with other glass forming oxides (Al_2O_3). Popular fiber optical amplifiers include ytterbium-doped fiber amplifiers (YDFA) or EDFA, semiconductor optical amplifiers (SOA), and Raman fiber amplifiers. In an EDFA, erbium ions (Er^{3+}) are added to the fiber core material as a dopant (impurity). Typical dopant levels are a few hundred parts per million. The fiber remains highly transparent to the Er emission (amplification) wavelength, even when light from the pump laser diodes excites it. The resulting gain medium in the fiber functions as an amplifier. Popular pump laser wavelengths for EDFAs are 980 and 1480 nm. As Figure 4.32 illustrates, multiple stages of amplification are commonly utilized for boosting the output power to the desired levels.

The physical components of an EDFA include multiple pump diode lasers at 920, 980, or 1480 nm wavelengths (equipped with monitor photodiodes), biconical fused fiber couplers, polarization insensitive optical isolators (isolating pump laser from input signals and isolating multiple stages of amplification from each other), and optical filters for gain flattening.

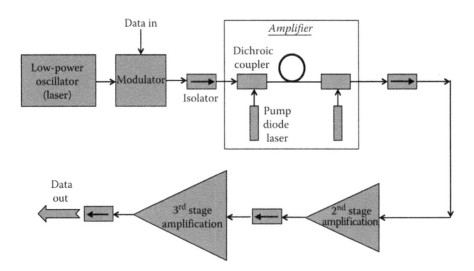

FIGURE 4.32 Schematic of a fiber amplifier showing some of the typical constituent components.

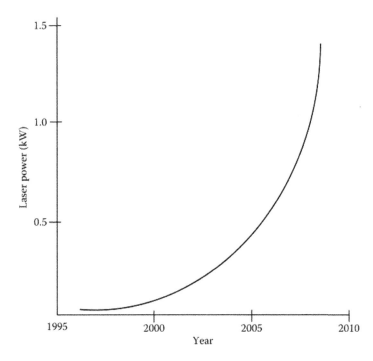

FIGURE 4.33 Exponential growth of single-mode core fiber amplifier output power as a function of time.

Figure 4.33 illustrates that the output power of fiber amplifier devices is increasing exponentially as a function of time. Therefore, communication systems are not limited by the lack of available transmitter power. However, the wall-plug efficiency enhancements of these sources are lagging behind but improving due to several concentrated efforts [58].

Raman fiberoptic amplifiers take advantage of the phenomenon that a small part of light scattered by the fiber has certain frequencies removed from the frequency of the incident beam by a quantity equal to vibration frequencies of the fiber material's scattering characteristics. When the input monochromatic beam is adequately intense (above a certain threshold), photons at Raman frequencies become amplified. The strong buildup of the field has the characteristics of stimulated emission. Thus, it is referred to as the stimulated or coherent Raman effect.

Currently, double-clad doped fibers provide the highest powers and efficiencies. In a double-clad fiber, a large multimode fiber surrounds the single-mode core [59]. Maintaining single-mode behavior is important for propagating a diffraction-limited Gaussian beam. Such amplifiers can be end-pumped or side-pumped with broad area high-power laser diodes, while the fiber is fairly insensitive to pump misalignment [60, 61]. Following multiple total internal reflections, pumped light is absorbed into the doped core. A drawback of double-clad fibers is that the fiber length must be increased to ensure absorption of the pump power. Large mode area fibers ease the thermo-optical burden on the fiber, thereby significantly reducing the fiber core power density.

A disadvantage of large mode area fibers is the potential for propagation of higher-order transverse modes, whereas a diffraction-limited quality beam is desired for most applications. To select out a single transverse mode utilized at the 1000- to 1600-nm wavelength range, a core diameter of less than 15 μm would be required for the step index fibers. Typical large mode area fibers have a core diameter of 30 μm or greater. One of the techniques identified to preferentially select a single mode is to wind the fiber tightly (within the allowable bending radius of the fiber). This technique is found to reject the higher-order modes [62].

Photonic crystal fibers (PCF), or holey fibers, offer advantages over conventional fibers, including substantial design flexibility and strictly single mode of operation over a large wavelength range. PCF cladding consists of a triangular array of air holes [63]. Single-mode Yb-doped cores, 50 μm in diameter, have been demonstrated. Double-clad PCFs are also achievable. Rare-Earth-doped PCFs offer several unique properties due to very high pump core numerical aperture and expansion of mode area while maintaining single-mode operation. This allows improved performance scaling (compared to conventional fiber lasers and amplifiers). Similar to more conventional fibers, nonlinear effects, damage threshold, and thermal loading drive the fiber amplifier design.

To generate higher output powers, multi-stages of amplification are often utilized, where booster stages with differing optical gains are chained together and typically isolated from back reflection by optical isolators. Table 4.8 points to the major advantages of fiber amplifiers. Many different fiber amplifiers have been demonstrated [54]. Table 4.9 compares the merits of some of the more popular fiber-based amplifier sources.

Yb-doped fibers provide amplification over the 975- to 1200-nm wavelength range [64]. The Er-doped fibers operate at the 1460- to 1560-nm range. Table 4.10 trades the advantages and disadvantages of each option.

The atmospheric, beam effects, and background light effects of the laser wavelength, assuming an aperture diameter d, are compared in Table 4.11.

Neglecting the current (nearly a factor of two or higher) overall efficiency of 1000-nm fiber amplifiers relative to 1550-nm fiber amplifiers, link analysis shows the difference between the influences of the two wavelengths on the link margin is minor.

TABLE 4.10
Benefits and Drawbacks of Different Fiber Amplifiers

Architecture	Characteristics
Diode-pumped Nd	High powers demonstrated
Diode-pumped Yb	975–1200 nm amplification range; high powers demonstrated
	Free from competing radiative processes encountered by other doped fibers. Relatively high efficiency
Diode-pumped Er	Parts commercially available from multiple sources
	Multiple stages often required
Nd-fiber-pumped Yb/Er	High powers demonstrated
	Robust pumping
Yb-fiber-pumped Yb/Er	Potential for highest efficiency. Robust pumping

TABLE 4.11

Comparison of the Properties of 1000- and 1500-nm Wavelength Fiber Amplifiers

Wavelength (nm)	Pros	Cons
1000	More efficient due to lower quantum defect; corresponding detectors slightly more efficient	Lower commercial availability of parts
	Lower diffraction losses (e.g., less space loss)	Higher quantum noise
1500	Compatible with fiberoptics components	Higher diffraction losses

The number of suppliers for passive and active components is more prevalent at the 1500-nm wavelength. Table 4.12 summarizes the properties of some of the commercially available high-power fiber amplifiers.

4.2.2.2.1 Key Advantages of Transmitters Based on Fiber Amplifiers

Laser transmitters that are based on fiber amplifiers offer major advantages over conventional solid-state laser technologies, e.g., diode-pumped bulk crystal solid-state lasers and semiconductor optical amplifiers (SOA). Some of the key advantages are listed below.

4.2.2.2.1.1 High Overall Efficiency As more efficient pump diode lasers with electrical-to-optical conversion efficiency exceeding 65% become commercially available, optical-to-optical conversion efficiencies greater than 70% have been demonstrated in Yb-doped fiber amplifiers [65, 66]. Due to the quantum defect of converting a 980-nm input to a 1550-nm output, Er-doped fiber amplifiers are less efficient by approximately a factor of 2, relative to Yb-doped fibers.

4.2.2.2.1.2 High Beam Quality and Beam Pointing Stability The fiber amplifier systems typically utilize a single-mode fiber. Therefore, high stability and quality of a single spatial mode output beam are expected. The mode quality is independent of mechanical jitter and thermal lensing, or other thermally induced aberrations

TABLE 4.12

Atmospheric Effects, Beam Effects, and Background Light Effects of Two Primary Fiber Laser Wavelengths

Parameters	1000 nm	1500 nm
Background light	1000×2.3	1
Transmittance	1	1500×1.08
Transmitted beam divergence (space loss effect)	1	$1500 \times (1.5)^2 d$

typically associated with diode-pumped solid-state lasers. For laser communications, a very low beam pointing jitter for the laser is desired. Again, in contrast to the diode-pumped bulk-crystal lasers with time-varying and temperature sensitive beam pointing, the output beam pointing of fiber-based systems is typically dominated by the stability of the final connector at the end of the fiber.

4.2.2.2.1.3 Suitable for Flight Use Optomechanical requirements for a spacecraft-worthy laser demand a source that is rugged, compact, and lightweight. Fiber-based laser transmitters can meet all of these requirements, simultaneously. Fiber amplifiers are typically free of the requirement for a major mechanical structure, traditionally associated with other laser sources. The output beam may be piped through difficult turns, allowing for placement of the pump diode lasers where they could be conveniently heat sunk. Relative to other flight-compatible lasers of comparable output power, fiber-based sources are considerably more compact. Also, due to elimination of the traditional mechanical bench, overall weight of the system is low.

4.2.2.2.1.4 Thermal Management With diode-pumped bulk crystals, the requirement for thermal management of the bulk crystal and control to a few degrees centigrade of pump lasers, typically poses major challenges. In a fiber-based system, thermal load is distributed across the fiber, and active cooling is generally not required. Pump diode lasers may be located away from the amplification medium (typically on a radiator) to remove excess heat. Unlike other diode-pumped solid-state lasers, pump diode lasers may experience temperature variations of as much as 40°C, without much noticeable effect on the performance of the fiber amplifier. This significantly contributes to improvements in wall-plug efficiency, size, and mass of the laser transmitter.

4.2.2.2.1.5 Flight Qualification In a fiber-based system, multiple optical elements are replaced with connectors, and the bulk crystal is replaced with fiberoptics. Thermal vacuum and vibration sensitivity are significantly reduced with fiber-based sources, whereas radiation susceptibility moderately increases. Significant progress has been made in development of radiation hard fibers and fiberoptic components [67]. Chapter 10 discusses the reliability issues in more detail.

4.2.2.2.1.6 Redundancy Implementation Fiber-based sources allow simpler architectures for implementing redundancy of the laser transmitter. Typically, pump diode lasers have the shortest life expectancy. The redundancy implementation task is made easier with a fiber-based laser system, where inclusion of multiple pump sources with the fiber is more easily permitted.

4.2.2.3 SOA Transmitters

SOAs are essentially Fabry-Perot laser diodes, where the end mirrors are replaced with nonreflecting fiberoptics. SOAs are highly compact, electrically pumped amplifiers that operate (amplify signals) over the same wavelength ranges as laser diodes. Current SOAs are inferior to conventional doped fiber amplifiers, as indicated by higher noise figure and coupling loss, limited output power, electrical-to-optical power efficiency, high nonlinearity with fast transient time, and moderate

FIGURE 4.34 Picture of an SOA. (From CIP Technologies; with permission. Available at www.ciphotonics.com.)

polarization dependence. Figure 4.34 illustrates pictures of a commercial SOA (www.ciphotonics.com).

SOAs with multi-watt average output power typically have a discrete semiconductor MOPA structure that allows separate optimization of the oscillator and amplifier [68]. More importantly, optical amplifier feedback into the oscillator can be avoided more effectively with this architecture. The amplified spontaneous emission (ASE) degrades SOA efficiency during the off pulse due to lack of energy storage in the amplifier material. An important consideration for use in telecommunications is that due to the common presence of a substantial amount of ASE, the SOA modulation extinction ratio is on the order of 10 dB (low compared to 20 dB for fiber amplifiers and approximately 40 dB for bulk solid-state devices). Figure 4.35 illustrates how the gain, noise figure, and polarization dependent gain of a typical SOA vary with the wavelength of the oscillator.

4.2.2.4 Raman Amplifier Transmitters

Fiberoptic Raman amplifiers (FRAs) utilize stimulated Raman scattering, a nonlinear interaction between pump laser(s) and the signal, to create a wide (up to 50 nm)

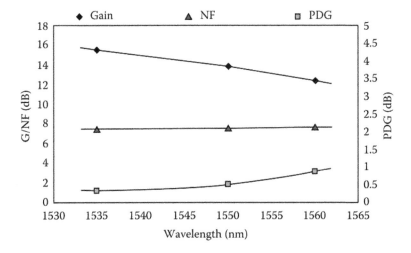

FIGURE 4.35 Gain (G), noise figure (NF), and polarization dependent gain (PDG) characteristics of an SOA as a function of wavelength (at 20°C). (From CIP Technologies; with permission. Available at www.ciphotonics.com.)

optical gain bandwidth [69]. The power, wavelength, and wavelength separation of pump lasers determine the output wavelength, gain, and noise of the amplifier. In a distributed Raman amplifier, the transmission fiber itself functions as the gain medium. In a lumped Raman amplifier, a dedicated shorter length of highly nonlinear fiber is utilized to provide amplification.

Advantages of FRAs over EDFAs include

- Lower noise buildup
- Simpler design, the optical fiber itself is used as the amplification medium
- Broader gain bandwidth
- More flexible signal frequency assignment
- Gain flattened bandwidth when combined with other amplifiers

Disadvantages of FRAs relative to EDFAs include

- Higher pump power lasers at specific wavelengths are required
- Possibility of cross talk among different channels
- Possibility of higher amplitude jitter for amplified signal

4.2.2.5   Erbium-Doped Waveguide Amplifier Transmitters

The erbium-doped waveguide amplifiers (EDWAs) operate on the same basic principle as doped fiber amplifiers and utilize a bulk waveguide, as opposed to a fiber waveguide used in EDFAs. This amplifier is capable of providing comparable gains to EDFAs [70]. Gain clamping that utilizes fiber Bragg grating or tunable band-pass filters are among the options commonly applied to EDWAs. Integration of multiple functions onto a mass producible photonic integrated circuit is one of the main attractions of EDWAs.

Wavelength, spectral bandwidth, and noise figure characteristics similar to EDFAs are generally measured for EDWAs. EDWAs utilize a higher (10–20 times) density of impurity ions than fiber amplifiers and have shorter amplification path length. EDWAs have traditionally shown inferior optical-to-optical pump power conversion efficiency and the signal coupling efficiency relative to doped fiber amplifiers (at this time). Due to lack of splicing capability, more free-space components are used with EDWAs than with fiber amplifiers.

4.2.2.6   Polarization Maintaining Transmitters

In some scenarios (e.g., background signal evasion, polarization multiplexing, and control), it is advantageous to utilize a polarization maintaining fiber for the amplifying medium. Maintenance of the polarization extinction ratio (typically <30 dB) under the airborne or space-borne environment is often a challenge [71]. Multiple architectures and fiber mediums have been devised to generate amplified output in a single polarization [72, 73].

4.2.3   Amplified Spontaneous Emission

Forward ASE, i.e., the spontaneous emission that is amplified within the fiber, is a key noise source in optical amplifiers. Since the emission spectrum of ASE closely matches the amplifier gain spectrum, backward ASE can result in reduction of the amplifier gain. ASE filters are effective in minimizing negative effects of this undesired emission.

4.2.4 MODULATION

Direct modulation of high-output power (mW) amplifiers is often difficult, due to the unavailability of modulators that can handle those power levels. Fortunately, MOPAs facilitate this requirement, whereby the low-power master oscillator is modulated, and the amplifiers preserve modulation characteristics of the input beam.

Popular methods of modulation include direct modulation of the semiconductor laser current and external modulation of the low-power beam. The direct modulation technique is simple, low-cost, and compact. However, due to carrier induced chirp and temperature variations from carrier modulation, the output frequency may shift with the drive signal. Due to the preference for nonreturn to zero modulation, the modulation extinction ratio is limited. External modulation techniques, including electrooptical modulation and electro-absorption, enable higher speed, larger extinction ratio, lower chirp, and lower modulation distortion. The electro-absorption modulators are highly compact (as they are integrated with the chip), but signal chirp may still occur. The major disadvantage of external modulation is about 6 dB of insertion loss. External modulators with bandwidth of 40 GHz are now commercially available.

The electro-absorption modulator is a semiconductor device, wherein an applied electric field is used to change the absorption spectrum. Usually, carriers are not injected into the active region but are generated due to light absorption. Waveguide type electro-absorption type modulators result in a higher extinction ratio relative to transmission type modulators.

4.3 SUMMARY

In summary, only certain semiconductor lasers, solid-state lasers, and fiber amplifiers are practical for near-Earth laser communications. Fiber amplifiers provide high-power levels, wide bandwidth, extremely high modulation rate, and reliability needed for most demanding laser communication applications. External modulation provides cleaner pulses with, higher extinction ratio, but introduces higher insertion loss (relative to direct modulation). In high power amplifiers, noise, nonlinear effects, and ASE should be avoided as much as possible.

YDFAs and EDFAs can offer a combination of ideal characteristics, including over 50 dB of gain, greater than 30 dBm of output power and (for a well-compensated amplifier) about 3 dB of noise (Table 4.13). Challenges associated with these amplifiers include polarization dependent gain, polarization hole burning (causing gain anisotropy in saturated amplifiers), finite gain spectrum, and nonuniform spectral gain shape. Gain equalizers can mitigate the nonuniformity of the amplifier gain spectrum. A well-designed amplifier minimizes the nonlinear effects in the high-power fiber amplifier. These effects include stimulated Raman and Brillouin scattering, self and cross-phase modulation, and four-wave mixing.

Figure 4.36 illustrates the behavior of fiber amplifier gain and noise figure as a function of the length of the fiber and as a function of copropagating and

TABLE 4.13
Examples of Commercially Available High-Power Fiber Amplifier Sources

Manufacturer: Parameters	IPG Photonics	Keopsys	Emcore/Ortel
Fiber type	Single mode	Single mode	Single mode
Wavelength range (nm)	1540–1605	1540–1565	1545–1562
Saturated output power (dBm)	>40	40	35
Optical input power (dBm)	3	5–15	12
Typical noise figure (dB)	<7	<9	<6
Data-rate capability (Gbps)	>40	>40	>40
Output power stability (dB)	<0.2 (over 8 h)		±0.25
Polarization	Random	Polarized	Polarized
Input/output optical isolation (dB)	40/25		40/40
Mean time between failure (MTTF) of pump diodes (hours at 20°C)	100,000		

counterpropagation pump sources. Amplifier gain, under the condition of counter-propagating pump laser, increases more rapidly (relative to copropagating fibers) as the fiber length increases. Conversely, noise figure increases dramatically with fiber length for the case of counterpropagating pump. Figure 4.37 shows the typical effect of input signal wavelength on the gain and noise figure of the amplifier. The noise figure shows a slight decrease with increase in signal wavelength.

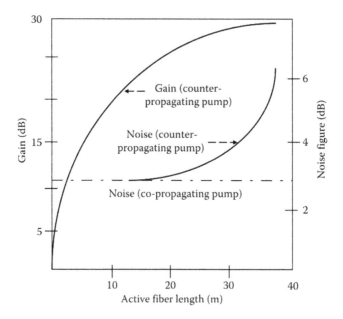

FIGURE 4.36 Gain and noise figure example measured for a particular EDFA as function of the fiber amplifier's length (results vary for amplifiers, e.g., depending on Er dopant concentration, and pump power).

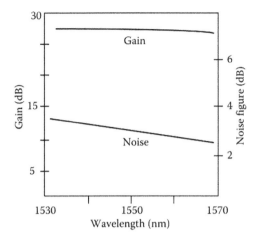

FIGURE 4.37 Gain and noise figure example measured for a particular EDFA as function of the signal wavelength (results vary from amplifier to amplifier).

Future technology development of high bandwidth lasercom will focus on enhancing the true overall efficiency and reliability, including laser lifetime. Wavelength division multiplexing (WDM) commonly utilized in fiberoptics industry is applicable to free-space communications over long ranges, as well as extending the data rate capability to hundreds of Gbps. Some of the applications that require multi-gigabit data rates include cross-link between satellite (such as multi- and hyperspectral imaging) and synthetic aperture radar (SAR) imaging. Researchers are now extending the output wavelength of fiber lasers and amplifiers (Figure 4.38).

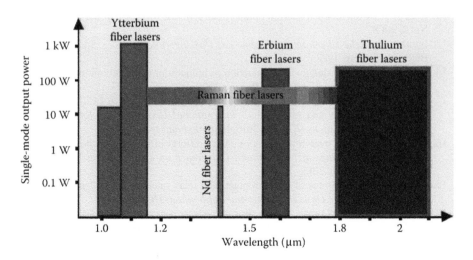

FIGURE 4.38 Wavelength span of current fiber amplifiers based on differing fibers and dopants. (From ipgphotonics.com; with permission.)

REFERENCES

1. Smutny B. et al. "In-orbit verification of optical inter-satellite communication links based on homodyne BPSK." *Free-Space Laser Communication Technologies, XX*, Vol. 687702, 2008.
2. Sodnik Z. et al. "Optical intersatellite communication." *IEEE Journal of Selected Topics in Quantum Electronics*, Vol. 16, No. 5, 2010 p 1051–1057
3. ESA. *EDRS (European Data Relay Satellite) Constellation/SpaceDataHighway.* Brochure. 2019. https://directory.eoportal.org/web/eoportal/satellite-missions/e/edrs
4. Fuchs C. et al. "SOTA optical downlinks to DLR's optical ground stations." *International Conference on Space Optics (ICSO)*, 2016, Biarritz, France.
5. Dreischer T. et al. "Functional system verification of the OPTEL-µ laser downlink system for small satellites in LEO." *Proceedings of International Conference on Space Optical Systems and Applications (ICSOS)*, 2014, Kobe, Japan.
6. Chishiki Y. et al. "Overview of optical data relay system in JAXA." *Free-Space Laser Communication and Atmospheric Propagation XXVIII, SPIE*, Vol. 9739, 2016, San Francisco, California, United States.
7. Moision B. et al. "Demonstration of free-space optical communication for long-range data links between balloons on Project Loon." *Free-Space Laser Communication and Atmospheric Propagation XXIX, SPIE*, Vol. 10096, 2017, San Francisco, California, United States.
8. Chen C. et al. "Demonstration of a bidirectional coherent air-to-ground optical link." *Free-Space Laser Communication and Atmospheric Propagation XXX, SPIE* Vol. 10524, 2018, San Francisco, California, United States
9. Horwath J. et al. "Experimental verification of optical backhaul links for high-altitude platform networks: Atmospheric turbulence and downlink availability." *International Journal of Satellite Communications and Networking,* Vol. 25, No. 5, 2007.
10. Perlot N. et al. "Optical GEO feeder link design." *Future Network & MobileSummit*, 2012, http://ieeexplore.ieee.org/document/6294234, Berlin, Germany.
11. Camboulives A. R. "Compensation des effets de la turbulence atmosphrique sur un lien optique montant sol-satellite geostationnaire: Impact sur la architecture du terminal sol." Ph.D. thesis. l'Universite Paris-Saclay, 2018.
12. Ip E. et al. "Coherent detection in optical fiber systems." *Optics Express 753*, Vol. 16, No. 2, 2008.
13. Franz J. *Optische Übertragungssysteme mit Überlagerungsempfang.* Springer-Verlag Berlin Heidelberg, 1988.
14. Kahn J. E. et al. "Modulation and detection techniques for optical communication systems." *Optics InfoBase Conference Papers*, 2006.
15. Sodnik Z. et al. "The ESA's optical ground station for the EDRS-A LCT in-orbit test campaign: Upgrades and test results." *International Conference on Space Optical Systems and Applications (ICSOS)*, 2016, Biarritz, France.
16. Edwards B. L. "An overview of NASA's latest efforts in near earth optical communications." *Proceedings of International Conference on Space Optical Systems and Application (ICSOS)*, 2015.
17. Witting M. et al. "Status of the European data relay satellite system." *International Conference on Space Optical Systems and Applications (ICSOS)*, 2012.
18. Luzhansky E. et al. "Overview and status of the laser communication relay demonstration." *Free-Space Laser Communication and Atmospheric Propagation XXVIII, SPIE*, Vol. 9739, 2016.
19. Liu Xiang et al. "M-ary pulse-position modulation and frequency-shift keying with additional polarization/phase modulation for high-sensitivity optical transmission." *Optics Express*, Vol. 19, No. 26, 2011.

20. Caplan D. O. et al. "Multi-rate DPSK optical transceivers for free-space applications." *Free-Space Laser Communication and Atmospheric Propagation XXVI, SPIE*, Vol. 8971, 2015.
21. Araki T. "Digital coherent receiver technique for onboard receiver of future optical data relay system." *Free-Space Laser Communication and Atmospheric Propagation XXVIII, SPIE*, Vol. 9739, 2016.
22. MacDougall J. et al. "Transmission and pump laser modules for space applications." *Free-Space Laser Communication and Atmospheric Propagation XXIX*, Vol. 100960I, 2017.
23. Kaminski A. A. *Laser Crystals.* Springer-Verlag Berlin Heidelberg, 1990.
24. Rose T. S. et al. "Performance and spectroscopic characterization of irradiated Nd:YAG." *Inorganic Crystals for Optics, Electro-Optics, and Frequency Conversion, SPIE*, Vol. 1561, 1991.
25. Johann U. A. et al. "Diode pumped Nd:Host laser transmitter for intersatellite optical communications." *Free-Space Laser Communication Technologies II, Proceedings of SPIE*, Vol. 1218, 1990.
26. Kane T. J. et al. "Monolithic, unidirectional single-mode Nd:YAG ring laser." *Optics Letters*, Vol. 10, No. 2, pp. 65–67, 1985.
27. Schmitt N. P. et al. *Diodengepumpte Festkörperlaser.* Springer-Verlag, Berlin, 1993.
28. Schmitt N. P. *Abstimmbare Mikrokristall-Laser.* Shaker Verlag, Aachen, 1995.
29. Pribil K. et al. "High data rate inter-satellite-link communication system SOLACOS." *CRL International Workshop on Space Laser Communications, Tokyo, Japan*, 1998.
30. Pribil K., Leeb W. R., and Scholtz A. L. "140 Mbit/s heterodyne ASK data transmission with diode pumped Nd:YAG lasers." *Free-Space Laser Communication Technologies II, SPIE*, Vol. 1218, 1990.
31. Koechner W. *Solid State Laser Engineering.* Springer Verlag, New York, 1989.
32. Scheps R. "Single-mode operation of a standing wave miniature Nd laser pumped by laser diodes." *Applied Optics*, Vol. 28, No. 24, 1989.
33. Rasch A. et al. "Optical carrier modulation by integrated optical devices in lithium niobate." *Optical Space Communication II, SPIE*, Vol. 1522, 1991.
34. EOspace. *Advanced products 2019.* Short-form product catalog. 2019. www.eospace.com
35. IXBlue. *Modulators & Modbox.* Brochure. 2019. www.photonics.ixblue.com
36. Fickenscher M. et al. "Characteristics of an integrated optical phase modulator at high optical energy densities." *Laser in Research and Engineering—Proceedings of the 11th International Congress Laser 93.* 1993.
37. Wang C. et al. "Integrated lithium niobate electro-optic modulators operating at CMOS- compatible voltages." *Nature*, Vol. 562, 2018.
38. Axenic. "aXMD2150 optical modulator." Data sheet. 2019. www.axenic.co.uk
39. Caplan D. O. et al. "High-sensitivity multi-channel single-interferometer DPSK receiver." *Optics Express*, Vol. 14, No. 23, 2006.
40. Lu G. W. et al. "40-Gbaud 16-QAM transmitter using tandem IQ modulators with binary driving electronic signals." *Optics Express*, Vol. 18, No. 22, 2010.
41. Neophotonics. "High bandwidth coherent driver modulator (HB-CDM)." Short-form data sheet. 2019. www.neophotonics.com
42. Petsch T. "Electro-optic modulator for high-speed Nd:YAG laser communication." *Optical Space Communication II, SPIE*, Vol. 1522, 1991.
43. Kane T. J. et al. "Diode pumped Nd:YAG amplifier with 52 dB gain." *Free-Space Laser Communication Technologies VII, SPIE*, Vol. 2381, 1995.
44. Gapontsev Y. P. et al. "3W saturation power polarization maintaining 1064 nm ytterbium fiber amplifier." *Free-Space Laser Communication Technologies XI, SPIE*, Vol. 3615, 1999.

45. Baister G. C. et al. "The OPTEL terminal development programme-enabling technologies for future optical crosslink applications." *21st AIAA International Communication Satellite Systems Conference,* 2003, Long Beach, California, United States.

46. Kingsbury R. W. et al. "Compact optical transmitters for CubeSat free-space optical communications." *Free-Space Laser Communication and Atmospheric Propagation XXVI,* Vol. 9354, 2015.

47. Ladaci A. et al. "Optimization of rare-earth-doped amplifiers for space mission through a hardening-by-system strategy." *Free-Space Laser Communication and Atmospheric Propagation XXIX,* Vol. 10096, 100960F, 2017.

48. Stampoulidis L. et al. "Radiation-hard mid-power booster optical fiber amplifiers for high-speed digital and analogue satellite laser communication links." *Proceedings of International Conference on Space Optical Systems and Application (ICSOS),* 2014, Tenerife, Canary Islands, Spain.

49. IXBlue. *Rad Hard Space Grade Doped Fibers.* Brochure. 2019. www.photonics.ixblue. com

50. Girard G. et al. "Recent advances in radiation-hardened fiber-based technologies for space applications." *Journal of Optics,* Vol. 20, 2018.

51. Chan V. W. S. "Optical space communications." *CRL Workshop on Space Laser Communications, CRL Tokyo,* 1997.

52. Jost G. *Hochleistungslaserverstaerker. Shaker Verlag, Aachen,* 2000.

53. Walpole J. N. et al. "High-power, strained-layer amplifiers and lasers with tapered gain regions." *IEEE, Photonics Technology Letters,* Vol. 5, 1993.

54. D. J. F. Digonnet and M. J. F. Digonnet, *Rare Earth Doped Fiber Lasers and Amplifiers,* Marcel Dekker, New York, 2001.

55. IPG Photonics, http://www.ipgphotonics.com

56. H. Hemmati and D. Copeland, Laser transmitter assembly for optical communications demonstrator, *Proc. SPIE,* 2123, 283–291, 1994.

57. M. Pfennigbauer and W. R. Leeb, Optical satellite communications with erbium doped fiber amplifiers, *Space Commun.,* 19, 59–67, 2003.

58. C. M. Stickly, *An Overview of DARPA's SHEDS and ADHELS Programs,* SPIE, 6104–6106, 2006.

59. E. Snitzer, H. Po, F. Hakimi, R. Tumminelli, and B. C. McCollum, *Double-Clad, Offset Core Nd Fiber Laser,* OFC, 1997.

60. P. Glas, M. Naumann, A. Shirramacher, and T. Petsch, The multi-core fiber—a novel design for a diode pumped fiber laser, *Optics Commun.,* 151, 187–195, 1998.

61. J. P. Koplow, L. Goldberg, and D. A. V. Kliner, Compact 1-W Yb-doped double cladding fiber amplifier using V-groove side-pumping, *Photon Technol. Lett.,* 10, 793–795, 1998.

62. P. Koplow, L. Goldberg, R. P. Moeller, and D. A. V. Kliner, Single-mode operation of a coiled multimode fiber amplifier, *Opt. Lett.,* 25, 442, 2000.

63. J. C. Knight, Photonic crystal fibers, *Nature,* 424, 847, 2003; P. Russell, Photonic crystal fibers, *Science,* 299, 358, 2003.

64. R. Paschotta, J. Nilsson, A. Tropper, and D. C. Hanna, Ytterbium-doped fiber amplifiers, *IEEE J. Quant. Elec.,* 33, 1049–1056, 1997.

65. Y. Jeong, J. K. Sahu, D. N. Payne, and J. Nilsson, Ytterbium-doped large-core fiber laser with 1.36 kW continuous-wave output power, *Opt. Exp.,* 12(25), 6088–6092, 2004.

66. J. Limpert, N. Deguil-Robin, S. Petit, I. Manek-Honninger, F. Salin, P. Rigail, C. Honninger, and E. Mottay, High power Q-switched Yb-doped photonic crystal fiber laser producing sub-10 ns pulses, *Appl. Phys. B,* 81, 19–21, 2005.

67. C. Barnes, M. Ott, H. Becker, M. Wright, A. Johnson, C. Marshall, H. Shaw, P. Marshall, K. LaBel, and D. Franzen, NASA electronics parts and packaging (NEPP) program assurance research on optoelectronics, *Proc. SPIE,* 589707, 2005.

68. M. W. Wright, Characterization of high-speed, high power semiconductor master-oscillator power amplifier (MOPA) laser as a free-space transmitter, *JPL TMO Progr. Rep.*, 42–142, 1–11, 2000.

69. S. Namiki and Y. Emori, Ultra-band Raman amplifiers pumped and gain equalized by wavelength-division-multiplexed high-power laser diodes, *IEEE J. Sel. Top. Quant. Electron.*, 7, 3–16, 2001.

70. D. Zimmerman and L. H. Spiekman, Amplifiers for the masses: EDFA, EDWA, and SOA amplets for metro and access applications, *J. Lightwave Technol.*, 22, 63–70, 2004.

71. P. Wysocki, T. Wood, A. Grant, D. Holcomb, K. Chang, M. Santo, L. Braun, and G. Johnson, High reliability 49 dB gain, 13W PM fiber amplifier at 1550 nm with 30 dB PER and record efficiency, *Proceedings of OFC (Optical Fiber Communications) Conference, paper PDP17*, Anaheim, CA, 2006.

72. D. A. V. Kliner, J. P. Koplow, L. Goldberg, A. L. G. Carter, and J. A. Digweed, Polarization-maintaining amplifier employing double-clad bow-tie fiber, *Optics Lett.*, 26, 184–186, 2001.

73. Chen C. et al. "High-speed optical links for UAV applications." *Free-Space Laser Communication and Atmospheric Propagation XXIX, SPIE*, Vol. 10096, 2017.

5 Flight Optomechanical Assembly

Hamid Hemmati

CONTENTS

5.1   INTRODUCTION

The optomechanical assembly (OMA) of a flight optical transceiver typically consists of a transmit/receive aperture and an aft optics assembly that route the transmit and receive signals through the system. This assembly conditions the input laser transmitter beam to achieve the designed output beam divergence. An OMA typically comprises the following elements:

- Main transmit and receive aperture (telescope)
- Aft optics that route the incoming and outgoing beams
- Acquisition and tracking sensors, and associated actuators for beam pointing
- Mechanical structure that integrates the optics
- Coarse-pointing mechanism that compensates for the platform's attitude
- Inertial sensors and isolation mechanisms (in some systems)

Additional functions associated with this assembly include solar light rejection (such as filtering and baffling), thermal management of the optics, protection against contaminants, and interface with the host platform. One critical function of a laser communications transceiver is to deliver the signal reliably while maintaining a precise coalignment of the mutual line of sight (LOS). The quality and stability of the flight optical system must be preserved to satisfy the stringent requirements characteristic of laser communication from space, e.g., vibration and temperature gradients, and fluctuations. Figure 5.1 shows a block diagram of a generic laser communication terminal.

The optical train may include a common transmit/receive aperture, followed by data transmit, data receive, acquisition and tracking, and reference channels. The transmit laser signal reflects from a two-axis fine-steering mirror (FSM) before traveling through the telescope. The beacon signal from a remote receiver (on Earth or at an intermediate relay location) can be imaged onto a focal plane detector array for spatial acquisition and tracking. The receive path may accommodate both data and beacon signals. The beacon signal aids with precision LOS pointing, acquisition, and tracking (PAT). A small portion of the transmit signal can also be imaged onto the focal plane in order to keep track of the outgoing beam [1, 2]. An alternative approach that has the potential to simplify the optomechanical system development uses a fiber-bundle receiver, where a fiber array leads to a tracker

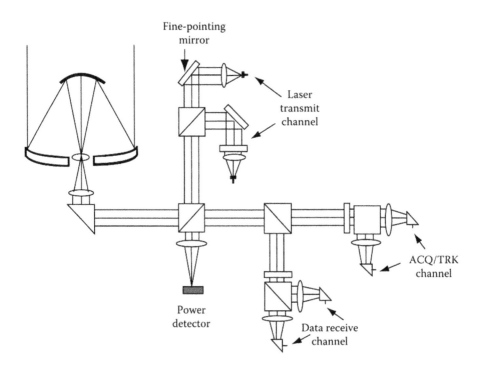

FIGURE 5.1 Example of an optical train for a lasercom transceiver, utilizing a single aperture to transmit and receive, with redundant transmitter and receivers. Drawing not to scale.

focal plane (or a detector array) and is bundled in the aft optics segment of the optical assembly [3].

5.2 GENERIC REQUIREMENTS

The generic requirements discussed below apply to both direct and coherent optical signal detection schemes with an emphasis on direct detection. Optomechanical alignment tolerance requirements for a lasercom system operating in the coherent detection mode can be more stringent than that based on direct detection. Key requirements on the optics assembly include high thermo-mechanical stability, immunity to (launch) shock and vibrations, and maintaining the overall tight optical and mechanical tolerances that are typically imposed by the assembly's design to enable micro-radian level laser beam pointing accuracy from space.

These requirements are typically derived from a link budget. For example, an optical aperture diameter sets minimum diffraction-limited divergence angle and therefore antenna gain, but there are also losses from wavefront error, throughput (absorption and scattering), and pointing alignment and jitter. These are all high-level terms in a link budget that must be decomposed based on the selected architecture, and allocated to the various constituent terms in order to ensure that the overall system meets its requirements at the end.

In addition to requirements imposed by the link budget, certain requirements are dictated by the overall link geometry and initial acquisition scheme. For example, the coarse-pointing assembly (CPA) field of regard is dictated by the orbital trajectories. Both CPA accuracy and the acquisition sensor field of view (FOV) have to be sized in consideration of a spacecraft attitude knowledge uncertainty (e.g. the star tracker accuracy) to ensure a timely and successful spatial acquisition.

- A structure with high fundamental frequency and high temporal and thermal stability to meet the sub-microradian laser pointing accuracy requirement.
- Minimum mass and swept volume.
- An optical system capable of supporting multiple optical wavelengths.
- The transmit optical path must have an FOV sufficient to provide steering capability to control the downlink over the entire system FOV and must maintain good wavefront quality over the entire FOV.
- A receive optical path to collect the incident photons from the input aperture and direct the signal to the data detector.
- Excellent background light rejection. The receive optical path must provide narrow band filtering to reduce the amount of background radiation and must provide the wide FOV necessary to cover the spacecraft's angular vibration excursions.
- For some missions there is a need for point-ahead angle compensation. This will also vary based on orbital trajectory—for nearby systems it may be ≤ 12 μrad, for farther-away and/or faster-moving targets, it could be ≥ 50 μrad.

Examples of critical specifications and requirements, driving the optomechanical subsystem design, include:

Aperture diameter	On-axis or off-axis telescope
Background light rejection and baffling	Obstructions secondary optics, baffles
Beam divergence	Operational wavelengths
Beam walk control	Operational link ranges
Beam size ratio between aft optics, telescope	Optical efficiency, surface smoothness
Bidirectional reflectance distribution function (BRDF)	Point-to-point or point-to-multipoint
Combined or separate apertures	Pointing, acquisition, tracking concept
Detector type and efficiency	Polarized light accommodation
Effective focal length	Radiation hardness of refractive optics
Field of view of detectors/focal planes	Redundancy
Field of regard of course pointer	Solar filtering
FSM(s) accommodation	Stray light control
Focal or afocal telescope	Strehl of optics
Interfaces (mechanical, electrical, and thermal)	Stability (temporal and thermal)
Interfaces (data and command handling)	Stops and shutters
Laser beam clipping	Thermal extremes and cycling
Laser feedback (minimized)	Throughput
Line-of-sight coalignment	Transmit/receive isolation
Mass, volume, input power, and cost	Unidirectional or bidirectional link
On- or off-axis vignetting loss	Wavefront quality of optics

Some of the major design drivers and options for the flight system optics design are discussed below.

5.3   OPTICAL DESIGN DRIVERS

Among the multiple design drivers on the flight transceiver are communication range between transceivers, the sun angles, propagation through the atmosphere, maximum throughput of near diffraction-limited output beam quality, and optomechanics that maintain allocated tolerances in space [4]. The telescope's principle functions are to serve as the aperture that receives the data or beacon signal from a remote terminal, and to serve as the aperture for transmission of the outgoing laser beam to a remote terminal. Two general options are available: (1) a single aperture to perform both functions, in which case the often stringent transmit and receive isolation issues have to be addressed; or (2) separate apertures for transmission and reception. In the latter case, angular mechanical drift between the apertures will have to be addressed as it relates to precision tracking and pointing. Both refractive and reflective telescope systems represent viable options.

5.3.1   TELESCOPE SELECTION

It is important to select an optical design that can accommodate diverging requirements of a lasercom transceiver without a significant deterioration of optical

TABLE 5.1

Generalized Characteristics of Certain Telescope Architecture Suitable for Flight

Characteristics	Cassegrain	Catadioptric Cassegrain	Maksutov Cassegrain	Schmidt Cassegrain	Off-Axis Gregorian	Off-Axis Newtonian
Field of view	Narrow to moderate	Moderate	Large	Large	Narrow	Narrow to moderate
Off-axis light rejection	Good	Good	Fair (due to corrector)	Fair (due to corrector)	Excellent	Very good
Weight (relative)	Light to medium	Medium	Heavy	Medium	Medium	Medium
Length (relative)	Short	Short	Short to medium	Short to medium	Long	Long

performance. Table 5.1 shows examples summarizing the merits of multiple options available for flight optics design. The actual choice of the telescope architecture depends on requirements such as size (e.g., diameter), mass, spectral range, FOV, obscuration, alignment tolerances, and the thermal environment. A refractive system may be a practical choice for smaller diameter transceivers. Otherwise, reflective systems are generally chosen with larger lightweight mirrors to reduce the contribution to total system mass.

The telescope may be afocal, generating a collimated beam path after the telescope, or focal. An afocal telescope (essentially used in a beam-expanding mode) is suitable for single apertures in a transceiver configuration. Relative to focal telescopes, an afocal mode of operation allows for easier incorporation of beam-matching optics and FSMs in the aft optics area of the telescope. The optical magnification of the telescope (i.e., the beam expansion ratio) may be selected to best match the requirements for the FSM.

Design drivers that affect the telescope/beam expander magnification selection include the following:

1. LOS jitter angles are "shrunk" proportional to the magnification of the telescope (a consequence of the Lagrange Invariant) when transitioning from the small beam space to the output space. Therefore, if a beam coming from the aft optics assembly is jittering around with 10-μrad root mean square (RMS) jitter, it goes through an afocal telescope with 10× lateral magnification, and it will only have 1-μrad RMS jitter in output space. In this sense, a larger telescope magnification helps shrink open-loop LOS jitter.

2. The converse of the previous statement is that, any FSM mechanisms "upstream" of the telescope (i.e., in the aft optics) will have to work harder/travel a larger physical distance to impart the same amount of output-space beam pointing. For example, using an FSM that needs to apply a 50-μrad point-ahead angle in output space, going from a telescope magnification of 5 to 10 will double the required travel from of the FSM mechanism.

3. For a fixed output aperture size, having a larger telescope magnification will mean a smaller beam size in the small beam space. However, as a consequence of the same effect mentioned above, off-axis rays with a certain divergence/FOV in output space will have a proportionally larger FOV/cone angle in small beam space. These two effects will counteract each other, so basically the takeaway is that having a larger telescope magnification does not necessarily mean smaller back-end optics (since it depends on FOV as well).

Variants of the Cassegrain telescope include Classical Cassegrain (CC), Ritchey Chretien (RC), Schmidt Cassegrain (SC), and Cassegrain with a corrective refracting (CR) optics between the secondary mirror and the focal plane [5]. The CC telescope consists of a paraboloidal primary mirror and a hyperboloid secondary mirror. This configuration provides significantly higher imaging performance and a larger FOV (approximately ±2 mrad) than a single paraboloid mirror. The RC configuration consists of aspheric primary and secondary mirrors with slightly weakened zonal curvatures (relative to CC telescope mirrors). The SC configuration incorporates a refractive plate having a toroidal aspheric figure to correct for spherical aberrations. The CR design in which refractive optics is added to the converging beam (in the aft section of the telescope) affords additional aberration correction and delivers good image quality over a large FOV. A Schiefspiegler telescope is a variant of the Cassegrain design with off-axis mirrors to avoid the secondary obscuration.

In the aft optics section of the telescope, the signal to be downlinked to Earth is moved between the laser signal generator unit and the main aperture (the telescope) via an optical train consisting of a combination of fiberoptics, mirrors, and lenses. Here, signal energy is generated or detected, and the optical beam is shaped, split, magnified, controlled, and sampled. The aft optics may consist of the transmit and receive channel, the acquisition and tracking channel, and channels that accommodate the laser beam and alignment, calibration, or inertial sensors with an optical output.

The transmit signal optics relay the optical beam from laser transmitter output to the exit aperture of the optics. The transmit optical path must provide steering capability to control the downlink beam direction over the entire system FOV and must maintain a good wavefront quality over the entire FOV.

The receive optical path collects the incident photons from the input aperture and directs them to the data detector. The receive optical path must provide narrow band filtering to reduce the amount of background radiation and provide the wide FOV necessary to cover the spacecraft's range of angular excursions. The receive optical path must also provide sufficient isolation such that the effect of signal feedback from the transmit path is minimized. Wavefront quality is not critical for the receive optical path so long as it can provide a small enough received signal spot size to fit on a receive detector small enough to provide adequate signal-to-noise ratio. However, matching a wide FOV to a small diameter detector could become a significant optical design challenge. The transmit path places the most critical demand on the image quality of the optical system. As dictated by the link budget, typically telescope and imaging optics combined are required to maintain a wavefront on the order of

TABLE 5.2
Example: Wavefront Quality of Transmit Channel

Component	Wavefront Error (λ/)	RSS (at 1 μm) (λ/)
Transmit optical path		14.6
Laser beam shaper optics	20	
Point-ahead mirror	50	
Lens and mirrors	30	
Quarter wave plate	60	
Other elements	50	
Common optical path		14.4
Telescope	30	
Power optics	50	
Dichroic	50	
Fast-steering mirror	50	
Flat fold mirror	20	
ω = wavefront error (RMS) λ/		10.3
Strehl $\sim e^{-(2\pi\sigma)2}$		0.69

RSS = Root Mean Square

λ/15 RMS (at 500 nm). That means, the individual reflectors need to have a wavefront error better than λ/20 (RMS at 500 nm).

The Strehl ratio is a measure of the wavefront quality and is defined by ratio of ideal on-axis irradiance for a perfectly unaberrated beam vs. actual on-axis irradiance for an aberrated beam transmittance/reflectance (throughput) of each optical component in the system which determines the overall system losses [6]. Highest wavefront quality (highest Strehl, approaching 1.0) is required to minimize the losses. The system wavefront quality is calculated by root mean squaring (RMSing) the wavefront contributions of individual optical elements in a given path. The example in Table 5.2 demonstrates the influence of individual optical elements wavefront contributions. Highest throughput is accomplished by utilizing lowest loss optical coatings.

Following the completion of initial optics design, thermal and structural analysis of the OMA should be conducted to identify the anticipated thermal gradients and expected range of component motions. Performance of the perturbed optical system (including wavefront budgets and alignment budgets) can then be calculated by feeding these findings back into the optical model.

5.4 REJECTION OF UNDESIRED PHOTONS

5.4.1 BACKGROUND LIGHT REJECTION/FILTERING

The thermal management of a laser communication terminal becomes more manageable if the entrance aperture of the terminal is equipped with a window that has been coated for optical band-pass characteristics that reject wavelengths not utilized in the terminal's optical communication. If a common transmit/receive aperture is used, the filter needs to be highly transparent at all incoming and outgoing wavelengths. This window will substantially help in maintaining the often stringent requirements on the environment's cleanliness for the purpose of alleviating the undesired stray light caused by scattering from contaminant particles on optics.

Scattering from the optics and (in particular) the primary mirror could potentially be a major source of stray light. Scattering from the optics will arise due to dust or other optical contaminants, or from pitting of the primary mirror from collisions with micrometeorites and on-orbit debris. To control stray light, external and central baffles are often added. The central baffle may include a field stop to control the FOV of the optical system. Stray light due to scattering from vanes holding the secondary mirror may be alleviated by including a Lyot filter to the optical system. Field stops and conical baffles are also an effective means of stray light control. The baffles are typically painted black to minimize reflection. Several baffling schemes (such as stacked-type vanes) have been devised for this purpose [7].

MIL-Standard MIL-F-48616 is relevant to optical filters for flight use. Some of the key specifications for a filter include polarization effects, cosmic background light radiation effects, dimensional stability, substrate material, substrate flatness (specified in waves at a certain wavelength), surface quality, center wavelength, nominal bandwidth, minimum transmittance, out-of-band rejection characteristics, and operating temperature. To avoid signal loss, the spectral width of the narrow band-pass filter has to be wider than the line width of the transmitted laser plus the Doppler shift of the signal. Also, the transmission bandwidth of most band-pass filters varies as a function of incidence angle, and often as a function of temperature. A collimated incoming beam and temperature control may be required to obtain optimum performance. Table 5.3 characterizes some of the key characteristics of band-pass filters that have conventionally been utilized.

TABLE 5.3
Summary of Key Characteristics of Narrow-Band Optical Filters for Lasercom

Optical Filters in the Visible to Near-IR	Bandwidth (nm)	Transmission (%)	Tunability (nm)
Atomic resonance	0.01	10–50	<0.002
Fabry-Perot			
Interference	<0.2	>60	Several nanometers
Single cavity etalon	<0.1	>90	
Double cavity etalon	<0.1	>80	
Volume Bragg grating	>0.25	>90	~5
Birefringent			
CdS and quartz, Lyot and Solc	~0.2	<30	Hundreds of nm
Acousto-optic			
Quartz, TeO$_2$	>2	>90	>500

Note: Some data are typical while others are state of the art (2008).

5.4.1.1 Solar Rejection Filters

During telecom operations when the Sun is within the spacecraft FOV, depending on the flight telescope's diameter, a fraction of the approximately 1300 W/m^2 solar constant (above the atmosphere) enters through the open aperture, amounting to tens of watts of solar radiation over the aperture area. A simple bandpass filter with low wavefront distortion located at the entrance aperture of the telescope with high transmittance at both the outgoing laser and the incoming beacon signal wavelengths not only reduces the background light, but also minimizes the heat that can enter the telescope. Even with a prefilter at the entrance aperture of the telescope, a portion of the incident optical power is absorbed at the mirror surface, potentially causing thermal distortion. To prevent this thermal problem, the solar filter has to reject greater than approximately 99% of the incident light. Multifunctional use of the system (e.g., as an Earth image tracker or optical navigation system) would require a broad wavelength transmission through the filter.

An effective approach to mitigate the effects of stray sunlight, Earth IR, or other stray light in general uses a synchronous detection ("lock-in amplifier") approach to filter out the desired modulated signal from background noise at all other frequencies [8]. Another interesting approach to solar rejection is to operate the laser wavelength within one of the Fraunhofer lines [9]. These lines are caused by absorption of sunlight in solar atmospheric layers above the photosphere. A high-resolution spectrum shows hundreds of dark lines. Therefore, receiving signals through a Fraunhofer filter have the advantage of naturally rejecting a fraction of the background light. A Fraunhofer filter typically consists of a Fabry-Perot etalon coated with a multilayer dielectric interference filter. The etalon provides many narrow band-passes, with one centered on a Fraunhofer line. The dielectric filter blocks transmission at all the etalon passbands, except the one centered at the Fraunhofer line.

5.4.2 TRANSMIT/RECEIVE ISOLATION

Detecting an extremely faint signal at the data detector/tracking array can be challenging when the system is transmitting a high-power laser with several watts of average power and kilowatts of peak power. If not correctly compensated for (through adequate isolation and baffling) backscatter from the optics (due to imperfect coatings or contamination and leakage around optical elements), it will lead to the spurious laser signal overwhelming the data/tracking detector. Estimates of the required level of isolation may be arrived at through a knowledge of the transmit laser's power and estimates of the backscattered power. Table 5.4 provides an example of the level of isolation required for the proposed system. The assumed values are for end of mission life.

Typical requirement on optical isolation between the transmit-channel and the receive- channel is above 140 dB. Optical isolation methods include optical filters, polarization discrimination, and spatial/temporal rejection of undesired radiation. Table 5.5 summarizes the properties of some of the optical isolation techniques used today.

Figure 5.2 illustrates an example of such a design, developed by the Facebook, Inc., for a bidirectional 100 Gbps lasercom link where the optical assembly is

TABLE 5.4

Example: Optical Isolation Budget

Contributing Element	Signal Contribution (dB or dB/W)
Laser transmitter power (peak)	+3
Scattered power into single pixel	−70
Transmission through band-pass filter	−45
Transmission through dichroic beam splitter	−40
Transmission through other beam splitters	−4
Common path to dichroic beam splitter	−1
Backscatter at the data/tracking detector	−157
Received power at data/tracking detector (peak)	−57

incorporated within the CPA. A coudé path to relay the optical signal at the pupil plane on to the small folded-path optics bench follows the 7.5-cm aperture refracting telescope. The relay optics, the narrow and the wide FOV optical tracking sensors, optical filters, and a number of FSMs are housed within the small optics bench to launch the received beam on a single-mode fiber [10].

Micro-Electromechanical System (MEMS) common-mode FSMs within the bench stabilize the incoming LOS. Separation of the beacon signal from the communications signal is accomplished via a tracking beam splitter. One of the FSMs along with a fiber-focusing lens steers the communications beam onto a single-mode communications fiber, thereby accomplishing the goal of signal emitted out of the transmitter's fiber and launching it into the receive aperture's fiber, tens to thousands of kilometers apart (Figure 5.3). Thermally induced fine position adjustments for both focusing lenses are compensated via miniaturized motorized translation stages [11].

TABLE 5.5

Characteristics of Some of the Known Optical Isolation Techniques

Characteristic	Wavelength	Polarization	Spatial	Temporal
Data rate applicability	Unlimited	Unlimited	Unlimited	Low (Mb/s)
Degree of isolation (dB)	110–120	60–80	≥120	High
Single wavelength applicability	No	Yes	Yes	Yes
Implemented in hardware?	Yes	Yes	Yes	Yes
Mechanism for interference	Backscatter off optical elements	Backscatter off optical elements	Scatter off nearby structure	Signal EMI

Note: EMI, electromagnetic interference.

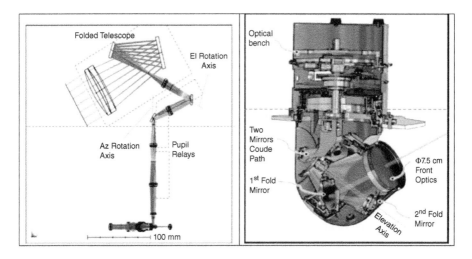

FIGURE 5.2 The optics layout and schematic of a laser communications transceiver optics developed for 100 Gbps bidirectional coherent links between airplanes and atmosphere. (Credit Facebook, Inc.)

5.6 ALIGNMENT TOLERANCES

A telescope for flight must be tolerant of the thermal and mechanical environments experienced during ground operation, launch, and in-flight operations. Thermal excursions and gradients, material relaxation and creep, and vibrations within the spacecraft and from mechanisms imparted into the transceiver represent the primary cause of optomechanical misalignments. Thermal and mechanical stability must be adequate to ensure near diffraction-limited operation in-flight. Much information about alignment sensitivities can be obtained from optical design software tools. This includes the effects of tilt, spacing, and centering error between the telescope's

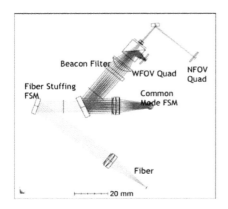

FIGURE 5.3 The small optics bench for the transceiver shown in Figure 5.2 incorporating a large number of components into a highly compact area. (Credit: Facebook, Inc.)

primary and secondary mirrors. A thorough knowledge of optical alignment tolerances and wavefront errors can be budgeted more intelligently after comprehensive thermal and structural models are developed, identifying the degree of mechanical motion over the environmental conditions. Another design driver is tolerance requirement implications on the initial assembly. Tight tolerances make initial assembly more challenging, which could take more time, increase cost, require more complex mounts or adjustment features, etc.

Requirements on alignment tolerances, placed on proper (stable) thermal and mechanical design, may be estimated quantitatively once the RMS wavefront error (represented as a fraction of wavelength) is specified for the optical assembly. Misalignments may be put into three categories: (1) pure axial misalignment (introducing defocus), (2) pure lateral misalignment (e.g., introducing third-order axial coma), and (3) pure tilt misalignment (also often introducing coma). In defocus, the optical axes of primary and secondary mirrors remain aligned, while the mirror separation changes. In the lateral misalignment, optical axes are no longer overlapped, while the mirror separation remains fixed. The latter will take the form of a tilt or a decenter of the secondary mirror axis relative to the primary mirror axis.

Temperature variations (ΔT) introduce local nonelastic strains in the mirror material proportional to $\alpha \Delta T$, where α is the linear coefficient of thermal expansion (CTE). A soaked (homogeneous) temperature change across the entire volume of the optical assembly causes the mirror focal length (f) to adjust by an amount Δf, according to $\Delta f = f \cdot \alpha \cdot \Delta T$ [12]. This equation does not apply to refractive telescopes where the dn/dT of the refractive material has to be taken into account. A localized (gradient) temperature change results in localized surface deformations that lead to wavefront aberrations.

5.7   OPTICAL SYSTEM STABILITY ASSESSMENT AND MITIGATION

Minute motion, on the order of a few microns, within the optomechanical system is enough to degrade the image quality to less than that of a diffraction-limited system. It is challenging to assess the effect of thermal gradients on the optical system without a comprehensive thermal model of the system. Proper heat sinking is required to mitigate locally generated thermal loads. Since the entire optical assembly may experience significant temperature variations as the spacecraft orbits the Earth (or another object in our solar-system) sun-shields and heaters are frequently added to maintain the system temperature near constant. The larger the telescope diameter, the more thermal gradient one can expect across its clear aperture, since it becomes more difficult to reach an equilibrium across the aperture. Driven by the requirement for unhindered access to space for signal transmission and reception, larger optical apertures are more difficult to thermally passivate.

Maintenance of primary-secondary mirror separation and overlap is a key design driver affected by temperature variations. Inclusion of an active focusing element, though not desirable due to additional risks that it introduces into flight optics, is one method to mitigate optical misalignments. Also, utilizing the same material with high thermal conductivity for the mirror substrates and the structure also minimizes thermally induced axial misalignments. In this case, the solar heat load on the structures and the reflective optics is conducted away promptly. A third option for reducing the thermal effects is to select materials with small CTE, such as ULE, CER-VIT, CFRP, ZERODUR, and SiC [13]. For example, use of an all SiC material telescope is a means of athermalizing the system, i.e., making it less sensitive to thermal variations and gradients [14]. The residual differential thermal expansion resulting from this approach can cause defocusing. This may be compensated by the use of primary to secondary mirror connecting spacers that are made from a material with proper CTE (such as a specific invar metering rod). SiC has a very good thermal figure of merit (conductivity vs. CTE), which is why this telescope is insensitive to gradients.

5.8 TELESCOPE MATERIALS

Selection of telescope materials is driven by (1) mechanical stability (resistant to mechanical deformations and stress), (2) thermal stability (resistant to soaked and localized temperature changes), (3) low density (varies nonlinearly with mirror diameter), (4) affordability (cost increases approximately as square of the diameter), and possibly (5) space radiation resistance.

Use of low-density (lightweight) and high modulus (stiff) structures will reduce the flight transceiver mass and provide structural integrity during launch and operations. Isotropic materials and those with a highly uniform CTE demonstrate high dimensional stability as their thermal environment varies. To be useful as a mirror, the material must be machinable, polishable to RMS surface roughness of approximately 2 nm, and adhere to either metallic or dielectric coatings.

Reflective mirrors and the telescope structure may be electroformed, fabricated entirely from aluminum, or made from a variety of silicon carbide compounds. These materials, along with beryllium and low thermal expansion glass, are excellent candidates for mirror substrates [15]. Figure 5.4 shows the picture of a 30-cm aperture all-silicon-carbide lightweight (<5 kg) telescope developed by L3-Communications SSG, and a 20-cm diameter lightweight telescope fabricated by Media Lario through high accuracy nickel electroforming replication technology.

FIGURE 5.4 An all-silicon-carbide telescope (left) and an electroformed telescope (right).

TABLE 5.6

Approximate Values for Properties of Selected Telescope Mirror and Structure Materials

Material Property	Al 6061-T6	Be (BeO-50)	ULE (Si) (7971)	Zerodur (Si)	β SiC (CVD)	SiC (RC)	Al/SiC (SXA)	GEC
Density (10^3 kg/m³)	2.78	1.85	2.20	2.55	3.21	3.1	2.97	1.63
Specific heat (J/kg K)	899	1880	708	820	700	733	847	589
Thermal conductivity (W/m·K)	237	216	1.3	6.0	200	198	160	47.7
CTE (10^{-6}/K)	24	12	0.03	0.05	2.2	4.5	10.8	0.25
Elastic modulus (GPa)	76	303	67	90	466	467	145	93.8
Poisson's ratio	0.35	0.05	0.17	0.24	0.21	0.14	0.33	0.33
Thermal distortion (Wm^{-1} × 10^7)	0.95	1.9	4.3	4.0	10.4	0.23	0.68	0.05
Dynamic response (kg m⁷)	23.8	282		971		1500		
Fundamental frequency (N m$^{7/2}$)$^{-1}$ × 10^{-3}	60.6	224		7.2		115		
Surface finish (nm RMS)	10	6–8	<0.5	<0.5	<0.5	0.5	<0.5	>10
Figures of merit								
Thermal stability (K × 10^{-8} m/W)	11	7.1	56	55	1.3	1.21	9.1	
Stiffness/weight (10^7 N m/kg)	2.6	15.4	3.3	3.3	14.5	12.2	4.6	5.7
NASA readiness level (TRL)	9	9	8	8	5	7	7	6

Note: RC, reaction bonded; SXA, Al/SiC composite; GEC, graphite epoxy composite.

Table 5.6 summarizes some of the key characteristics of popular telescope mirrors and optomechanical structure materials.

High CTE of a mirror material is of concern due to induced aberrations as the result of thermal dynamics. Variable Emittance Coatings (VEC) technique offers an approach to influence radiative heat loss based on maximum heat loading requirements [16]. VEC makes use of a mechanism which only provides low and high emissive states. With VEC, electronically controls the thermal emittance of the mirror surface allowing for rigorous control of its temperature, and possibly for mirror's surface-figure regulation.

5.9   MECHANICAL ACTUATION

To reduce risk, minimizing the use of optical mechanisms (actuation elements) within the flight transceiver is advisable. Use of certain mechanisms such as two-axis gimbals for coarse pointing or FSMs for fine beam-pointing is typically unavoidable. Other moving elements may include adjustable lens systems for beamwidth reduction or enhancement, pop-up mirrors (sliding or flip mirrors) to select beam paths, redundant channel switchers, and shutters to avoid excessive power levels on sensors or optics, and variable field stops.

5.9.1 FINE-STEERING MIRRORS

The PAT subsystem consists of optics and associated optomechanical structures, electro-optical detectors for acquisition and tracking, FSM(s) for the fast steering of transmit laser's beam and when the cross velocity between the two ends of the links is large enough to require point-ahead angle compensation.

The system must be articulatable to allow acquisition and lock of transmitter/ receiver beam paths. The receive optics should provide intentional image blurring to spread the tracking signal over a few pixels. However, the optical signal spot must be void of coma and other asymmetric aberration patterns. A tracking reference path may be included to image the transmit signal (after beam steering mirror) and measure the instantaneous position of the downlink signal.

FSMs stabilize the transmit and receive beam in the presence of spacecraft jitter, and also provide the proper point-ahead (or point behind) angle for accurate pointing to the receive station. Typically, a collimated beam path is used to provide an image of the FSM at the telescope primary mirror. This minimizes beam walk in the telescope by way of angular motion of the FSM. The FSM's construction may include flexure suspension, electromagnetic (e.g., voice coil) actuators, piezoelectric transducer (PZT) actuators, proximity sensors (e.g. from Kaman or Blue Line), or optical encoders. The performance of the FSM is almost entirely inseparable from the performance of its control electronics. The control bandwidth of the FSM is dominated by the control bandwidth of the entire loop consisting of FSM + track sensor + feedback control algorithm implemented in electronics and/or software or firmware.

5.9.2 COARSE-POINTING MECHANISMS

The CPA is incorporated with the flight transceiver as an alternative to simply body pointing/steering the spacecraft, which may be a preferred approach on some platforms (e.g., CubeSat). In mobile laser communication terminals, suitable beam-pointing angles often vary quickly, in that case coarse beam pointing becomes an essential subsystem. The coarse beam pointer is a large angular motion, low-bandwidth mechanism for initial open-loop pointing, acquisition searching, and variable rate slewing. As compared in Table 5.7, options for independent optical terminal pointing include a gimbaled telescope and gimbaled mirror. Each option offers specific advantages. The term "gimbaled flat mirror" refers to an actuated mirror located in front of the telescope aperture to manipulate the transmit beam.

Typical requirements and design drivers for CPA include field of regard (the area that can be covered by the actuated optics assembly), open-loop pointing accuracy, motion control smoothness, slew rate, contamination control, flex wraps, and rotary

TABLE 5.7

Comparison of Transceiver Coarse-Pointing Options

Parameters	Gimbaled Flat Mirror	Gimbaled Telescope (Only)	Gimbaled Transceiver
Total mass	Moderate	Moderate	Higher
Useful orbit range	MEO and GEO	MEO and GEO	LEO and MEO
Sensitivity to jitter	High	Moderate	Moderate
Relative size	Large	Moderate–large	Moderate
Field of regard	Small	Large	Large
Coudé path	Not possible	Possible	Possible
Output beamwidth	Narrow (internal FSM)	Narrow (internal FSM)	Wide (no FSM)
Exposure to Sun	Medium	Low	Low
Mirror contamination	Exposed	Protected	Protected

electrical interfaces (including optics). A variety of CPA options are in use including heliostat, coelostat, gimbaled telescope, coudé system [17]. A CPA's requirements are primarily driven by acquisition scheme/spacecraft uncertainty vs. acquisition sensor FOV.

Nonmechanical coarse-pointing and fine beam steering technology is advancing rapidly, and gradually becoming mature enough for field implementation [18]. These devices are based primarily on the optical phased array technology.

5.9.3 BEAM DIVERGENCE/FOCUS CONTROL MECHANISMS

A variety of mechanisms have been devised to control beam divergence of laser beams that traverse the laser communications terminal. Table 5.8 trades the properties of a few of these mechanisms.

TABLE 5.8

Characteristics of Some of the Known Beam Divergence Control Mechanisms

Characteristics	Continuous Zoom Lens(es)	Decentered Pupil	Pop-In Refractors	Electro-Optical
Beam divergence	Continuous	Intermittent	Continuous	Continuous
Acceptance angle (relative)	Wide	Wide	Wide	Narrow
Accuracy	Fair	Good	Fair	Excellent
Divergence	Limited	Wide	Limited	Wide
Encoder requirement	Yes	Yes, for multisteps	Yes, for multisteps	No
Switching time (ms)	1000	50	10	1

(The Method columns span Decentered Pupil, Pop-In Refractors, and Electro-Optical)

5.10   CLEANLINESS CONTROL

Optics contamination control is driven by the bidirectional reflectance distribution function (BRDF) specifications set for the system [19]. Primary sources of contamination leading to stray light within the optical system are ground handling, storage, and spacecraft launch. Stray light requirements are frequently a driving factor in optical terminal design. Examples include uniform distribution of stray light over the tracking detector FOV, and internal stray light levels not to exceeding 1 µW/nm sr (measured relative to telescope pupil). An optical system of class 300 or better cleanliness is frequently employed to minimize the scattered light that can find its way to any one of the system sensors. For this purpose, flight transceivers are often integrated and tested in a class 100 clean room. A sealed cover at the exit aperture of the telescope (that can be blown-off shortly after the launch) can be utilized to protect the system from the environmental, structural and assembly (e.g., soldering) contaminants. Similarly, use of materials with a low outgassing coefficient, ground handling covers, and gas purging the late-stage integration and test are among the precautions that has to be taken to manage the contamination control.

5.11   MECHANICAL PACKAGING

The requirement for stress-free mounting of optical elements places stringent demands on mechanical packaging. Mounting stresses degrade surface quality of reflective elements and induce index of refraction changes in refractive elements. Mirrors contribute significantly more to wavefront error than do refractive optics, given the same figure error values [20]. Good mechanical packaging practices include three-point kinematic seating or lapped multipoint pads, compressive loads on refracting optics instead of tensile loads, sufficient clearance between cell and optics for thermal excursions, and bonding with elastomeric agents [21].

The optomechanical assembly is typically designed concurrently with the optical system design. Typically, the primary mirror is bonded to a central hub mount by applying a flight-certified room temperature vulcanizing (RTV) bonding material. This mount is then structurally connected to a flange in the main housing through three sets of blade flexures, evenly spaced about the optical axis 120° apart. Three flexure assemblies at the end of the main housing support the secondary mirror housing.

The spacing between the primary and secondary mirrors critically influences the alignment of the optical system. Therefore, three invar rods whose CTE is selected based on the type of glass mirror substrate and mirror-housing material used control this spacing. The invar rods connect the flange inside the main housing with the secondary spider.

To ensure the optical assembly meets the dynamic thermal and structural environment of the spacecraft, it is important to set up the mechanical system design work product such that it is easily transferable to a structural analysis software tool to study the thermal and dynamic frequency responses of the optical assembly.

Figure 5.5 shows a form, fit, and function prototype of a lasercom terminal developed by the NASA-Jet Propulsion Laboratory (JPL) [22]. Specifically selected

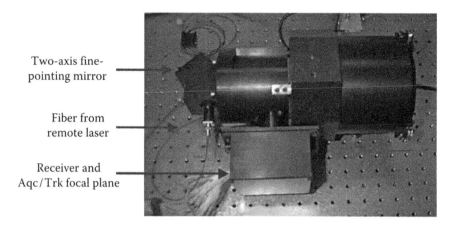

Two-axis fine-
pointing mirror

Fiber from
remote laser

Receiver and
Aqc / Trk focal plane

FIGURE 5.5 Example of a lasercom assembly's optics module (called optical communications demonstrator) comprising the transmit/receive aperture, aft optics, two-axis FSM, a fiber input for laser transmitter, and a receiver and focal plane array for beacon beam acquisition and tracking.

stainless steel metering rods connect the primary mirror to the secondary mirror. To avoid thermal gradients, only the critical assemblies that had to be collocated with the OMA were incorporated into this part of the system. Besides the optical assembly, major components include two FSMs and a focal plane array for laser acquisition and tracking of the incoming beacon signal. The laser source that is transmitted is located away from the telescope in a location convenient for its thermal management. The laser beam is then piped in via a single-mode fiber.

5.12   ACTIVE AND PASSIVE MECHANICAL ISOLATION

Addition of an isolation platform to the flight transceiver helps to lower the FSM closed-loop bandwidth requirements, and conceivably use a lower performance FSM. Platform disturbance causes LOS jitter and directly affects the PAT process. PAT control design requires a proper understanding of the spectrum and amplitude of vibrational disturbances of the host platform to which the optical transceiver will be mounted. These values are typically not well known until the platform assembly is mostly designed and analyzed. Platform vibration frequencies are in the range of 0.1 Hz to hundreds of Hertz, with amplitudes typically much larger than the transmit laser's beamwidth. Passive isolators and inertially referenced stabilized platforms utilizing angle sensors and/or accelerometers can adequately reject frequencies greater than tens of Hertz [23]. Inertial reference unit options include AC-coupled devices, such as, angle sensors (e.g., the magnetohydrodynamic type) and accelerometers, and DC-coupled sensors, like fiber-coupled gyros.

Platform disturbance mitigation techniques may be classified as passive and active isolation mechanisms. Passive mechanisms include viscoelastic materials (e.g., particular rubber with low outgassing), sealed fluid, and Eddy current [24]. Active isolation mechanisms include voice coil, magnetic piezoelectric actuators, and smart materials.

FIGURE 5.6 An active isolation bipod developed by Honeywell consisting of two isolators with integral launch lock and accelerometer payload sensors.

As an example, Honeywell, Inc., has developed a miniaturized active vibration isolation system (MVIS II) specifically for isolation of flight optical systems [25]. The expected in-flight capability is greater than 20 dB RMS active jitter suppression over the 1 to 200 Hz range, shown in Figure 5.6 [26].

The need for a fast-tracking subsystem may be substantially circumvented by mounting the (lightweight) lasercom transceiver on a platform that can effectively isolate the transceiver from the host platform, such as the disturbance-free platform (DFP) [27]. Some of the advantages of this architecture include the following:

- Reduction in the amount of LOS jitter coupled to the optical element results in much longer integration time at the focal plane array to detect the signal. This enables the use of a faint beacon as a pointing reference. Long integration time also has the benefit of making the pointing concept relatively insensitive to scintillation-induced fades of the uplink beacon.
- Besides mechanical isolation, the DFP also provides coarse and fine-pointing adjustment, thereby relaxing the spacecraft pointing requirement.
- Elimination of the high bandwidth LOS control components (such as an FSM) results in significant simplification, mass, and power reduction of the optical package.

REFERENCES

1. N. A. Page, Design of the optical communications demonstrator instrument optical system, *Proceedings of SPIE*, 2123, 498, 1994.
2. N. A. Page and H. Hemmati, Preliminary opto-mechanical design for the X2000 transceiver, *Proceedings of SPIE*, 3615, 206, 1999.

3. C. Fuchs, H. Henniger, and D. Giggenbach, Fiber-bundle receiver: A new concept for high-speed and high-sensitivity tracking in optical transceivers, *Proceedings of SPIE*, 6709, 67090V, 2007.

4. M. Toyoshima, N. Takahashi, T. Jono, T. Yamawaki, K. Nakagawa, and A. Yamamoto, Mutual alignment errors due to the variation of wave-front aberrations in a free-space laser communication link, *Optics Express*, 9, 592–602, 2001.

5. R. Czichy, Optical design and technologies for space instrumentation, *Proceedings of SPIE*, 2210, 420, 1994.

6. J. C. Wyant and K Creath, Basic wavefront aberration theory for optical metrology, *Applied Optics and Optical Engineering*, R. R. Shannon and J. C. Wyant (Eds.), Academic Press, New York, 1992, Vol. XI, pp. 1–53.

7. F. Grundahl and A. N. Sorenson, Detection of scattered light in telescopes, *Astronomy and Astrophysics Supplement Series*, 116, 367–371, 1996.

8. E. Miller, K. Birnbaum, C. Chen, A. Grier, M. Hunwardsen, and D. Jandrain, Fine Pointing Concepts for Optical Intersatellite Links, *Proceedings of International Conference on Space Optical Systems (ICSOS)*, November 2017.

9. E. L. Kerr, Fraunhofer filters to reduce solar background for optical communications, *Optical Engineering*, 28, 963–968, 1989.

10. C.-C. Chen, A. Grier, M. Malfa, E. Booen, H. Harding, C. Xia, M. Hunwardsen et al., Demonstration of bi-directional coherent air-to-ground optical link, *Proceedings of SPIE Photonics West*, 10524, 2018.

11. K. Balmer, Telescope, *Encyclopedia of Physics*, S. Flugge (Ed.), Springer-Verlag, New York, 1967, p. 266.

12. H. Bach and D. Krause (Eds.), *Low Thermal Expansion Glass Ceramics*, Schott Series on Glass and Glass Ceramics, 2nd ed., Springer, Secaucus, NJ, 2005.

13. M. Schwalm and B. Gority, Silicon carbide telescopes for space based lasercom, *IEEE Aerospace Conference Proceedings*, 3, 79–83, 2000.

14. D. Vukobratovich, Ultra-lightweight optics for laser communications, *Proceedings of SPIE*, 1218, 178–192, 1990.

15. B. R. Burg, S. Dubowsky, J. H. Lienhard V., and D. Poulikakos, Thermal control architecture for a planetary and lunar surface exploration micro-robot, *AIP Conference Proceedings*, 880, 43–50, 2007.

16. B. Burg, Thermal control architectures for planetary and lunar (sub-) surface exploration robots, Master Thesis, ETH Zurich, 2006.

17. E. D. Miller, M. DeSpenza, I. Gavrilyuk, G. Nelson, B. Erickson, B. Edwards, E. Davis, T. Truscott, A prototype coarse pointing assembly for laser communications, *Proceedings of SPIE Photonics West*, V. 10096, 2017.

18. P. McManamon, An overview of optical phased array technology and status, *Proceedings of SPIE*, 5947, 59470I, 2005.

19. P. R. Spyak and W. L Wolfe, Scatter from particulate-contaminated mirrors. Part 1: Theory and experiment for polystyrene spheres polystyrene spheres and lambda = 0.6328 μm, *Optical Engineering*, 31, 1746, 1992.

20. M. Born and E. Wolf, *Principles of Optics: Electromagnetic Theory of Propagation, Interference and Diffraction of Light*, Cambridge University Press, New York, 1999.

21. P. R. Yoder, Jr., Parametric investigations of mounting-induced axial contact stresses in individual lens elements, *Proceedings of SPIE*, 1998, 8–20, 1993.

22. V. Lossberg and B. Von Lossberg, Mechanical design and design processes for the telescope optical assembly of the optical communications demonstrator, *Proceedings of SPIE*, 2123, 505–517, 1994.

23. G. Ortiz, A. Portillo, S. Lee, and J. Ceniceros, Functional demonstration of accelerometer-assisted beacon tracking, *Proceedings of SPIE*, 4772, 112–117, 2001.

24. G. Baister and P. V. Gatenby, Small optical terminal designs with softmount interface, *Proceedings of SPIE*, 2990, 172, 1997.

25. T. T. Hyde and L. P. Davis, Optimization of multi-axis passive isolation systems, *Proceedings of SPIE*, 3327, 399–410, 1998.

26. M. B. McMickell, T. Kreider, E. Hansen, T. Davis, and M. Gonzalez, Optical payload isolation using the miniature vibration isolation system, *Proceedings of SPIE*, 6527, 652703, 2007.

27. C. C. Chen, H. Hemmati, A. Biswas, G. Ortiz, W. Farr, and N. Pedrerio, A simplified lasercom system architecture using a disturbance-free platform, *Proceedings of SPIE*, 6105, 610505, 2006.

6 Coding and Modulation for Free-Space Optical Communications

Bruce Moision and Jon Hamkins

CONTENTS

6.1   INTRODUCTION

Previous chapters have presented methods to transmit and reliably reproduce a set of optical symbols. For example, the symbols could be distinct phase, polarization, wavelength, or amplitudes of the optical carrier. In the simplest and most common case, one transmits either the presence or the absence of an optical pulse. In order to use the optical channel to transmit information, one requires a *modulation*—a mapping of user bits to the physical symbols. On reception, an estimate is made of the transmitted symbol and mapped back to estimates of the transmitted bits. The reliability of this operation may be dramatically improved by first encoding the user bits with an *error-correction code* (ECC). An ECC adds redundant information to the user bits to enable error correction at the receiving end, and results in more efficient use of power and bandwidth. The limit on the gain achievable with an ECC is given by the *capacity* of the channel. In this chapter, we discuss modulation and coding for, and capacity of, the optical channel.

When designing a coded modulation scheme, one wants to maximize the rate at which the communication system can transmit data error-free.[1] The system has certain resources at its disposal: power, bandwidth, time (as delay or latency), and complexity (as computational power). We want to use these resources efficiently. In any given system some resources will be more scarce, or constrained. For a downlink from a spacecraft at Mars to a Earth-based receiver, latency and bandwidth may be relatively plentiful, while power efficiency is the dominant constraint, and complexity is asymmetric (less constrained at the receiver than at the transmitter). At the other end of the spectrum, a terrestrial ethernet backhaul link may be primarily bandwidth and latency constrained. All these constraints impact system design. In this chapter, we describe methods to quantify the trade-offs between resources in the design of coded modulations.

For our purposes, the optical communications channel may be reduced to the block diagram shown in Figure 6.1. A block of information bits **u** is mapped via

FIGURE 6.1   A coded optical channel.

[1] By which we mean with an error rate arbitrarily close to zero.

encoding and modulation to a codeword **x**, which is transmitted over an optical channel. The optical detector and receiver (by which we mean the device that maps detector outputs to a quantized signal, such as a photon count) are considered to be part of the channel model. The output of the detector is quantized and synchronized, providing either symbol estimates or likelihoods **y** to the decoder. Using the channel output information, the decoder, which may also perform the function of demodulation, generates estimates of the transmitted user data. The system may also incorporate an interleaver to compensate for fading and, on bidirectional links with moderate round-trip-travel times, the entire system may be wrapped with a automatic repeat request (ARQ) system.

The fidelity of the system is measured in terms of the bit error rate (P_b) and codeword error rate (P_w) at the output of the decoder, with a P_b target typically on the order of 10^{-6} and a P_w target on the order of 10^{-4}. Codeword errors may be detected with high reliability, and when detected typically an entire codeword in error will be discarded.

The chapter is organized as follows. In Section 6.2 we define three modulations that will serve as examples in the chapter. In Section 6.3 we present statistical models of the detector output. In Section 6.4 we show channel capacities that result from various combinations of modulations and statistical channel models. In Section 6.5 we take one example, the Poisson PPM capacity, and use it to illustrate some trade-offs and rules of thumb in channel design. In Sections 6.6 and 6.7 we examine two major impediments to optical communications: fading and jitter, illustrating their impact on the channel capacity. Finally, in Section 6.8, we discuss the application of ECC to the channel and illustrate the gains achievable with different classes of codes.

6.1.1 NOTATION, ASSUMPTIONS

Throughout we assume source bits are equally likely 0 and 1 and independent. We refer to the signal at the receiver, assuming transmission losses have been factored out. In particular, average signal power, denoted P_{av}, always refers to received power, not transmitted power.

We use lowercase u, x, c, y to denote realizations of corresponding random variables U, X, C, Y. Boldface **x** refers to a vector, $\mathbf{x} = \{x_0, x_1, \ldots, x_n\}$. The notation $f_Y(y)$ is used to denote the probability density or mass function of random variable Y evaluated at y. When the random variable is clear from context, we simply write $f(y)$ for $f_Y(y)$. Similarly $f_{Y|X}$ denotes a probability mass function of Y conditioned on X. $F_Y(y)$ denotes the cumulative probability density or mass function.

6.2 MODULATION

A modulation is a physical representation of information—a representation that allows the information to be transmitted and retrieved. In the free-space optical communication channel, information is encoded into physical properties of light, e.g., the phase, intensity, polarization, and wavelength. For our purposes, the light wave encoding the information may be represented as

$$a(t) = \mathrm{R}\{x(t)e^{j\omega_0 t}\}$$

where R is the real part of the signal, $x(t)$ is a modulating waveform, and ω_0 is the optical carrier wave frequency. The average power of the signal is

$$P_{av} = \lim_{T \to \infty} \frac{1}{T} \int_0^T |a(t)|^2 \, dt$$

For example, a system could encode information bits into the intensity of pulses of light from a source, with a one bit encoded into a pulse and a zero bit encoded into the absence of a pulse. The choice of a modulation depends on the ability to modulate the source and detect the resulting modulation and has a large impact on the efficiency of the link. We describe several frequently used modulations that will be explored further throughout this chapter.

6.2.1   ON-OFF KEYING

The modulation described above, wherein a zero is represented by the absence of a pulse and a one is represented by a pulse, is referred to as On-Off Keying (OOK). Photodetectors inherently measure intensity and OOK, a binary modulation of the intensity is perhaps the most straightforward modulation to detect. It is also power efficient in certain photon flux regimes.

Let $x_k \in \{0,1\}$ be the k-th bit transmitted by the source. In OOK x_k is mapped to a pulse of light of duration T_s seconds

$$a(t) = R\{2x_k \sqrt{P_{av}} e^{j\omega_0 t}\} \, , t \in ((k-1)T_s, kT_s]$$

6.2.2   PULSE-POSITION MODULATION

In pulse-position modulation (PPM) a modulation symbol consists of M time slots, each of duration T_s. A set of $\log_2(M)$ bits,

$$\mathbf{x}_k = \{x_{k,0}, x_{k,1}, \ldots, x_{k,\log_2(M)-1}\},$$

are mapped to the position of a single pulse in one of the M slots. No signal is transmitted in the other $M-1$ slots. Let $\mathbf{x}_k \in \{0,1,\ldots,M-1\}$ also denote the integer represented by the M-bit vector. In PPM, $\mathbf{x}_k$ is mapped to

$$a(t) = R\{x(t)\sqrt{2MP_{av}} e^{j\omega_0 t}\} \, , t \in ((k-1)MT_s, kMT_s]$$

where on $((k-1)MT_s, kMT_s]$

$$x(t) = \begin{cases} 1 & , t \in (((k-1)M + \mathbf{x}_k)T_s, ((k-1)M + \mathbf{x}_k + 1)Ts], \\ 0 & , \text{otherwise} \end{cases}$$

By using one pulse to represent $\log_2(M)$ bits, PPM is generally more power efficient (greater bits/Joule) than OOK. PPM provides a method to signal at a high peak-to-average power ratio, and the optical link achieves high power efficiencies at high peak to average power ratios [1–3]. The price paid for this increase in power efficiency is a reduction in bandwidth efficiency. We can approximate the bandwidth occupancy of an OOK or PPM signal as $1/Ts$. PPM has a maximum bandwidth efficiency of $\log_2(M)/M$ (bits/s/Hz), while OOK has a maximum efficiency of 1 (bits/s/Hz). This is a fundamental trade-off: an increase in power efficiency typically comes with a reduction in bandwidth efficiency (and vise versa).

6.2.3 Phase Shift Keying and Differential Phase Shift Keying

In Phase Shift Keying (PSK), information is encoded into the phase of the carrier. We'll consider the simplest case, binary PSK (BPSK). Let $x_k \in \{0,1\}$ be the k-th bit transmitted by the source. In BPSK, x_k is mapped to a pulse of light of duration T_s seconds

$$a(t) = R\{\sqrt{2P_{av}}\, e^{j\omega_0(t+x_k\pi)}\}\,, t \in ((k-1)MT_s, kMT_s]$$

In BPSK bits, one and zero are encoded into the carrier with phase offsets of π radians. A homodyne or a heterodyne receiver may be used to extract this information. Higher-order encodings into the phase and amplitude of the signal allow extensions to bandwidth-efficient signals representing multiple bits per pulse, e.g., quadrature amplitude modulation (QAM) and 8-PSK. Higher-order bandwidth-efficient PSK signaling has not been as widely deployed for optical links as it has been for radio frequency links. This is due in part to the large bandwidth available in the optical range, emphasizing power efficiency over bandwidth efficiency, and in part to the higher complexity of a high-order PSK receiver. However, we expect this to change as bandwidth demands start to outstrip even the large bandwidth available in the optical regime.

In applications for which coherent reception is difficult, Differential Phase Shift Keying (DPSK) may be used. In DPSK, the k-th transmitted bit is first differentially encoded and then modulated with BPSK. At the receiving end, the phases of two consecutive bits are compared, using a delay-line interferometer. One advantage of this approach is that the receiver need not explicitly track the phase of the incoming signal.

6.3 STATISTICAL CHANNEL MODELS

In this section, we define statistical models of the optical channel. This will allow us to quantify differences in the efficiencies of various coded modulation schemes. We'll focus on two models that represent ideal implementations of receivers to detect intensity and phase, respectively.

6.3.1 Poisson

Suppose we transmit a two-level intensity modulated signal, such as OOK or PPM, and the signal is received with an ideal photodetector. An ideal photodetector produces a count of the number of photons incident on the detector over the interval T_s. The mean count is proportional to the incident signal power. Let n_s denote the mean detected signal photons per pulse. We have

$$n_s = \begin{cases} 2P_{av}T_s / E & \text{,OOK} \\ MP_{av}T_s / E & \text{,PPM} \end{cases}$$

where E is the energy per photon. In general, there is also background light of power P_n incident on the detector. We similarly let $n_b = P_nT_s / E$ denote the mean detected background photons per pulse.

Let $x \in \{0,1\}$ denote whether the slot was pulsed or non-pulsed, and $y \in \{0,1,2,\ldots\}$ denote the photon count. The random number of detected photons may be modeled as Poisson with mean n_b for a non-pulsed slot and $n_s + n_b$ for a pulsed slot:

$$f_{Y|X}(y \mid 0) = \frac{e^{-n_b} n_b^k}{y!} \tag{6.1}$$

$$f_{Y|X}(y \mid 1) = \frac{e^{-(n_s+n_b)}(n_s + n_b)^y}{y!}. \tag{6.2}$$

A Poisson model follows from the assumptions that (1) a single event occurring in an epoch of duration t is proportional to t, (2) the probability of more than one event happening in an epoch of duration Δt goes to zero as Δt approaches zero, and (3) events in nonoverlapping time intervals are statistically independent. A single mode of coherent radiation can be described by a quantum state, and the number of photons in the state meets all three of these assumptions.

Realizable photodetectors convert the incident light wave to a photocurrent, with each detected photon producing a current impulse. The current may be edge detected or integrated over a slot and thresholded to produce an estimated photon count. Highly sensitive superconducting nanowire receivers, see, e.g., Ref. [4], approach the behavior of an ideal photodetector. Noise in the detection process may be accounted for by using more accurate models for specific classes of detectors, e.g., the Webb-Gaussian model for avalanche photodiode detectors [5] or the Polya model for photomultiplier tubes (PMTs) [6].

6.3.2 GAUSSIAN

For some receivers the sampled output of the photodetector is well modeled as Gaussian. This may occur when the incident photon rate is large, in which case the current output is well approximated as Gaussian. In addition thermal noise in the detection process is Gaussian and may dominate (dark noise may also be modeled as Gaussian but tends not to dominate).

For example, for a BPSK signal detected by the heterodyne receiver, the signal is effectively amplified by mixing it with a local oscillator prior to photodetection. The local oscillator power is typically chosen sufficiently large so that the detected signal is well approximated as Gaussian [7]. Furthermore, when the local oscillator power is sufficiently large, shot noise dominates other noise sources. In this case, the sampled output of the detector may be modeled as

$$f_{Y|X}(y \mid 0) = \frac{1}{\sqrt{2\pi}} e^{-(y+\sqrt{2n_s})^2/2} \tag{6.3}$$

$$f_{Y|X}(y \mid 1) = \frac{1}{\sqrt{2\pi}} e^{-(y-\sqrt{2n_s})^2/2} \tag{6.4}$$

where $n_s = P_{av} T_s / E$, the mean signal photons per pulse.

In other types of receivers, the detector output may be modeled as Gaussian, but with a variance that depends on the information being sent. For example, with OOK or PPM transmission being received by an avalanche photodiode detector (APD), the output may be approximiately Gaussian, with both the mean and the variance proportional to the received signal flux. That is, there is a higher variance when a "1" is sent, compared to a "0". In such systems, it may not be possible to express system performance in terms of a single signal-to-noise ratio parameter as is commonly done for other Gaussian communications models.

6.3.3 HARD DECISIONS

In some receivers, the photodetector output is thresholded and immediately converted to a sequence of estimates of the corresponding transmitted bits **x**. We refer to this as a hard-decision channel (as opposed to a soft-decision channel where there is no preliminary decision on the quantized output and the decoder receives bit-likelihoods rather than bits). This may be done to reduce the complexity of the receiver and in most cases results in a loss in performance.

We'll model the decision as part of the statistical channel model since it determines limits on the achievable data rate. The mapping of signal power to error rates is a function of the modulation and the statistics at the input to the decision. The resulting model depends on the type and reliability of the decisions. We consider several extensions of the prior channels to hard-decision channels.

6.3.3.1 Binary Asymmetric Channel

Suppose the Poisson OOK output, Eqs. (6.1) and (6.2), is mapped to binary decisions. The decision that minimizes the bit error rate is [8, p. 251]

$$\hat{x} = \begin{cases} 0 & , y \leq \tau \\ 1 & , y > \tau \end{cases}$$

$$\tau = \frac{n_s}{\ln\left(1 + \dfrac{n_s}{n_b}\right)}$$

The probability of a bit error is asymmetric:

$$p_{01} = P(\hat{x} = 0 \mid x = 1) = F_{Y|X}(\lfloor \tau \rfloor \mid 1)$$

$$p_{10} = P(\hat{x} = 1 \mid x = 0) = 1 - F_{Y|X}(\lfloor \tau \rfloor \mid 0)$$

where $\lfloor . \rfloor$ is the floor function. The bit error rate is

$$P_b = \frac{1}{2} p_{01} + \frac{1}{2} p_{10}$$

In the special case $n_b = 0$, we have $\tau = 0$,

$$p_{01} = e^{-n_s}$$

$$p_{10} = 0$$

$$P_b = \frac{1}{2} e^{-n_s}$$

which is referred to as a Z-channel.[2]

For the Gaussian BPSK channel, Eqs. (6.3) and (6.4), mapped to binary decisions the error rates are symmetric and the channel is characterized by the bit error rate:

$$P_b = \frac{1}{2} \mathrm{erfc}\left(\sqrt{n_s}\right) \tag{6.5}$$

6.3.3.2   Poisson PPM Channel

With the PPM channel, the hard decision can be with respect to the PPM symbol or the bits. The symbol error rate (P_s) (the probability of choosing the wrong PPM symbol) of a Poisson PPM channel Eq. (6.1), Eq. (6.2), may be accurately computed as [9]:

$$P_s = \begin{cases} \dfrac{(M-1)e^{-n_s}}{M} & , n_b = 0 \\[2ex] \displaystyle\sum_{k=0}^{N} f_{Y|X}(k \mid 1)\left(1 - \dfrac{F_{Y|X}(k \mid 0)^M - F_{Y|X}(k-1 \mid 0)^M}{M f_{Y|X}(k \mid 0)}\right) & \\[2ex] \qquad + \displaystyle\sum_{k=N+1}^{\infty} f_{Y|X}(k \mid 1)(M-1)g(k) & , n_b > 0 \end{cases} \tag{6.6}$$

where $g(k) = 1 - F_{Y|X}(k-1 \mid 0)$.

Bit decisions on the PPM channel produced by mapping the PPM symbol estimates to bit estimates have bit error rate

$$P_b = \frac{M}{2(M-1)} P_s$$

---

[2] A drawing of the transition probabilities forms the letter Z.

6.3.3.3 Ties and Erasures

At small n_b the Poisson PPM channel has a non-negligible probability of producing two or more slots with the same photon count—a tie. Inherent in Eq. (6.6) is the assumption that the receiver resolves ties when selecting a maximum. Consider an alternative receiver that, rather than making a decision, produces an erasure symbol when there is a tie. The probability of an erasure, a correct decision, and an error for the Poisson PPM channel is

$$P_e = \sum_{k=0}^{\infty} f_{Y|1}(k) F_0^{M-1}(k-1) \sum_{l=1}^{M-1} \binom{M-1}{l} \left(\frac{f_{Y|0}(k)}{F_{Y|0}(k-1)} \right)^l \tag{6.7}$$

$$P_c = \sum_{k=1}^{\infty} f_{Y|1}(k) F_{Y|0}(k-1)^{M-1} \tag{6.8}$$

$$P_s = 1 - P_e - P_c \tag{6.9}$$

We'll see that including an erasure output improves the performance of the channel at small n_b.

6.4 OPTICAL CHANNEL CAPACITY

Recall our goal is to choose a modulation and ECC that, for a given signal power and bandwidth, yield the largest possible data rate with a low error rate. A number of practical constraints will guide and limit this decision. It's useful to determine, for a given modulation and ECC overhead, what data rates are achievable. The *capacity* of the channel is the least upper bound on the data rate that may be transmitted at arbitrarily small error rates [10]. The capacity may serve as a guide in parameter selection and a benchmark for assessing the performance of a particular design. In this section, we discuss the computation of the channel capacity and some of its properties.

The capacity is a function of the modulation, the statistical channel model, and the operating point (signal and noise powers). Inherent in the channel model, as described in Section 6.3.3, is an assumption about the type of information made available by the detection method. In one case, the receiver makes estimates of each bit or PPM symbol, passing these estimates, or *hard* decisions, to the decoder. In this case, the capacity may be expressed as a function of P_b or P_s. In the second case, the receiver makes no explicit symbol decision, but passes on the probability of each symbol conditioned on the observables directly to the decoder. These *soft* decisions yield more information about the transmitted signal and hence result in a greater capacity than the hard-decision case. The capacity in this case may be expressed as a function of the channel statistic $f_{Y|X}$.

Each channel use takes T_s seconds. We refer to this interval as a *slot*. To factor out the slot-width of a particular implementation, capacities (data rates) use normalized units of bits/slot. Similarly we use units of photons/slot for average power:

$$1\,\text{Watt} = 1\,\text{Joule} / \sec = (T_s / h\nu)\,\text{photons} / \text{slot}$$

We also assume the channel is memoryless, i.e., that the received statistics of a slot, conditioned on the slot being pulsed or not, are independent of other slots (we address estimating the capacity of a channel with memory in Section 6.7). We present expressions without derivation, which may be found in the literature, see, e.g., Refs. [8, 10, 11].

6.4.1 BINARY ASYMMETRIC CHANNEL

The binary asymmetric channel (BAC) has capacity

$$C = 1 + h\left(1 - p_{01} + p_{10}\right) - \frac{1}{2}\left(h(p_{01}) + h(p_{10})\right)$$

where

$$h(p) = -p\log_2(p) - (1-p)\log_2(1-p)$$

is the binary entropy function. In the case of the Poisson OOK channel with $n_b = 0$, we have $p_{10} = 0$, resulting in a Z-channel with

$$C = 1 + h(1 - e^{-n_s}) - \frac{1}{2}h(e^{-n_s})$$

For the hard-decision Gaussian BPSK and binary-decision Poisson PPM channels we have $p_{01} = p_{10}$. In this case, the BAC reduces to a binary symmetric channel (BSC) with capacity

$$C = 1 - h(P_b) \tag{6.10}$$

The Poisson OOK channel at moderate to large n_b has $p_{01} \approx p_{10}$ and is also well approximated as a BSC.

6.4.2 GAUSSIAN BPSK CHANNEL

The capacity of a binary-input continuous-output channel is

$$C = 1 - \sum_{x \in \{0,1\}} \int \frac{1}{2} f_{Y|X}(y \mid x)\log_2\left(\frac{f_{Y|X}(y \mid 0) + f_{Y|X}(y \mid 1)}{f_{Y|X}(y \mid x)}\right)dy \tag{6.11}$$

Substituting the expressions (6.3) and (6.4) for the Gaussian BPSK channel with a heterodyne receiver yields

$$C = 1 - \int \frac{1}{\sqrt{2\pi}} e^{-y^2/2\sigma^2} \log_2\left(1 + \exp\left(y\sqrt{8n_s} - 4n_s\right)\right)dy \tag{6.12}$$

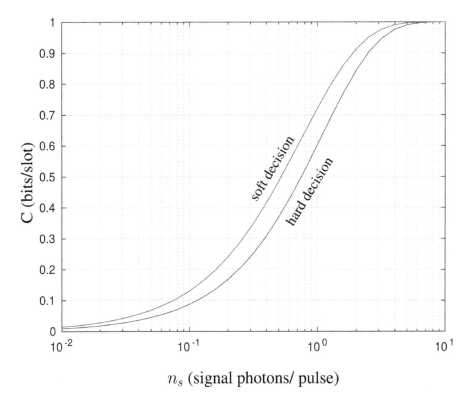

FIGURE 6.2 Heterodyne BPSK (Gaussian) channel capacity, hard and soft decisions.

which may be evaluated efficiently as a sample mean. Figure 6.2 illustrates the Gaussian BPSK capacity and the BSC capacity for a BSC derived by thresholding the Gaussian BPSK signal [Eq. (6.10) evaluated at Eq. (6.5)]. This illustrates the loss in thresholding the output. For example, a hard-decision channel requires an increase in n_s by 1.56 dB relative to a soft-decision channel to achieve $C = 0.5$ bits/slot.

6.4.3 HARD-DECISION POISSON PPM CHANNEL

For the Poisson PPM errors and erasures channel, the channel capacity is

$$C_{h,e} = \frac{1}{M}\left((1 - P_e)\log_2\left(\frac{M}{1 - P_e}\right) + P_c \log_2(P_c) + P_s \log_2\left(\frac{P_s}{M - 1}\right) \right) \quad (6.13)$$

where P_e, P_c, P_s are given by Eqs. (6.7), (6.8), (6.9), respectively. When $n_b = 0$, only signal photons are detected. If any signal photons are detected, the corresponding symbol is known exactly, hence $P_s = 0$, $P_e = e^{-n_s}$, and

$$C_{h,e} = \frac{\log_2(M)}{M}(1 - e^{-n_s}) \quad (6.14)$$

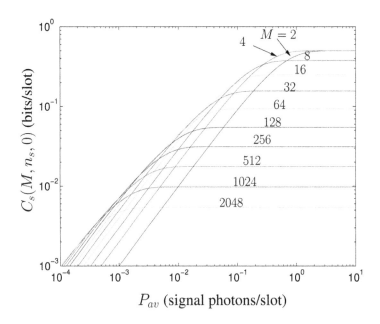

FIGURE 6.3 Capacity for Poisson PPM channel, $C_{h,e}$, $n_b = 0$, $M \in \{2,4,\ldots,2048\}$.

The capacity for $n_b = 0$ is illustrated in Figure 6.3 as a function of average power $P_{av} = n_s / M$ for a range of M.

When ties are resolved we have $P_e = 0$ and the channel is an M-ary input, M-ary output symmetric channel with capacity

$$C_h = \frac{1}{M}\left(\log_2 M + (1 - P_s)\log_2(1 - P_s) + P_s \log_2\left(\frac{P_s}{M-1}\right) \right) \qquad (6.15)$$

where P_s is given by Eq. (6.6).

6.4.4 SOFT-DECISION POISSON PPM CHANNEL

For $n_b = 0$, the soft-decision Poisson PPM capacity is equivalent to that for the hard-decision channel, given by Eq. (6.14). We can see this since the symbol is determined exactly for any nonzero photon count and is unknown for a zero photon count. Hence, the number of photons detected conveys no information.

For $n_b > 0$ the capacity is parameterized by the triple (M, n_s, n_b) [11]:

$$C_s(M, n_s, n_b) = \frac{\log_2(M)}{M}\left(1 - \frac{1}{\log_2 M} E_{Y_1,\ldots,Y_M} \log_2\left[\sum_{i=1}^{M}\left(1 + \frac{n_s}{n_b}\right)^{Y_j - Y_1} \right] \right) \qquad (6.16)$$

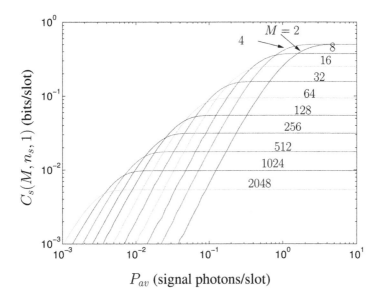

FIGURE 6.4 Capacity for Poisson PPM channel, $C_s(M,n_s,n_b)$, $n_b = 1.0$, $M \in \{2,4,...,2048\}$.

where the expectation is over $Y_1,...,Y_n$, with Y_j having density $f_{Y|X}(y\,|\,1)$ for $j = 1$ and density $f_{Y|X}(y\,|\,0)$ otherwise. The M-fold integration in Eq. (6.16) is straightforward to approximate by a sample mean.

The case $n_b = 1$ is illustrated in Figure 6.4 as a function of average power $P_{av} = n_s\,/\,M$ for a range of M. In Figures 6.3 and 6.4 each curve has a horizontal asymptote at $\log_2 M\,/\,M$, the maximum bits/slot for M-ary PPM.

6.4.4.1 Hard vs. Soft Decision

The difference between hard- and soft-decision capacities for the Poisson PPM channel depends on the noise level and whether ties are resolved. Figure 6.5 illustrates the required average power to achieve $C_s = 1\,/\,8$ given by Eq. (6.16), $C_h = 1\,/\,8$ given by Eq. (6.15), and $C_{h,e} = 1\,/\,8$ given by Eq. (6.13), for an $M = 16$ channel as a function of the noise background n_b. This is the capacity corresponding to a rate $1/2$ code and corresponds to the $C = 0.5$ threshold for Gaussian BPSK where we see a 1.56 dB loss.

At high-noise backgrounds, the gap between C_s and C_h is approximately 1 dB. At low-noise backgrounds, the gap between C_s and C_h grows to approximately 2.2 dB. As n_b approaches 0, the soft-decision capacity approaches the $n_b = 0$ limit of -1.6 dB. At $n_b < 0.1$ we see a gain in the hard-decision case by declaring erasures, and we see $C_{h,e}$ approach C_s at $n_b < 0.001$. This illustrates regimes where there are gains in using different receivers as well as regimes where more complex receivers provide no gain.

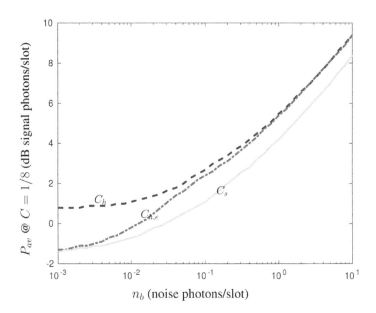

FIGURE 6.5   Average power thresholds in dB photons/slot to achieve $C = 1/8$ bits/slot on a Poisson PPM channel with $M = 16$ PPM for hard decision C_h, hard decision with erasure $C_{h,e}$, and soft decision C_s receivers.

6.5   TRADE-OFFS FOR THE POISSON PPM CHANNEL

In this section, we take the example of the Poisson PPM channel and examine in more detail a number of trade-offs and rules-of-thumb that are illustrated by the channel capacity.

6.5.1   SIGNAL POWER VS. DATA RATE FOR FIXED NOISE POWER

One can show [11] the asymptotic slope of C_s for fixed M and small $P_{av} = n_s / M$ in the log-log domain is 1 for $n_b = 0$ and 2 for $n_b > 0$, as shown in Figures 6.3 and 6.4. Hence, over a broad range, for a fixed PPM order, each 1 dB increase in average power increases the achievable throughput by 1 dB on the noiseless channel and by 2 dB on the noisy channel.

Let

$$C_s(n_s, n_b) = \max_m C_s(2^m, n_s, n_b),$$

the capacity maximized over order $M = 2^m$. $C_s(n_s, 0)$ and $C_s(n_s, 1)$ are the upper shells of the functions in Figures 6.3 and 6.4 and are illustrated in Figure 6.6 along with $C_s(n_s, n_b)$ for $n_b \in \{0.01, 0.1, 10\}$. The slope of the shell is ≈ 1, hence, when allowed to choose an optimum PPM order, each 1 dB increase in average power increases the achievable throughput by ≈ 1 dB.

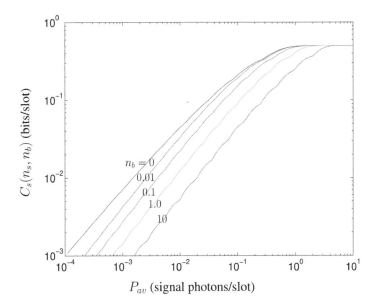

FIGURE 6.6 Capacity maximized over order, $C_s(n_s,n_b)$, $n_b \in \{0,0.01,0.1,1,10\}$.

6.5.2 SMALL n_s BEHAVIOR—SIGNAL AND NOISE POWER TRADE-OFFS

We may obtain an estimate of C_s for small n_s by expanding $C_s(M,n_s,n_b)$ in a Taylor series about $n_s = 0$ [11],

$$C_s(M,n_s,n_b) = \left(\frac{M-1}{2M\ln 2}\right)\frac{n_s^2}{n_b} + O(n_s^3) \tag{6.17}$$

Hence, for small n_s, the capacity goes as n_s^2 / n_b. In this region, each 1 dB increase in signal power compensates for a 2 dB increase in noise power. Let $\lambda_s = n_s / MT_s$ and $\lambda_b = n_b / T_s$ denote the signal and noise photon rates in photons/sec. The dependence on slot-width is shown by substituting for n_s, n_b in Eq. (6.17), and converting units to bits/sec,

$$C_s(M,n_s,n_b) \approx \left(\frac{M-1}{2M\ln 2}\right)\frac{(\lambda_s MT_s)^2}{\lambda_b T_s}\frac{1}{T_s}\text{(bits / second)}$$

$$= \left(\frac{M(M-1)}{2\ln 2}\right)\frac{\lambda_s^2}{\lambda_b}\text{(bits / second)}.$$

Hence the capacity is approximately invariant to slot-width for small n_s. This shows, e.g., that one may increase the slot-width for fixed M to simplify receiver implementation while paying a negligible penalty in capacity.

6.5.3   PEAK AND AVERAGE POWER CONSTRAINTS

The laser transmitter will have peak and average power constraints imposed by physical limitations and available resources. Let P_{pk} denote the peak power constraint. To
satisfy a joint peak and average power constraint we must have

$$\frac{n_s}{M} \le P_{av}$$

$$n_s \le P_{pk},$$

which imply $n_s \le \min\{MP_{av}, P_{pk}\}$. Hence the peak-to-average power ratio $P_{pa} = P_{pk} / P_{av}$
divides PPM orders into two regions. For $M \le P_{pa}$ the channel is average power constrained. For $M > P_{pa}$ the channel is peak power constrained.

One can show [12] that for fixed n_s and n_b the capacity is monotonically decreasing
in the PPM order. It follows that peak and average power constraints impose an effective constraint on the optimum order of $M \le P_{pa}$. This can be seen as follows. If there
is a signaling scheme with average photons per pulse n_s and duty cycle $1 / M$ that meets
the peak and average power constraints, then $n_s \le \min\{MP_{av}, P_{pk}\}$. If $M > P_{pa}$ (and P_{pa}
is a power of 2), then we can increase the capacity by reducing M to $M = P_{pa}$, without
violating the power constraints. Hence, in the presence of peak and average power
constraints, the PPM order should be set so that the average power is met with equality
(or as close as possible), even if the peak constraint cannot be met with equality.

6.5.4   THE IMPACT OF DEAD TIME

Certain lasers have a required dead time after the transmission of a pulse during
which another pulse cannot be transmitted. Suppose this dead time is a an integer d
multiple of the slot-width. To satisfy the dead time constraint with PPM signaling,
we can impose a period of d slots after each symbol during which a pulse cannot be
transmitted. Dead time may also be added intentionally [13] to provide a tone to aid
in synchronization, or to vary the duty cycle without changing the PPM order. How
does this added dead time impact the capacity of the channel?

Suppose we can close an optical link of PPM order M with average power
$P_{av} = n_s / M$. Adding a dead time of d slots while retaining n_s photons/pulse reduces
the capacity by a factor of $M / (M + d)$. However, the average power is also decreased
by the same factor when keeping n_s fixed. Hence, for each point

$$(\text{photons / slot}, \text{bits / slot}) = (P_{av}, C_s(M, P_{av}M, n_b))$$

achievable with PPM there exists a family of points

$$(\text{photons / slot}, \text{bits / slot}) = (P_{av}M / (M + d), C_s(M, P_{av}M, n_b)M / (M + d))$$

achievable by adding dead time. In a log-log domain plot of average power vs. capacity, this is represented by extending each point on the C_s vs. P_{av} curve down and to
the left with a line of slope 1.

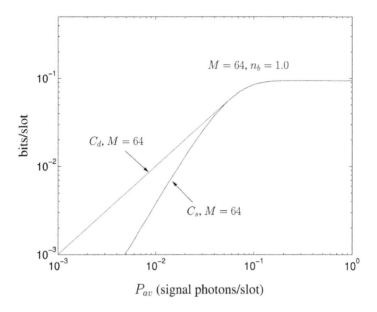

FIGURE 6.7   Achievable rates for $M = 64$, $n_b = 1$, with and without dead time.

Let

$$C_d(M, P_{av}, n_b) = \max_d \left\{ C_s(M, n_s, n_b) \frac{M}{M+d} \middle| n_s = P_{av}(M+d) \right\},$$

the capacity maximized over dead time d, illustrated in Figure 6.7 for $M = 64$. C_d is equal to C_s for P_{av} above the point where the tangent of C_s has slope one, and equal to that tangent below that point. Comparing this to the $n_b = 1$ shell of Figure 6.6 we see it is an efficient way to approach the capacity at small duty cycles.

6.5.5   SUBOPTIMALITY OF PPM

PPM is an efficient method to achieve a high peak-to-average power ratio when the ratio is large. At smaller duty cycles, PPM becomes less efficient (relative to other modulations with the same duty cycle). Here we quantify the loss in efficiency, illustrating regions where non-PPM modulations might be pursued.

Suppose we are utilizing a slotted, binary modulation with duty cycle $1/M$, but are not restricted to PPM. The capacity of a binary modulation with duty cycle $1/M$ on the Poisson channel is given by[3]

$$C_{OOK}(M, n_s, n_b) = \frac{1}{M} E_{Y|1} \log \frac{f_{Y|X}(Y|1)}{f_Y(Y)} + \frac{M-1}{M} E_{Y|0} \log \frac{f_{Y|X}(Y|0)}{f_Y(Y)}$$

---

[3] We use the subscript OOK, and think of this as a generalization of OOK to arbitrary duty cycles.

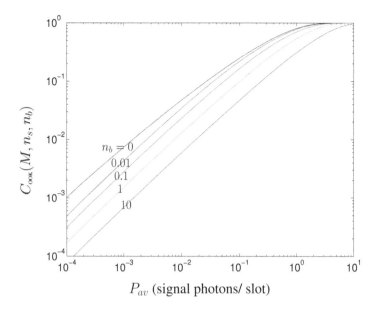

FIGURE 6.8 Duty cycle constrained capacity maximized over order, $C_{OOK}(M,n_s,n_b)$, $n_b \in \{0,0.01,0.1,1,10\}$.

where $f_{Y|X}(y|0), f_{Y|X}(y|0)$ are given by Eqs. (6.1) and (6.2), and

$$f_Y(y) = \frac{1}{M} f_{Y|X}(y|1) + \frac{M-1}{M} f_{Y|X}(y|0)$$

is the probability mass function for a randomly chosen slot.

Recall $h(p)$ is the binary entropy function, $h(p) = p\log_2(1/p) + (1-p)\log_2(1/(1-p))$. $C_{OOK}(M,n_s,n_b)$ has a horizontal asymptote at $h(1/M)$, the limit imposed by restricting the input to duty cycle $1/M$. Let

$$C_{OOK}(n_s,n_b) = \max_M C_{OOK}(M,n_s,n_b),$$

the capacity maximized over real-valued duty cycle $1/M$, illustrated in Figure 6.8 for $n_b \in \{0,0.01,0.1,1,10\}$.

Figure 6.9 illustrates the ratio $C_{OOK}(n_s,n_b)/C_s(n_s,n_b)$, reflecting the potential gain in using an arbitrary duty cycle constraint relative to PPM. The gains are larger for high average power, corresponding to small PPM orders, and for smaller background noise levels. We can potentially double the capacity for moderate to high average power. This requires a modulation that efficiently implements an arbitrary duty cycle. There are systematic methods to design such modulations, see, e.g., Ref. [14].

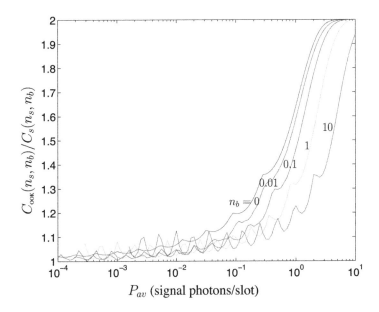

FIGURE 6.9 Relative loss due to using PPM, $C_{OOK}(n_s, n_b) / C_s(n_s, n_b)$, $n_b \in \{0, 0.01, 0.1, 1.0, 10\}$.

6.6 FADING

The received power over a free-space optical communications link fluctuates with time, or fades, due to changes in the atmospheric path from the transmitter to the receiver and time-varying pointing errors. This can be a significant impediment for a free-space optical channel. In this section, we characterize losses due to fading and illustrate methods to mitigate fading.

Let $v(t)$ be the fading waveform, such that power $v(t)P_{av}$ is received at the aperture at time t. Let $f_V(v)$ be the probability density function (PDF) of a sample $v_k = v(t_k)$ ($v(t)$ is assumed stationary). Let $l = log(v)$, and f_L be the density of l. In many cases atmospheric fading is well modeled by a lognormal distribution [15, 16]

$$f_V(v) = \frac{1}{\sqrt{2\pi\sigma_l^2}} \frac{1}{v} \exp\left(\frac{-(\log v + \sigma_l^2/2)^2}{2\sigma_l^2}\right) \tag{6.18}$$

$$f_L(l) = \frac{1}{\sqrt{2\pi\sigma_l^2}} \exp\left(\frac{-(l + \sigma_l^2/2)^2}{2\sigma_l^2}\right)$$

Note that the fading process v_k has been normalized to have mean 1.

Figure 6.10 illustrates a sample of an observed fading time series over a 45-km free-space link [15]. The observed histogram of samples v_k, and a best fit to $f_V(v)$ (shown as both a histogram of bin volumes corresponding to observations and a continuous function) given by Eq. (6.18) is illustrated in Figure 6.11. We see that the lognormal model is a good fit to observations. Numerical results in the remainder of the section will use the lognormal model.

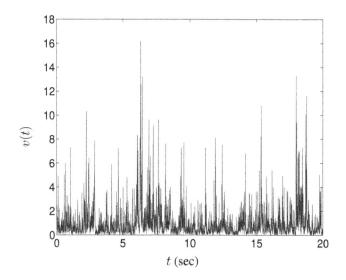

FIGURE 6.10 Fading time series.

Other sources of fading may be characterized by replacing $f_V(v)$ with an appropriate model. For example, the waveform for pointing-error-dominated fading may be modeled as [17]

$$v(t) = \exp\left(-\frac{w_x^2 + w_y^2}{2}\right),\tag{6.19}$$

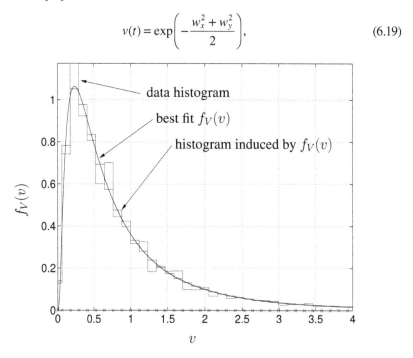

FIGURE 6.11 Histogram of fading samples and a lognormal probability density function (PDF) fit $f_V(v)$: $\hat{\sigma}_l^2 = 0.98$.

where w_x and w_y are jointly Gaussian random variables. Certain atmospheric fading processes may also be modeled by Eq. (6.19). It is straightforward to extend the methods described here to characterize and mitigate fading processes to other models.

6.6.1 COHERENCE TIME

Samples of the fading process separated by a sufficiently long interval will be uncorrelated. The *coherence time*, T_{coh}, of the fading process is the duration such that samples separated by more than the coherence time are (approximately) uncorrelated. T_{coh} is roughly the reciprocal of the bandwidth of the spectral density of the fading process. Let $S(f)$ be the power spectral density of $v(t)$ and, for $\xi \in [0,1]$, define the $100\xi\%$ bandwidth as

$$W(\xi) = \min\left\{ 2B \mid \int_{-B}^{B} S(f)df = \xi \int_{-\infty}^{\infty} S(f)df \right\}$$

In our analysis, we put

$$T_{coh} = \frac{1}{W(0.90)}$$

or the reciprocal of the 90% bandwidth of the process.

Figure 6.12 illustrates the power spectral density of the sequence illustrated in Figure 6.10. The 90% bandwidth is 117 Hz, yielding a coherence time of 8.5 msec, typical of an atmospheric link.

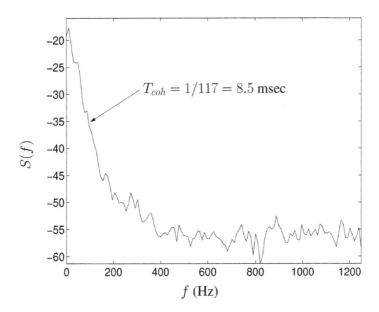

FIGURE 6.12 Power spectral density of sample fading time series with $T_{coh} = 8.5$ msec.

6.6.2 LOSSES DUE TO FADING

The fading process produces losses relative to a received signal in the absence of fading. We follow the approach in Ref. [17] to quantify the losses.

Let $C(P_{av})$ be the capacity of the channel with power P_{av} (in the absence of fading). The *fading capacity* of the channel is given by

$$C_{fad}(P_{av}) = \int_0^\infty C(vP_{av})f_v(v)\,dv$$

The fading capacity is the limit on the achievable throughput. The performance for a finite block length ECC is well predicted by the *outage probability*. For positive integer N_f, let

$$\bar{C} = \frac{1}{N_f}\sum_{i=1}^{N_f} C(v_i P_{av})$$

the *instantaneous capacity* of the channel, where v_i are independent realizations of the process $\{v_k\}$. For a PPM-encoded system with ECC rate R and block length chosen such that there are N_f independent fades observed per codeword, the outage probability is given by

$$P_{out}(R,P_{av}) = Pr\left(\bar{C} < \frac{R\log_2 M}{M}\right) \tag{6.20}$$

The losses due to fading may be divided into three components. The first, the *static loss*, is the reduction in average received power due to the fading process, and is non-recoverable. The static loss can be significant, but here we want to focus on losses due to power fluctuations, and we presume the fading process has been normalized so that the static loss is zero. The second, the *capacity loss*, is the difference P_{av} / P'_{av} where P_{av} and P'_{av} are the average powers satisfying $C(P_{av}) = C_{fad}(P'_{av})$, and is also non-recoverable. The third loss, the *dynamic loss*, is the difference between the fading capacity $(P_{av} \mid C_{fad}(P_{av}) = R\log_2 M / M)$ and the power required to achieve a desired outage probability. In the next section, we discuss methods to recover the dynamic loss.

6.6.3 MITIGATING FADING LOSSES

Fading may be mitigated at a system level by introducing diversity into the communication system. For example, to mitigate pointing errors, one can utilize multiple transmit beams with separate pointing control systems such that the instantaneous pointing errors are uncorrelated. To mitigate scintillation one can use a larger aperture, averaging over many fading processes. Here we discuss methods not to reduce the variance of the fluctuations, but to mitigate the losses due to those fluctuations in signal processing. Our running example will be a system transmitting M-ary PPM with a rate R ECC and slot-width T_s. However, the approaches can be applied to any coded modulation.

6.6.3.1   Adaptive Coding and Modulation

The most robust method to mitigate fading losses is to adapt the coding and modulation to the instantaneous capacity of the channel. Reliable communication is possible only if the throughput is less than the capacity:

$$\frac{R\log_2(M)}{MT_s} \leq C(v(t)P_{av}) \tag{6.21}$$

Suppose the transmitter can learn $v(t)P_{av}$. Then the channel can achieve the fading capacity (up to the efficiency of the ECC) by varying R such that Eq. (6.21) holds. More generally, the transmitter can adapt any of the signaling parameters: (R,M,T_s). In this case, the relevant instantaneous capacity is the maximum over the allowable pairs (M,T_s).

Adaptive coding and modulation is common in wireless radio frequency links. It requires knowledge of the received power at the transmitter, and the ability to change signaling parameters (baud rate, ECC rate, modulation). The received power may be obtained from the receiver via a bidirectional communication link. Hence, this is typically constrained to links where the received power is changing much more slowly than the round-trip time (RTT).

6.6.3.2   Channel Interleaving

An interleaver, inserted after the ECC in the signaling, mitigates losses by introducing temporal diversity. The interleaver spreads the bits of one ECC codeword over a duration of time sufficiently long that the codeword sees N_f independent realizations of the fading process, effectively increasing N_f. Interleaving does not require knowledge of the received power and can be implemented on links with a long RTT and without bidirectional communication.

Interleaving may be implemented via a convolutional interleaver, illustrated in Figure 6.13. A convolutional interleaver consists of N rows of delays (shift registers) of length $0, B, 2B, \ldots, (N-1)B$. Each register contains the observations corresponding to one PPM symbol. The number of registers required to implement the interleaver or de-interleaver is $N(N-1)B/2$.

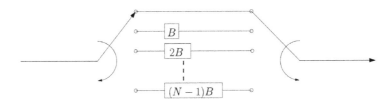

FIGURE 6.13   A convolutional interleaver.

Letting N_f be the number of uncorrelated fades per codeword, we have

$$N_f \approx \frac{N(N-1)B}{N_{coh}}$$

where N_{coh} is the number of PPM symbols per coherence time. Assume the receiver quantizes slot measurements to 4 bits. Each register, corresponding to one PPM symbol, requires $M/2$ bytes, hence the required de-interleaver memory to induce N_f fades/codeword is

$$N_b = \frac{MN(N-1)B}{4} \, bytes$$

$$\approx \frac{MN_fN_{coh}}{4} \, bytes$$

Figure 6.14 illustrates capacity and fading capacity thresholds at which $C = C_{fad} = 1/4$ and P_{out} for lognormal fading with $\sigma_l = 0.8$, $M = 4$, $R = 1/2$, and $n_b = 1.0$ photons/slot. There is a capacity loss (the gap between $P_{av} \mid C(P_{av}) = \log_2 M /(2M)$ and $P_{av} \mid C_{fad}(P_{av}) = \log_2 M /(2M)$) of 0.6 dB. The 13 dB loss (measured at $P_{out} = 10^{-4}$) that would be incurred due to fading in the absence of interleaving can be reduced to 1.1 dB with an interleaver that produces an $N_f = 128$. For a system with slot-width $T_s = 1$ nsec, coherence time $T_{coh} = 8.5$ msec, $M = 4$ and 4 bits/slot, an interleaving depth of $N_f = 128$ could be implemented with a 272-Mbyte de-interleaver.

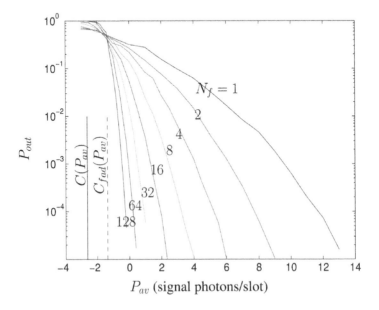

FIGURE 6.14   Capacity and outage probabilities, $\sigma_l = 0.8, M = 4, R = 1/2, n_b = 1$.

The dynamic loss may be completely recovered by a sufficiently long interleaver, since

$$C_{fad} = \lim_{N_f \to \infty} \bar{C}$$

However, this comes at a cost in latency from the interleaving and de-interleaving steps. Note that this is less robust than adaptive coding and modulation, since adaptation can achieve the fading capacity as the fading capacity changes. The methods are not exclusive, and a system can adopt both.

6.6.3.3 Automatic Repeat Request (ARQ)

Interleaving is an efficient method to mitigate fading losses, but it comes at a cost in latency from the interleaving and de-interleaving processes. On bidirectional links with a latency constraint that prohibits interleaving, fading losses may be efficiently mitigated with an ARQ mechanism, see, e.g., Ref. [18]. In an ARQ scheme, transmitted codewords are stored at the transmitter. The receiver transmits acknowledgments (ACKs) of correctly received codewords back to the transmitter. Once the transmitter receives an acknowledgment, it drops the codeword from its memory. When the transmitter determines a codeword was lost, it retransmits it. The transmitter may infer that a codeword has been lost if an ACK has not been received within a RTT, or it may respond to an explicit message (NACK) from the receiver.

There are many variants on the implementation of an ARQ scheme. We can think of ARQ as a method to allow the transmitter to monitor when the link is available and to only transmit data when the link is up. It adaptively trades codeword error rate (which it can drive to zero) for latency. Consider an ideal ARQ system where the transmitter has perfect knowledge of the channel state (it knows perfectly which codewords were lost or, equivalently, when the channel is up). Let R_t be the channel transmission rate (information bits/sec), R_a be the rate of arriving traffic (information bits/sec), and the outage rate be P_{out}. Suppose further that we are in a regime where outage events are nonoverlapping, so that the transmitter buffer is cleared between outage events. Then the transmitter can drive the error rate to zero by transmitting packets only when the channel is up, supporting an average throughput of $R_a \leq R_t(1 - P_{out})$ at an approximate cost in mean latency (the increase in mean latency over the time-of-flight) of

$$E[\tau] \approx \frac{T_{coh} P_{out}}{2} \frac{1}{1 - \dfrac{R_a}{R_t(1 - P_{out})}}$$

One can show practical ARQ schemes, with $RTT < 2T_{coh}$, can approach the performance of this ideal ARQ performance.

6.7 JITTER

In the results to this point we made an implicit assumption that there is no inter-symbol(slot) interference (ISI), i.e., that the signal energy in a pulsed slot is not detected in other slots. This assumption can break down in a practical system as one pushes the bandwidth limits. Slot boundaries are approximated by clock recovery circuits, and errors in clock recovery can lead to ISI. Practical photodetectors also have minimum reproducible pulse widths due to detector jitter. In this section, we examine the degradation due to ISI caused by detector jitter and discuss methods to compensate for the jitter.

Suppose the receiver uses a photon-counting detector, i.e., a detector that produces a detectable current pulse in response to a single incident photon. In any such detector, there is a random delay from the time a photon is incident on the detector to the time a pulse is produced in response to that photon. This delay, which we refer to as detector jitter or simply jitter, causes degradation in performance relative to a system with no jitter. The standard deviation of the jitter for a photon detector may range between a few hundredths of a nanosecond up to a nanosecond. The degradation is negligible for systems where the slot-width is much larger than the jitter standard deviation, but becomes significant as the slot-width is narrowed to achieve higher data rates.

Let $\{s_j\}_{j=1,2,\ldots}$ be the collection of photon arrival times at the detector. Detector jitter is modeled by presuming that pulses are produced at the output of the detector at times $t_j = s_j + \delta_j$, where δ_j are assumed to be independent of one another as well as of the arrival times $\{s_j\}$, and identically distributed.

Figure 6.15 illustrates measured jitter histograms (normalized to integrate to one) for an InGaAsP PMT, a niobium nitride superconducting single photon detector (NbN SSPD), and an InGaAsP Geiger-mode avalanche-photodiode (APD). The histograms were fit to a weighted sum of Gaussians of the form [19]

$$f_\delta(\delta) = \sum_{k=1}^{K} \gamma_k \frac{1}{\sqrt{2\pi\sigma_k^2}} e^{-(\delta-\mu_k)^2/(2\sigma_k^2)} \tag{6.22}$$

using the Levenberg-Marquardt algorithm, see, e.g., Ref. [20], and normalized to have zero mean. The smallest value of K yielding a good fit was used in each case. The fitted models are overlaid on the histograms. Table 6.1 lists the parameters of the fitted models, the standard deviation,

$$\sigma = \sqrt{\sum_{k=1}^{K} \gamma_k (\sigma_k^2 + \mu_k^2)}$$

and the photon detection efficiency (β), the probability that an incident photon produces a pulse. This allows one to generate deviates modeling the jitter process by choosing $k \in \{1,2,\ldots,K\}$ according to the weights $\{\gamma_k\}$, i.e., $P(k=i) = \gamma_i$, and generating a Gaussian deviate with mean and variance (μ_k, σ_k).

6.7.1 CAPACITY

On a channel with jitter, signal photons from a pulse in one symbol may extend beyond the symbol boundaries, creating not only inter-slot-interference but also inter-symbol-interference. We examine this loss for a PPM-modulated system.

TABLE 6.1

Parameters for Detector Jitter Models. Units for σ, σ_k, μ_k are nsec.

Detector	γ_k	μ_k	σ_k
InGaAsP PMT ($\sigma = 0.9154$, $\beta = 0.09$)	0.7655	−0.0930	0.4401
	0.1940	0.8080	0.4511
	0.02723	−4.0780	0.7241
	0.01331	1.917	0.2751
NbN SSPD ($\sigma = 0.02591$, $\beta = 0.40$)	0.6550	2.838e-3	0.01890
	0.3450	−5.389e-3	0.03498
InGaAsP GM-APD ($\sigma = 0.2939$, $\beta = 0.45$)	0.4360	6.041e-2	0.4200
	0.2731	−0.1086	0.06024
	0.2909	1.141e-2	0.1172

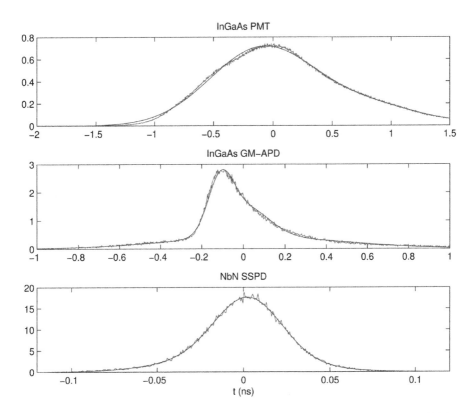

FIGURE 6.15 Detector jitter histograms and Gaussian fits for InGaAsP PMT, NbN SSPD, and InGaAsP GM-APD.

Suppose that in a single channel use we transmit n-PPM symbols, and there is no interference between channel uses. Let C_n denote the capacity of this channel with normalized units of bits/sec. The n-symbol capacity satisfies $C_1 \geq C_2 \geq \cdots \geq C_n$, yielding a sequence of nondecreasing upper bounds on $C_\infty = \lim_{n \to \infty} C_n$. We will see that over a broad range, C_1 is a good approximation to C_∞, the true channel capacity, as noted in Ref. [21].

We represent the transmission of n-PPM symbols as the transmission of an integer $x \in \{0, 1, \ldots, M^n - 1\}$. Expanding each x as a l-vector over base M, $x = \Sigma_{l=0}^{n-1} x_l M^l$, we let x_l select the l-th PPM symbol, i.e., the x_l-th slot of the l-th PPM symbol is pulsed. Let N be the number of detected photons when X is transmitted and $T = \{t_0, t_1, \ldots, t_{N-1}\}$ be the random N-vector of photon arrival times. The capacity of the n-symbol channel with equally likely transmitted PPM symbols is given by [10]

$$C_n(X; T, N) = \frac{1}{nMT_s} E_{T,N,X} \log_2 \frac{M^n f_{T,N|X}(T, N \mid X)}{\sum_{i=1}^{M^n} f_{T,N|X}(T, N \mid X = i)} \text{ bits / sec} \qquad (6.23)$$

where the expectation is over (T, N, X) and we have explicitly introduced the slot-width to examine the degradation as the slot-width is varied for fixed jitter statistics. Hence, in order to compute the channel capacity requires a determination of the conditional channel likelihoods.

6.7.2 CHANNEL LIKELIHOODS

As described previously, the input is a PPM signal and photon arrivals are modeled as a Poisson point process. Signal photons arrive at a rate of λ_s photons/sec and noise, or background, photons arrive at a rate of λ_b photons/sec. In the l-th symbol period, one of the M PPM symbols is transmitted by transmitting a pulse in the x_l-th slot of the M slot word, with slot-widths of T_s seconds. Let $\lambda_x(t) = M\lambda_s T_s \Sigma_{l=0}^{n-1} p(t - (x_l + Ml)T_s) + \lambda_b$, the incident photon intensity function when x is transmitted, where $p(t)$ is a pulse with amplitude $1/T_s$ on $[0, T_s]$ and zero elsewhere (a more accurate pulse shape may be substituted without loss of generality in the following analysis).

The distribution of detected signal photons for a pulse transmitted in the first slot is given by $f(t) = (p * f_\delta)(t)$. Substituting Eq. (6.22) for f_δ yields

$$f(t) = \frac{1}{2T_s} \sum_{k=1}^{K} \gamma_k \left(\text{erfc}\left(\frac{t - T_s - \mu_k}{\sqrt{2\sigma_k^2}} \right) - \text{erfc}\left(\frac{t - \mu_k}{\sqrt{2\sigma_k^2}} \right) \right)$$

The *detected* intensity (as opposed to the incident) when x is transmitted is given by

$$\lambda_x'(t) = \beta \left(\lambda_b + M\lambda_s T_s \sum_{l=0}^{n-1} f(t - (x_l + Ml)T_s) \right)$$

The probability density of the detected signal, conditioned on the transmitted signal, is given by [22]

$$f_{T,N|X}(\{t_j\}, N \mid x) = e^{\int_T \lambda_x'(t)dt} \prod_{j=1}^{N} \lambda_x'(t_j) \qquad (6.24)$$

where T is the interval over which there is a nonzero probability of detected signal photons, and N is the number of detected photons over T, which is random with Poisson probability mass function. In practice, the accuracy of measuring the detection time is limited. Approximations to the likelihoods given by Eq. (6.24) may be used, trading-off performance for complexity [23].

6.7.3 RESULTS

In a typical system link design, the noise rate λ_b, the detector characteristics (f_δ and β) and certain system parameters, e.g., an ECC rate R and PPM order M, are fixed, and the designer determines the signal power required to support a desired throughput bits/sec. That is the approach we take in this section. In numerical results Eq. (6.23) is estimated via a sample mean. The parameters M, λ_b, f_δ, and R are presumed fixed and the minimum signal power λ_s required to close the link, supporting a throughput of

$$\frac{R \log_2 M}{MT_s} \text{ bits / sec}$$

is determined as a function of the slot-width T_s. We presume the use of a rate $R = 1/2$ ECC and consider various fixed PPM orders. Let

$$\lambda_s' = \left\{ \lambda_s \mid C_n = \frac{\log_2 M}{2MT_s} \right\}$$

the theoretical minimum required incident power to close the link with a rate 1/2 ECC. Define the loss due to jitter $loss_n$, as the ratio of λ_s' in the presence of jitter to λ_s' in the absence of jitter for the n-symbol channel and let $\tilde{\sigma} = \sigma / T_s$, the normalized jitter standard deviation.

Figure 6.16 illustrates jitter loss for Gaussian jitter with $\sigma = 1.0$ nsec (the case $K = 1$, $\mu_1 = 0$, $\sigma_1 = 1.0$). PPM orders $M = 4$ and $M = 8$ are illustrated for n-symbol capacities of $n \in \{1,2,3,4\}$. The loss is well fit as quadratic in $\tilde{\sigma}$, with a "knee," the point where the slope of loss vs. $\tilde{\sigma}$ is 1, around $\tilde{\sigma} \approx 1.0$. We see that for small $\tilde{\sigma}$, $C_1 \approx C_2 \approx C_3$, in agreement with Ref. [21], and that this approximation is tighter for larger M, since the impact of inter-PPM-symbol interference is less. This justifies using C_1 and $loss_1$ as proxies for C_∞ and $loss_\infty$.

Figure 6.17 illustrates the required signal power λ_s to achieve a specified throughput for the three detectors characterized in Table 6.1 and an ideal detector ($\beta = 1$, no jitter) with $M = 16$, $R = 1/2$, and $\lambda_b = 1$ photon/nsec. When normalized, we see

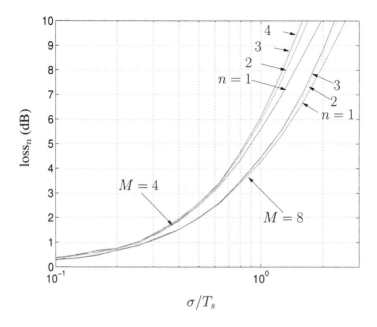

FIGURE 6.16 Loss as a function of normalized jitter variance for $n = 1, 2, 3, 4$, $M = 4$ and $M = 8$. Gaussian jitter, $\sigma = 1.0$ nsec, $\lambda_b = 1$ photon/nsec, $R = 1 / 2$ ECC, $\beta = 1.0$.

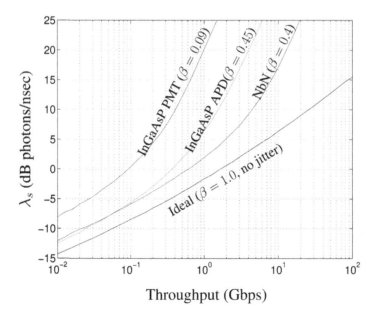

FIGURE 6.17 Required power to achieve specified throughput: $M = 16$, $R = 1 / 2$, ECC $\lambda_b = 1$ photon/nsec.

a similar loss as a function of $\tilde{\sigma}$ for all (nonideal) detectors, with a knee occurring at $\tilde{\sigma} \approx 1$. At low throughputs, the detector performance is differentiated only by the photon detection efficiency, with, e.g., the InGaAsP GM-APD demonstrating a 0.5 dB gain over the NbN SSPD. However, as the throughput is increased by narrowing the transmitted slot-width, the jitter loss begins to dominate, so that the NbN SSPD, with less jitter, demonstrates large gains at high throughputs.

6.8 ERROR-CORRECTION CODES

An ECC adds redundant information to the message to be transmitted in a systematic manner so that errors in the data may be corrected on reception. An (n, k) binary ECC maps each k information bits to n coded bits, and we say the rate of the code is $r = k / n$. In return for this added redundancy, the ECC provides large gains over an uncoded system. Figure 6.18 illustrates the bit error rate of an uncoded $M = 16$ PPM channel with $n_b = 1$, as well as bounds on the achievable bit error rate for hard- and soft-decision rate 1/2 codes. Error rate bounds are determined from the converse to the coding theorem

$$P_b \geq h^{-1}\left(1 - \frac{C(P_{av})}{R}\right)$$

see, e.g., Ref. [24]. Figure 6.18 illustrates that, in theory, a rate 1/2 code can provide gains up to 10.6 dB over an uncoded system (9.2 dB for hard decision), measured at $P_b = 10^{-9}$. We'll see that practical ECCs can perform within 1 dB of the channel capacity. Hence, at an error rate of 10^{-9}, a typical performance target, rate 1/2 soft- and hard-decision ECCs, can provide approximately 9.6 and 8.2 dB gain, respectively, over an uncoded system. This gain is reduced by 3 dB when normalized to a constant energy/information bit to account for the rate 1/2 code rate overhead, in which case ECC provides 5 to 6 dB gain over an uncoded system.

This section will primarily discuss ECCs for use with PPM. When a binary modulation such as BPSK or DPSK is used, ECCs originally designed for RF communications often work well. For example, NASA's Laser Communications Relay Demonstration (LCRD) uses DPSK together with low-density parity-check (LDPC) codes used in the second-generation Digital Video Broadcast standard [25].

In this section we illustrate performance for two types of ECCs: hard and soft decision. This distinction refers to the decoding algorithm used to decode the code, rather than the structure of the code itself. However, the design of the codes is intimately tied to the decoding algorithms, hence we use this common nomenclature to refer to the codes as well. Hard-decision codes accept symbol decisions, and their performance is bounded by the hard-decision capacity. Similarly, soft-decision codes accept symbol likelihoods, or photon counts, and their performance is bounded by soft-decision capacity. We focus on the class of Reed-Solomon (RS) hard-decision codes and the class of SCPPM soft-decision codes. These two classes represent the current state-of-the-art solutions for each case.

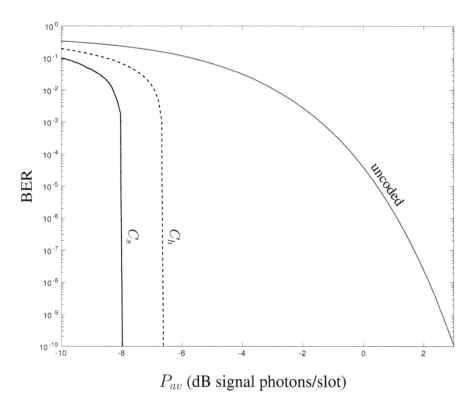

P_{av} (dB signal photons/slot)

FIGURE 6.18   Channel bit error rates and soft and hard capacity bounds for a rate 1/2 ECC, $M = 16, n_b = 1.0$.

6.8.1   REED-SOLOMON CODES

A RS code is a linear block code over a Galois field with q elements, denoted GF(q). The code maps blocks of k symbols to n symbols, each over GF(q), where $n = q - 1$. The field order, q, is commonly taken to be a power of 2, in which case the symbols have binary representations, and the code has an equivalent representation as a $(n \log_2 q, k \log_2 q)$ binary code.

The minimum number of symbol differences between two codewords of an RS(n, k) code is $d_{min} = n - k + 1$. This is the largest achievable minimum distance of any (n, k) linear block code over $GF(q)$, and RS codes are optimal in this sense. Note, however, this optimality is specific to the symbol distance. For example, the equivalent binary code does not provide the largest minimum distance over GF(2). A code with minimum distance d can correct any pattern of $(d_{min} - 1) / 2$ errors. Hence an RS(n, k) code can correct any pattern of $(n - k) / 2$ symbol errors. The encoding and decoding operations may be carried out efficiently in hardware using shift registers, see, e.g., Ref. [18]. The excellent distance properties and efficient encoding and decoding operations have led to the widespread use of RS codes in communications systems.

In Ref. [26], a RS code was proposed for use on the *noiseless*, that is, $n_b = 0$, Poisson PPM channel. The noiseless Poisson PPM channel is a symbol erasure channel, and an (n,k) RS code may be tailored to fit an M-ary PPM channel by choosing $n = M - 1$ and taking code symbols from the Galois field with M elements. However, strictly following the convention of choosing $n = M - 1$ yields codes for small orders with short block lengths, and codes for small n perform poorly and have less flexibility in choosing the rate (with $M = 4$, this convention would yield the small class of $(3,k)$RS codes). This can be generalized by grouping together β M-PPM symbols to form an element of $GF(M^\beta)$ [27]. The RS code is then taken to be an $(n,k) = (M^\beta - 1,k)$ code. The optimum choice of β will be a function of the target bit error rate. We refer to the concatenation of an RS code with PPM formed in this manner as an RSPPM(n,k,M) code.

Figure 6.19 shows the performance of rate $\approx 3/5$ RSPPM with $M = 64$ for $\beta \in \{1,2,3,4\}$, along with hard- and soft-decision capacities of rate 3/5 coded 64-PPM. We would conventionally use the RSPPM(63,37,64) code, which matches 64-PPM, but the RSPPM(262143,157285,64) has better performance at a target $P_b = 10^{-6}$. As the block length increases, the waterfall region is seen to move slightly to the right and get steeper, so that the optimal β is a function of the target P_b. At a target $P_b = 10^{-6}$, the RS(262143,157285,64) code is within 1.45 dB of hard-decision capacity and provides a gain of 5.23 over the uncoded system. However, the restriction to utilize hard decisions limits the achievable performance. A soft-decision-based channel has a potential gain of 1.24 dB over this. In the next section, we illustrate codes that close the gap to soft-decision capacity.

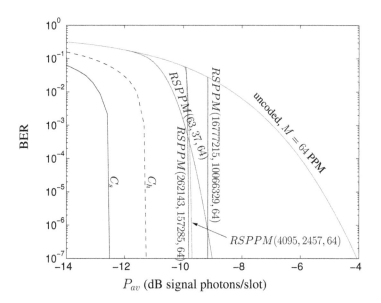

FIGURE 6.19   Performance of rate $\approx 3/5$ RSPPM, $M = 64$.

6.8.2   ITERATIVE SOFT-DECISION CODES

Convolutional codes (CCs) are linear codes whose codewords may be produced with (typically short) shift registers, allowing an efficient representation of the codewords as paths on a trellis. This trellis structure enables efficient decoding via the Viterbi algorithm, see, e.g., Ref. [18]. CCs were proposed for use on the noiseless PPM channel in Ref. [28], illustrating performance competitive with RS codes. However, the distinguishing characteristic of trellis codes is that the Viterbi algorithm accepts soft input information. Their use on the noisy PPM channel, accepting soft inputs, was explored in Ref. [29].

Turbo codes [30] achieved a breakthrough in code performance by combining a pair of CCs in parallel to form one large code. In a turbo code, a block of data is encoded by one CC. The same data is interleaved, and encoded by a second CC. A maximum likelihood decoder of the subsequent, typically long, block code would be prohibitively complex to implement. A key insight of turbo codes was to use a suboptimal decoder, decoding each constituent code independently, and passing information between them in an iterative manner. This decoding algorithm proves to have excellent performance in practice, achieving near capacity performance. Conventional turbo codes (parallel concatenated codes as described above) were applied to the PPM channel in Ref. [31–33], demonstrating large gains over RSPPM.

A variant on turbo codes utilizes a serial concatenation of the codes [34]. In this case, the output of the first code, as opposed to the input, is interleaved and encoded by a second code. This approach was applied to coded PPM in Ref. [35], utilizing a CC for the outer code and coded PPM as the inner code. In this approach, the modulation essentially becomes part of the ECC, providing large performance gains. These serially concatenated, convolutionally coded PPM (SCPPM) codes provide performance close to capacity for the PPM channel.

Figure 6.20 illustrates the performance of a rate 1/2 SCPPM code relative to $(n,k) = (4085, 2047)$ RS coded PPM for $M = 64$ and $n_b = 0.2$. The SCPPM code consists of a rate 1/2, memory 2 CC, a permutation polynomial interleaver, and a recursive accumulator—described in more detail in Ref. [35]. The RS block length was chosen to yield the best performance over all RS codes with the same rate associating an integer number of PPM symbols with each code symbol [11]. The code performs within 0.7 dB of soft-decision capacity and provides a 3 dB gain over the RS code.

6.8.3   COMPARISONS

Figure 6.21 illustrates achievable rates for $n_b = 1$ populated by points for a collection of SCPPM codes, a collection of RSPPM codes, and the uncoded PPM channel. Also illustrated are the shells $C_s(n_s, n_b)$ and $C_{OOK}(n_s, n_b)$, illustrating gaps to capacity and the efficiency of using PPM. Points correspond to the average power at which $P_b = 10^{-5}$. The coded channels are evaluated at a finite number of rates, connected to illustrate typical performance for the class. Note that a fixed bits/slot takes into account the overhead of the coded systems. The point RSPPM corresponding to $M = 64$ shows a cluster of points corresponding to the codes illustrated in Figure 6.19. Other RSPPM points use the convention $n = M - 1$.

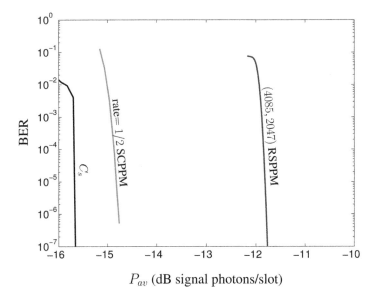

FIGURE 6.20 SCPPM and RSPPM performance, Poisson channel, $M = 64$, $n_b = 0.2$.

The class of SCPPM codes lies approximately 0.5 dB from capacity, while the class of RSPPM codes lies approximately 2.75 dB from capacity, and uncoded performance is 4.7 dB from capacity. These gaps will vary with n_b and the target P_b, but they provide an approximation over a range of expected background noise levels.

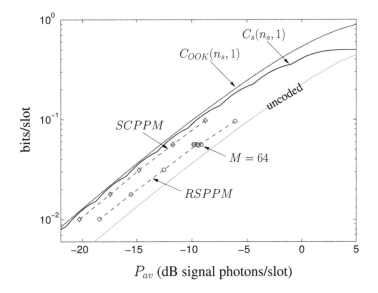

FIGURE 6.21 Sample operating points, $n_b = 1$.

For a conservative link budget calculation that does not require the design and evaluation of a specific code, one could assess a loss of 0.75 dB relative to capacity for iterative codes, 3.0 dB for RS codes, and 5.0 dB for uncoded. For the purpose of constructing link budgets, one might assess a conservative 1 dB loss for all operating points due to quantization and synchronization losses. A robust system design should also incorporate a margin above the minimum requirement, which we will take to be 3.0 dB. From these approximations, we may obtain achievable data rates as a function of average power and background noise, or the required average power for a specified data rate. To a rough approximation, the achievable data rate vs. average power curve (including receiver losses and margin) would be given by the capacity curve shifted by 4.75 dB for iterative codes, 7.0 dB for RS codes, and 9.0 dB for the uncoded case.

6.8.4 STANDARDS

As communications link designs mature, ECCs, as well as other aspects of modulation and signaling, are standardized in order to advance interoperability. A wealth of information and mature coded modulation designs may be found in standards documents. Standards for deep-space and near-Earth communication are captured in the CCSDS standards [36]. The SCPPM code appears in a current draft CCSDS standard [37], as well as a short block length family of LDPC codes targeted for uplink applications. Also of relevance for free-space optical communications are the Optical Transport Network standards, since free-space optical communication links may efficiently utilize technology developed for the fiber optic community. For example, RS codes appear in the first-generation OTN recommendations, and modern codes appear in the next generation standards [38].

REFERENCES

1. R. G. Lipes, "Pulse-position-modulation coding as near-optimum utilization of photon counting channel with bandwidth and power constraints," *DSN Progress Report*, vol. 42, pp. 108–113, Apr. 1980.
2. A. D. Wyner, "Capacity and error exponent for the direct detection photon channel–Part I," *IEEE Transactions on Information Theory*, vol. 34, pp. 1449–1461, Nov. 1988.
3. B. Moision and J. Hamkins, "Multipulse PPM on discrete memoryless channels," *IPN Progress Report*, vol. 42–160, Feb. 2005.
4. B. S. Robinson, A. J. Kerman, E. A. Dauler, R. J. Barron, D. O. Caplan, M. L. Stevens, J. J. Carney, S. A. Hamilton, J. K. Yang, and K. K. Berggren, "781 mbit/s photon-counting optical communications using a superconducting nanowire detector," *Optics Letters*, vol. 31, pp. 444–446, Feb. 2006.
5. S. Dolinar, D. Divsalar, J. Hamkins, and F. Pollara, "Capacity of pulse-position modulation (PPM) on Gaussian and Webb channels," *TMO Progress Report*, vol. 42–142, pp. 1–31, Aug. 2000.
6. A. Biswas and W. H. Farr, "Laboratory characterization and modeling of a near-infrared enhanced photomultiplier tube," *IPN Progress Report*, vol. 42–152, pp. 1–14, Feb. 2003.
7. J. R. Barry and E. A. Lee, "Performance of coherent optical receivers," *Proceedings of the IEEE*, vol. 78, pp. 1369–1394, Aug 1990.

8. H. Hemmati, ed., *Deep-Space Optical Communications, Ch. Coding and Modulation*, John Wiley & Sons, New York, 2004.

9. J. Hamkins, "Accurate computation of the performance of *M*-ary orthogonal signaling on a discrete memoryless channel," *IEEE Transactions on Communications*, vol. 52, pp. 1844–1845, Nov. 2004.

10. T. M. Cover and J. A. Thomas, *Elements of Information Theory*, Wiley, New York, 1991.

11. B. Moision and J. Hamkins, "Deep-space optical communications downlink budget: Modulation and coding," *IPN Progress Report*, vol. 42–154, Aug. 2003.

12. J. Hamkins, M. Klimesh, R. McEliece, and B. Moision, "Monotonicity of PPM capacity," tech. rep., JPL inter-office memorandum, October 2003.

13. K. J. Quirk, J. W. Gin, and M. Srinivasan, "Optical PPM synchronization for photon counting receivers," in *MILCOM 2008—2008 IEEE Military Communications Conference*, pp. 1–7, Nov. 2008.

14. B. H. Marcus, P. H. Siegel, and J. K. Wolf, "Finite-state modulation codes for data storage," *IEEE Journal Selected Areas Communications*, vol. 10, pp. 5–37, Jan. 1992.

15. A. Biswas and M. W. Wright, "Mountain-top-to-mountain-top optical link demonstration: Part I," *IPN Progress Report*, vol. 42–149, pp. 1–27, May 2002.

16. G. R. Osche, *Optical Detection Theory*, Wiley Series in Pure and Applied Optics, Wiley, New York, 2002.

17. R. J. Barron and D. M. Boroson, "Analysis of capacity and probability of outage for free-space optical channels with fading due to pointing and tracking error," in *Proceedings of SPIE* (G. S. Mecherle, ed.), vol. 6105, March 2006.

18. S. Lin, J. Daniel and J. Costello, *Error Control Coding: Fundamentals and Applications*, Prentice Hall, New Jersey, 1983.

19. B. Moision and W. Farr, "Communication limits due to photon detector jitter," *IEEE Photonics Technology Letters*, vol. 20, no. 9, pp. 715–717, 2008.

20. W. H. Press, S. A. Teukolsky, W. T. Vetterling, and B. P. Flannery, *Numerical Recipes in C*, 2nd ed., New York, Cambridge University Press, 1992.

21. A. Kachelmyer and D. M. Boroson, "Efficiency penalty of photon-counting with timing jitter," in *Proceedings of the SPIE* (San Diego, CA), Aug. 2007. preprint.

22. D. Snyder, *Random Point Processes*, Wiley & Sons, New York, 1975.

23. B. Moision, "Photon jitter mitigation for the optical channel," *IPN Progress Report*, vol. 11, p. 15, 2007.

24. R. G. Gallager, *Information Theory and Reliable Communication*, New York, John Wiley & Sons, 1968.

25. ETSI, "Digital Video Broadcasting(DVB): Second generation framing structure, channel coding and modulation systems for Broadcasting, Interactive Services, News Gathering and other broadband satellite applications," ETSI EN 302 307 v1.2.1 ed., 2009.

26. R. J. McEliece, "Practical codes for photon communication," *IEEE Transactions on Information Theory*, vol. IT-27, pp. 393–398, July 1981.

27. J. Hamkins and B. Moision, "Performance of long blocklength Reed-Solomon codes with low-order pulse position modulation." in *JPL Inter-Office Memorandum*, Nov. 2005.

28. J. L. Massey, "Capacity, cutoff rate, and coding for a direct-detection optical channel," *IEEE Transactions on Communications*, vol. COM-29, pp. 1615–1621, Nov. 1981.

29. E. Forestieri, R. Gangopadhyay, and G. Prati, "Performance of convolutional codes in a direct-detection optical PPM channel," in *IEEE Transactions on Communications*, vol. 37, no. 12, pp. 1303–1317, 1989.

30. C. Berrou and A. Glavieux, "Near optimum error correcting coding and decoding: Turbo-codes," *IEEE Transactions on Communications*, vol. 44, pp. 1261–1271, Oct. 1996.

31. J. Hamkins, "Performance of binary turbo coded 256-PPM," *TMO Progress Report*, vol. 42, pp. 1–15, Aug. 1999.

32. K. Kiasaleh, "Turbo-coded optical PPM communication systems," *Journal of Lightwave Technology*, vol. 16, pp. 18–26, Jan. 1998.

33. M. Peleg and S. Shamai, "Efficient communication over the discrete-time memoryless Rayleigh fading channel with turbo coding/decoding," *European Transactions on Telecommunications*, vol. 11, pp. 475–485, September-October 2000.

34. S. Benedetto, D. Divsalar, G. Montorsi, and F. Pollara, "Serial concatenation of interleaved codes: Performance analysis, design, and iterative decoding," *IEEE Transactions on Information Theory*, vol. 44, pp. 909–926, May 1998.

35. B. Moision and J. Hamkins, "Coded modulation for the deep space optical channel: Serially concatenated PPM," *IPN Progress Report*, vol. 42–161, 2005.

36. CCSDS.org, *The Consultative Committee for Space Data Systems(CCSDS)*. http://public.ccsds.org/.

37. Consultative Committee for Space Data Systems, *Optical Communications Coding & Synchronization*, CCSDS 141.1-R-1 ed., Draft Red Book, Dec. 2017.

38. ITU-T, "Forward error correction for high bit-rate DWDM submarine systems," Recommendation G.975.1, International Telecommunication Union, Series G: Transmission Systems and Media, Digital Systems and Networks, Feb. 2004.

7 Photodetectors and Receiver Architectures

Peter Winzer, Klaus Kudielka, and Werner Klaus

CONTENTS

7.1   REQUIREMENTS AND CHALLENGES

The optoelectronic receiver constitutes an important subunit of an optical inter-satellite terminal. Its position within the terminal and its interaction with other major subunits become clear from Figure 7.1. In the typical terminal structure shown in the figure, the signal is generated in the optical transmitter (TX) unit and passes the point ahead assembly (PAA), which is required whenever the angular separation between transmit and receive beam due to fast terminal motion is larger than the associated beam divergence, e.g., on geostationary platforms. The transmit signal then enters the input-output duplexer (DUP) unit. This unit assures sufficient isolation between the powerful transmit signal (e.g., on the order of 30 dBm) and the weak receive (RX) signal (e.g., on the order of −40 dBm). Isolation can be achieved through the use of orthogonal physical dimensions such as (i) different polarizations, (ii) different wavelengths, or (iii) disjoint apertures for transmission and reception. After passing the fine pointing assembly (FPA), which is used for beam tracking purposes, the transmit beam is directed to the telescope, acting as an optical antenna (ANT). In order to save terminal size and mass, transmit and receive paths typically share a single telescope (which eliminates the above-mentioned option (iii) for TX-RX isolation). In the receiving path, the incoming radiation received by the antenna passes the FPA and is directed to the RX by the duplexer. A small portion of the receive signal is typically diverted to the acquisition and tracking detector (ATD), which provides information for the FPA and the coarse pointing assembly (CPA) carrying the antenna. In this section, we deal only with the detection of user data, i.e., with the block shaded in gray in Figure 7.1, and not with concepts and optical sensors required for acquiring and tracking the counter terminal. These are covered in Chapter 3 on pointing, acquisition, and tracking.

As sketched in Figure 7.1, the optical power collected by the terminal's telescope is made available at the receiver input interface, either as a collimated free-space

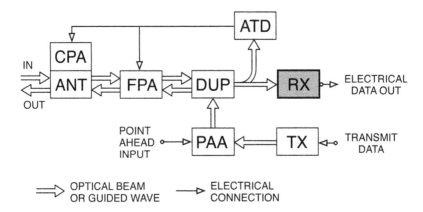

FIGURE 7.1   Basic block diagram of an optical transceiver for space-to-space laser links (ANT, antenna; CPA, coarse pointing assembly; FPA, fine pointing assembly; DUP, optical duplexer; ATD, acquisition and tracking detector; PAA, point ahead assembly; TX, transmitter; RX, receiver).

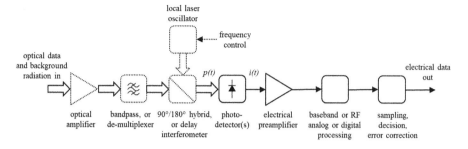

FIGURE 7.2 Generic block diagram of a digital optical receiver.

optical beam or as an optical wave guided in a fiber. The receiver has to convert the data carried by the optical input signal into an electrical output signal. As the optical input power is usually very low, a "high" *receiver sensitivity* is required. As discussed in depth in Section 7.4, for digitally modulated input data, this sensitivity is expressed either by the *number of photons per bit* or by the *optical power* (at a given data rate) needed to achieve a certain bit error ratio (BER) at the RX output. Optical receivers can also handle analog signals, typically characterized by their electrical signal-to-noise ratio (SNR). In this case, the linearity of the electronic circuits involved becomes a major issue that is not easy to cope with if the relative bandwidth is large.

Figure 7.2 shows the generic block diagram of a receiver for digitally modulated data, i.e., key elements contained within the RX block shown in Figure 7.1. In its simplest form, the receiver comes without the dashed subunits. The detection process is in general hampered by background radiation inadvertently received by the optical antenna. Hence, an optical bandpass filter centered at the carrier wavelength λ is usually implemented to reduce the adverse effects of background radiation on signal detection. The filter passes the optical data signal that is converted into an electrical current $i(t)$ using a photodetector. Electrical amplification and baseband processing is followed by sampling, decision, and forward error correction (FEC) decoding. This type of receiver, in which the information-bearing electrical signal current is linearly proportional to the optical signal power $p(t)$ at the detector, is called a *direct detection receiver*. As detailed in Section 7.4.6, one may achieve a considerable improvement in receiver sensitivity by including a low-noise *optical preamplifier*. Alternatively, one may use a *coherent receiver* in which the single-frequency radiation of a local oscillator (LO) laser is superimposed and mixed with the optical input signal upon square-law detection. The photodetector then generates—among other terms—an electrical signal directly proportional to the *optical field*[1] of the optical input signal. The electrical signal is centered at the frequency difference between the optical input signal and the local laser light. As discussed further in Sections 7.4.4 and 7.4.5, coherent receivers offer not only improved receiver sensitivity compared

---

[1] In optical sciences, the term "optical field" is often used to denote either one of the four electromagnetic field quantities observing the wave equation. It is usually expressed as a complex baseband quantity by eliminating the optical carrier frequency and is normalized such that its squared magnitude represents either the local optical intensity or the optical power received by a detector.

to direct detection without optical pre-amplification, but also the possibility to readily detect signals that are modulated in phase or frequency.

The photodetector, being at the heart of any optical receiver, is usually implemented in the form of a semiconductor photodiode. One of its main characteristics is the responsivity, S, defined as the current to optical power ratio i/p, measured in Amperes/Watt. An upper limit for S is reached if each photon is converted into an electron. The quantum efficiency η of the photodiode, introduced in Section 7.3.1, would then be 100%. This value can be approached to within ~10% in practice, leading to responsivities of typically $S = 0.5$ A/W at a wavelength of $\lambda = 0.8$ µm. Another important diode characteristic is the response time t_r (closely related to the diode's electrical bandwidth). In case of direct detection and non-return-to-zero coding (NRZ), the response time t_r obeys the relationship $t_r \leq 0.5/R$, where R is the data rate.

For coherent detection, we distinguish between three cases, as discussed in more detail in Section 7.5.4: If the LO laser and the received optical carrier have identical (optically phase-locked) wavelengths, we speak of *homodyne detection*. If the wavelengths differ, we speak of *heterodyne detection* for which the detector and RF circuitry bandwidth requirements are much higher. This concept lends itself well to applications with analog data modulation. *Intradyne detection* represents a middle ground between homodyne and heterodyne detection, avoiding the complexities of homodyne optical phase locking and heterodyne bandwidth requirements at the expense of a slightly more complex optical front end and significant digital signal processing (DSP).

For optical pre-amplification, optical amplifiers with considerable gain and low noise figures are required. The prime candidate are rare earth doped fiber amplifiers, most notably the Erbium-doped fiber amplifier (EDFA), as their slow gain relaxation time constants let them pass high-speed modulated signals without distortion. Semiconductor optical amplifiers (SOAs) are more problematic in this context, although more recent experiments show their renewed potential for high-speed coherent systems [1]. Using EDFAs, high data rate receivers that closely approach the quantum limit have been reported (see Section 7.4.6). Optical amplifiers typically ask for an optical single-mode input in the form of the fundamental transverse mode, but few-mode optical amplifiers have also been recently discussed for free-space communications [2], as discussed further in Section 7.4.9.

Background radiation originating from the Sun, Moon, planets, or Sun-lit Earth may be picked up by the receive antenna and directed to the photodetector, thus presenting "optical noise" to the receiver. If an optical booster amplifier or a pre-amplifier is used, its amplified spontaneous emission (ASE) constitutes an additional source of background radiation. Efficient filtering of background radiation is achieved by spatial and polarization filtering (allowing only the spatial and polarization modes of the background radiation that also contain signal energy to reach the detector), or optical bandpass filtering using bandwidths that do not exceed a few times the data rate [3].

The signal current generated by a pin photodiode in a direct detection receiver without optical pre-amplification is typically between 10^{-9} A and 10^{-5} A, so that low-noise electrical amplification is essential to achieve a good receiver sensitivity.

The role of electrical amplification is reduced if an avalanche photodiode (APD) is used, or in the case of a coherent receiver or an optically pre-amplified receiver, where the photocurrent is significantly higher. As discussed in Section 7.4, electronics noise is usually not the dominating noise source in these more advanced receiver architectures.

Of all the devices constituting the receiver, *space qualification* is an important consideration. For example, radiation may be an issue for EDFAs. The fiber of an EDFA is prone to darkening when subjected to hard radiation, but this effect may be avoided by proper material composition [4–6].

Before going into more detailed discussions on optical receivers in the following sections, we want to point out some basic differences between receivers for fiber-based systems and free-space links, which is important because the low-volume free-space communications market generally tries to leverage as much technology as possible from the significantly higher-volume fiber optics industry:

- In a free-space link, utmost receiver sensitivity, i.e., close to quantum-limited performance, is of prime importance for the following reasons: (a) there is no in-line amplification available, (b) receiver sensitivity can be traded against antenna diameter, and reduced antenna size allows one to reduce on-board mass.
- An optical inter-satellite terminal will usually consist of both TX and RX. Cross talk of the high-power transmit signal into the receiver subunit has to be kept at a negligible level, which asks for high signal isolation between the transmit path and the receive path. Establishing such isolation may imply that physical dimensions used for multiplexing in fiber optics (e.g., polarization) must be reserved to establish proper isolation in free-space optical terminals.
- If there is a relative movement between the two terminals to be linked, the received signal will be Doppler shifted. This effect may amount to several gigahertz and thus may require fast frequency tuning of the optical filter in case of an optically pre-amplified receiver, and/or of the local laser in case of a coherent receiver.
- The free-space optical channel is much simpler than the fiber-optic channel, devoid of chromatic dispersion and Kerr nonlinearity, which makes signal recovery much easier. However, random spatial distortions (speckle) induced by partial propagation through the atmosphere lead to fading effects that are not encountered in fiber optics.
- Terminal power, size, and weight are much more important in free-space communications than in fiber optics, especially when terminals are to be put onboard a satellite.
- Specifications concerning device reliability usually found in space systems may be more stringent than for undersea fiber-optic repeaters. As a result, subunits with critical reliability have to be devised redundant.

The remainder of this chapter is organized as follows: In Section 7.2, we will describe *fundamental aspects* of the optoelectronic detection process, including the impact of background radiation for various types of receivers. In Section 7.3, we will discuss important properties of state-of-the-art *receiver hardware*, in particular of pin and APDs, optical filters, optical amplifiers, local lasers, and coherent receiver components. In Section 7.4, we will study the *performance* of the various types of digital optical receivers and derive their sensitivities. Section 7.5 will review inter-satellite receivers in or close to operation, and Section 7.6 will provide a summary, point out critical aspects, and recall the choices that have to be made when designing a receiver for a free-space, high data rate link.

7.2 FUNDAMENTALS OF OPTOELECTRONIC DETECTION AND NOISE

7.2.1 SQUARE-LAW PHOTODETECTION

Photodetectors linearly convert the incident optical power $p(t)$, i.e., the squared magnitude of the optical field vector, $\vec{e}(t)$, into an electric current. The factor of proportionality S (A/W) is called the detector's responsivity. The photocurrent $i(t)$ then reads:

$$i(t) = S\, p(t) * h(t) = S\, |\vec{e}(t)|^2 * h(t) \tag{7.1}$$

where * denotes a convolution. The impulse response $h(t)$ represents the filtering action of the entire detection electronics, i.e., any filter elements within the receiver electronics as well as any inherent band limitation of the photodetector. The linear relation between optical power and electrical current holds for input powers below the detector's saturation power (cf. Section 7.3.1), which sets an upper limit to the receiver's dynamic range.

Since photodetectors are typically polarization-independent square-law detectors with respect to the optical field, they are fundamentally insensitive to any phase or polarization information contained in $\vec{e}(t)$. Thus, from a detection point of view, *optical intensity modulation* is the most straightforward way to establish an optical communication link. Intensity modulation can be analog, e.g., with many radio-frequency subcarriers sharing a common, intensity-modulated optical channel, or digital, either with two (on/off keying, OOK) or more intensity levels (e.g., M-ary pulse amplitude modulation, M-PAM), or even with information encoded on the arrival time of an optical pulse within the bit slot (pulse-position modulation, PPM). More sophisticated modulation formats (e.g., [differential] phase shift keying, (D) PSK, frequency shift keying, FSK, quadrature amplitude modulation, QAM, or polarization shift keying, Pol-SK), which also use the signal's phase or polarization to encode information, have to employ special *optical preprocessing* in order to make the phase or polarization information accessible to square-law photodetection [7]. Such preprocessing, which is described in more detail in Section 7.4, can be accomplished by means of a local laser source (coherent receiver) or by means of appropriate optical filtering (e.g., an optical delay line to demodulate DPSK, or a polarization beam splitter to demodulate Pol-SK).

7.2.2   COHERENT DETECTION

If two optical fields $\vec{e}_1(t)$ and $\vec{e}_2(t)$ are incident to a photodetector, it is the total optical power, i.e., the squared magnitude of the vector sum of all optical fields that generates the photo signal:

$$i(t) = S\,|\,\vec{e}_1(t) + \vec{e}_2(t)\,|^2 * h(t) = S\left\{|\,\vec{e}_1(t)\,|^2 + |\,\vec{e}_2(t)\,|^2 + 2\,\mathrm{Re}\{\vec{e}_1(t)\cdot\vec{e}_2^*(t)\}\right\} * h(t) \quad (7.2)$$

Here, Re{·} denotes the real part, and the asterisk stands for complex conjugation. As in Eq. (7.1), $h(t)$ represents the impulse response of the entire optoelectronic detection chain. The first two terms within the curly braces in Eq. (7.2) are the individual powers $p_1(t)$ and $p_2(t)$ of the two incident optical fields, as measured by the detector across its detection area in the absence of the respective other field. The third term, called *interference term, beat term,* or *heterodyne term,* reflects the fundamental property of two optical fields to interact upon detection. The strength of the interference term depends on the *inner product* of the two beating optical fields: The strongest beating is observed if the two fields are co-polarized, while no beating is found for orthogonally polarized fields. In Eq. (7.2), we simplistically assume that the two beating fields share the same transversal mode structure as well. If this is not the case, which is particularly true for coherent receivers using free-space optics, we have to calculate the *spatial overlap integral* of the two fields over the detection area A to determine the beat term, resulting in a quantity called *heterodyne efficiency,* μ [8, 9, 10]:

$$\mu = \frac{\left|\displaystyle\iint_A \vec{e}_1(\vec{r},t)\cdot\vec{e}_2^*(\vec{r},t)\,d\vec{r}\right|^2}{p_1(t)\,p_2(t)}. \quad (7.3)$$

With this definition, we can rewrite Eq. (7.2) in its common engineering notation:

$$i(t) = S\left\{p_1(t) + p_2(t) + 2\sqrt{\mu}\sqrt{p_1(t)\,p_2(t)}\,\cos\left(2\pi\Delta f + \phi(t)\right)\right\} * h(t) \quad (7.4)$$

As detailed further in Section 7.4.4, it is the *beat term* that is exploited in coherent receivers, where $p_1(t)$ takes the role of the weak, information-bearing signal, while $p_2(t) = P_2$ represents an unmodulated LO laser of substantial power. Under these conditions, $p_1(t)$ can be neglected compared to P_2, which itself only represents a temporally constant offset. The remaining beat term, oscillating at the difference frequency, Δf, between the signal's optical carrier and the LO, then contains full information of the signal field's *complex amplitude*, amplified by P_2, which includes the signal's *phase* $\phi(t)$. Therefore, coherent detection offers a straightforward way of decoding any optical modulation format.

7.2.3   THERMAL AND SHOT NOISE

Detection of optical radiation is corrupted by various sources of uncertainties that can be classified by looking at their dynamic behavior: We find uncertainties that are *slowly* varying compared to the bit duration (e.g., telescope vibrations, dynamic beam misalignment, or wavefront distortions due to atmospheric turbulence), as well as uncertainties that occur on the *bit scale* (e.g., noise within the detection electronics). While the latter class of uncertainties determines the probability at which bit errors may occur (bit-error ratio, BER), the slowly varying class of uncertainties collectively affect an extended string of information bits, and can therefore lead to complete system outage. Such fluctuations can be counteracted by allocating sufficient margins in the system's link budget if upper bounds for the penalties exist. Sometimes, however, there is only a small probability for the occurrence of exceedingly large penalties, which makes the allocation of excessive system margins impractical. In this case, one can use outage statistics[2] to describe the probability that these rare events will lead to a system outage.

As for detection noise, *electronics noise*, *shot noise*, and *beat noise* are the most relevant noise sources in optical communications. These three noise sources are independent in the sense that their variances can be added up to arrive at the total detection noise [11, 12].

The term *electronics noise* comprises the sum of all noise sources generated within the optoelectronic detection circuitry that are *independent* of the optical radiation incident to the detector. Examples are thermal noise, transistor shot noise, $1/f$-noise, or dark-current shot noise. The design of the receiver front-end electronics significantly impacts its noise performance, and is detailed in numerous texts [13]. On a system level, electronics noise is often characterized by an *equivalent noise current density*, i_n (A/√Hz), which is related to the variance of the photocurrent by:

$$\sigma_{elec}^2 = i_n^2 B_e \qquad (7.5)$$

Here, B_e denotes the noise-equivalent bandwidth [14] of the entire detection chain. If the photocurrent is converted into an electrical voltage signal by means of some (trans)impedance, R_T, the voltage noise variance is obtained by multiplying the photocurrent noise variance by R_T^2; this conversion also holds for all other noise variances encountered in this chapter.

If mostly thermal noise from the receiver's front end makes up for electronics noise, i_n can be approximated by[3]:

$$i_n^2 \approx kTC_D B_e, \qquad (7.6)$$

where $k = 1.38 \times 10^{-23}$ As/K is Boltzman's constant, T is the front-end temperature, and C_D is the capacitance of the photodiode and the electrical preamplifier. Note that the electronics noise current density increases with receiver bandwidth. In the several Gbit/s regime, i_n^2 typically amounts to 10 pA/√Hz. Alternatively to i_n^2, the

---

[2] The outage probability is defined as the probability that some slowly varying fluctuation perturbs the system by such a high amount that the margin allocated in the link budget to cope with that fluctuation is exceeded and communication errors occur. Outage statistics have been widely used in direct-detection fiber-optic communications in connection with polarization-mode dispersion (PMD) [8].

[3] This is a rule of thumb, based on a model with a single load resistor, but is roughly valid for transimpedance amplifiers as well.

electronics noise performance of receivers is often specified using the *noise equivalent power* (NEP) (W/√Hz), usually defined as the optical power per square root electrical receiver bandwidth that would be required at the photodiode to make the electrical signal power equal to the electronics noise variance:

$$\sigma_{elec}^2 = S^2 \text{NEP}^2 B_e. \tag{7.7}$$

While electronics noise can, at least in principle, be engineered to insignificance by cooling and circuit optimization, *shot noise* cannot. This fundamentally present noise source has its origin in the discrete nature of the interaction process between light and matter: Governed by the rules of quantum mechanics, such light-matter interactions can only take place in discrete energy quanta (photons). For laser light, the photonic interaction statistics are Poissonian. Thus, a discrete, random number of charge carriers (electron-hole pairs in the case of semiconductor photodiodes) is generated in the detector whenever light impinges on it. As a consequence, the photocurrent is composed of individual elementary impulses, each carrying an elementary charge of $e \approx 1.602 \times 10^{-19}$As, as visualized in Figure 7.3. This fine structure of the electrical signal is perceived as shot noise [11, 14, 15], and leads to amplitude fluctuations around the average photo signal, Eq. (7.1).

It can be shown [14] that the shot noise variance is given by:

$$\sigma_{shot}^2(t) = eS(p * h^2)(t) \approx 2eSp(t)B_e. \tag{7.8}$$

Note that shot noise is a *nonstationary* noise process: Its statistical parameters, most notably its variance, change with time. As evident from Eq. (7.8), the dynamics of shot noise are determined by convolving the optical power waveform incident on the detector with the *square* of the detection electronics' impulse response $h(t)$. If the optical power variations are much slower than the speed of the detection electronics, the expression for the shot noise variance simplifies to the frequently encountered

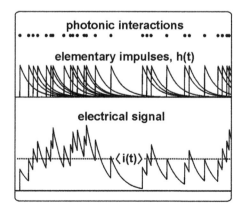

FIGURE 7.3 Photonic interactions are random and dictated by quantum statistics. Each photonic interaction produces an elementary electronic impulse. The impulses add up to produce the overall electrical signal, whose fluctuations are known as shot noise [14].

right-hand side of Eq. (7.8), with B_e being the detection electronics' noise equivalent bandwidth [14]. Accurate system performance analyses have to take into account the dynamic nature of shot noise, though, and even more accurate analyses its non-Gaussian probability density.

As discussed in more detail in Section 7.4.2, shot noise is usually negligible compared to electronics noise in direct-detection receivers, where optical signal power levels at the detector are low. In optically pre-amplified receivers, noise is typically dominated by beat noise. In practice, shot noise mostly plays a role in *coherent receivers* in the *absence of optical background radiation*, i.e., at high optical signal-to-noise ratios (OSNRs); in this case, shot noise is generated by the strong LO light. Shot noise can also be important in receivers that employ detectors with internal *avalanche multiplication* (e.g., APDs). As detailed further in Section 7.4.3, APD-based detectors make use of an internal charge carrier multiplication to produce, on average, an avalanche of M_{APD} charge carriers for each primary charge carrier generated by a photonic interaction. Thus, M_{APD} is also called the *avalanche gain*. Random fluctuations in the multiplication process exacerbate the fundamental shot noise fluctuations and lead to *multiplied shot noise* [14], quantitatively captured in the APD's *noise enhancement factor* $F_{APD} > 1$ via the multiplied shot noise expression:

$$\sigma^2_{mult.shot}(t) = eSM^2_{APD}F_{APD}(p*h^2)(t) \approx 2eSM^2_{APD}F_{APD}\,p(t)B_e, \qquad (7.9)$$

where the well-known approximation, again, holds for optical power variations slow compared to the detection electronics' speed. In addition to the multiplied shot noise, whose variance is proportional to the optical power at the detector, an APD also generates multiplied dark current shot noise through avalanche multiplication of dark current charge carriers (cf. Eq. [7.9] in Sect. 7.2.3). This noise term is stationary and independent of the optical power, and can thus be added to the electronics noise variance σ^2_{elec}.

7.2.4 BEAT NOISE

The third important noise source encountered in optical communications is called *beat noise*. It has to be considered whenever a substantial amount of *incoherent background radiation* (e.g., light from celestial bodies, stray light from the Earth's atmosphere, or ASE generated by optical amplifiers at the transmitter or at the receiver) impinges on the detector. Incoherent radiation is typically composed of a broadband and stationary random optical field whose complex amplitude obeys circularly symmetric Gaussian statistics [11, 15]. Most sources of background radiation are fully depolarized, i.e., the optical fields in any two polarizations are uncorrelated [16].

Background radiation is turned into *beat noise* by photodetection: With reference to Eq. (7.2), consider a well-defined optical signal field, $\vec{e}_1(t)$, and a random optical background field, $\vec{e}_2(t)$. The first term in the curly braces of Eq. (7.2) then constitutes the desired electrical signal (proportional to the received optical power), the second term is the randomly fluctuating instantaneous power of the background radiation, and the third term represents the beating between signal and background radiation. Since $|\vec{e}_2(t)|^2 = \vec{e}_2(t) \times \vec{e}_2^*(t)$, the second term can also be thought of as the beating of

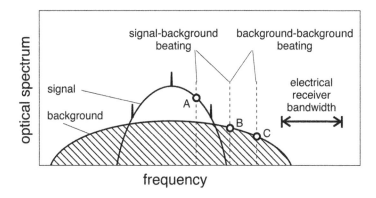

FIGURE 7.4 Signal-background beat noise is produced by those frequency components of the signal and the background light whose difference frequency falls within the electrical receiver bandwidth (e.g., by frequency components A and B whose difference frequency is smaller than the electrical receiver bandwidth indicated by the double arrow). Background-background beat noise is generated by the beating of background radiation with itself (e.g., by B and C).

background light with itself. The convolution with $h(t)$ in Eq. (7.2), or equivalently the multiplication with $H(f)$ in the frequency domain, makes sure that only beat components whose beat frequencies fall within the bandwidth of the detection electronics reach the receiver output. All other beat frequencies are filtered out by the band-limited nature of the detection chain. The process of beat noise generation is visualized in Figure 7.4, which shows a typical optical signal spectrum together with the spectrum of background radiation after having passed an optical bandpass filter. Each frequency component of the signal may beat with each frequency component of the background radiation, indicated by the circles denoted as A and B in Figure 7.4. The resulting beat frequency is given by the difference of the two beating spectral components. All beat frequencies that are smaller than the receiver bandwidth (indicated by the double arrow in Figure 7.4) will show up at the detector output. Since the background field is random in nature, the beating between each signal spectral component with each background spectral component will have a random amplitude and phase, thus the name *signal-background beat noise*. A similar process applies for the beating between any two spectral components of the background light (circles B and C in Figure 7.4), leading to *background-background beat noise*.[4]

Using Eq. (7.2) together with the statistics of background radiation, it can be shown [17] that the beat noise variances take the form:

$$\sigma^2_{sig-back}(t) = 2S^2 N_{back} \operatorname{Re}\left\{ \int_{-\infty}^{\infty}\!\!\int e_{sig}(\tau)e^*_{sig}(\tilde{\tau})r_{back}(\tau-\tilde{\tau})h(t-\tau)h(t-\tilde{\tau})\,d\tau\,d\tilde{\tau}\right\}, \quad (7.10)$$

---

[4] Beat noise has been extensively studied in the context of optical amplifiers, where ASE caused by the optical amplification process is an unavoidable source of background radiation. Therefore, the terms signal-ASE beat noise and ASE-ASE beat noise are commonly used in the vast literature available on optical amplifier noise.

and

$$\sigma^2_{back-back} = S^2 M_{pol} N^2_{back} \int_{-\infty}^{\infty} |r_{back}(\tau)|^2 \, r_h(\tau) \, d\tau. \qquad (7.11)$$

Here, N_{back} denotes the background radiation's peak power spectral density per (spatial and polarization) mode *at the detector*, and $r_{back}(\tau)$ is the background radiation's autocorrelation function (i.e., the inverse Fourier transform of the background radiation's power spectrum). The quantity M_{pol} gives the number of spatial and polarization modes of the background light that actually reach the detector, as discussed in more detail in Section 7.2.5. The signal field at the detector is denoted as $e_{sig}(t)$, and $r_h(\tau)$ is the electrical circuitry's autocorrelation function [17].

In the limit of rectangular filters and constant signal input power, P_{sig}, at the detector, the above relations simplify to [18]:

$$\sigma^2_{sig-back} \approx 4S^2 P_{sig} N_{back} B_e, \qquad (7.12)$$

and

$$\sigma^2_{back-back} \approx S^2 M_{pol} N^2_{back}(2B_o - B_e)B_e, \qquad (7.13)$$

where B_o denotes the optical filter bandwidth. While these relations lend themselves to valuable intuitive explanations, they are often too crude for quantitatively accurate predictions of receiver performance [19], especially in the presence of optical filters whose bandwidths are comparable to the optical signal bandwidth.

From Eqs. (7.9) and (7.11), we see that the signal-background beat noise, like shot noise, is *nonstationary* and its variance grows linearly with signal power. It is independent of B_o as long as the optical filter does not significantly influence the signal spectrum. The latter fact becomes intuitively clear when looking at Figure 7.4: Due to the limited spectral extent of the optical signal, increasing the spectral width of the background light does not produce more signal-background beat frequencies that can still fall within the limited electrical detection bandwidth. From Eqs. (7.10) and (7.12), we see that the background-background beat noise is stationary, and its variance grows linearly with both B_o and M_{pol}. Optical filtering (spectrally and spatially) therefore *only* reduces the background-background beat noise, but not the signal-background beat noise.

In the case of coherent detection of a signal with strong per-mode optical background radiation, the beating between LO and background radiation can become a non-negligible noise term. In general, the ratio of LO shot noise to LO-background beat noise is given by **[20]**:

$$\frac{\sigma^2_{LO-back}}{\sigma^2_{LO-shot}} = \frac{2N_{back}}{hf}. \qquad (7.14)$$

Therefore, coherent systems are only limited by LO-background beat noise instead of LO shot noise if the background power spectral density is significantly above

$hf / 2 \approx 10^{-19}$ W/Hz per spatial and polarization mode at near infrared wavelengths typically used in optical communications. As will be seen in Section 7.2.5, the Sun as the strongest natural source of background radiation barely reaches such levels. However, optical amplifiers typically produce significantly more background radiation (ASE), and hence almost always push coherent receivers into beat noise limited operation [20].

7.2.5 BACKGROUND RADIATION

Much more than microwave links, free-space optical transmission systems may suffer from the influence of background radiation. While a natural source like the Sun not only enlightens our daily life but is the basis for it, the radiation it and other celestial bodies emit make the life of the optical engineer troublesome. This broadband, incoherent, and depolarized radiation may represent a considerable source of noise when captured by the optical receiver's photodiode. As explained in Section 7.2.4, such background radiation manifests itself in the following, unwanted contributions to the photocurrent:

- a DC term proportional to the background power,
- a shot noise term related to the DC term,
- a beat noise term due to the beating of the background radiation with itself, and
- a beat noise term due to the beating of the background radiation with the data signal.

However, depending on the bit rate, filter bandwidths, and receiver design, not all of the above noise terms may have an impact on performance. If an optical power amplifier is implemented in the transmitter, or if an optical preamplifier is used in the receiver, the associated ASE (see Section 7.4.6) also has to be taken into account. It will affect the receiver performance in a similar way as the natural background sources mentioned above. In any case, the SNR in the receiver will be reduced, leading to reduced receiver sensitivity, and in the case of a digital system to an increased BER. In the following, we show how to calculate the background power, characterize the most important sources of background radiation in an optical receiver, and give examples of the receiver's sensitivity degradation [21].

7.2.5.1 Determination of Background Power

We first show how to determine the received background spectral density per mode, N_{back}, and the number of received background modes, M_{pol}. This allows us to quantitatively evaluate the noise terms caused by background radiation once the optical filter bandwidth is known (see also Section 7.2.5).

A background source is often characterized by its spectral radiance function, N_f, which gives the radiated power per emitting area into one steradian within a bandwidth of 1 Hz centered at frequency f (dimension [W/(m^{2}sr Hz)]). A receiver with an area A_{RX} and a field of view solid angle, Ω_{FV}, smaller than the solid angle, Ω_S, subtended by the background source at the receiver (see Figure 7.5a) receives a spectral power density N_{RX} (dimension [W/Hz]) of:

$$N_{RX} = N_f\, A_{RX}\, \Omega_{FV}. \tag{7.15}$$

We now employ the antenna theorem [22]:

$$\Omega_{DL} = \lambda^2/A_{RX}, \tag{7.16}$$

which relates, Ω_{DL}, the diffraction-limited value of the antenna's field of view at wavelength, $\lambda = c/f$, with the antenna area, A_{RX}. Combining Eqs. (7.2-41) and (7.2-42) yields:

$$N_{RX} = 2\cdot N_f \lambda^2 \Omega_{FV}\, /\, \Omega_{DL}. \tag{7.17}$$

In Eq. (7.17), the factor $2 \times \Omega_{FV}/\Omega_{DL}$ can be interpreted as the number of spatial and polarization modes, M_{pol}, of the background light that actually reach the detector, i.e.,

$$M_{pol} = 2\cdot\Omega_{FV}\, /\, \Omega_{DL}. \quad (\text{if } \Omega_{FV} < \Omega_S) \tag{7.18}$$

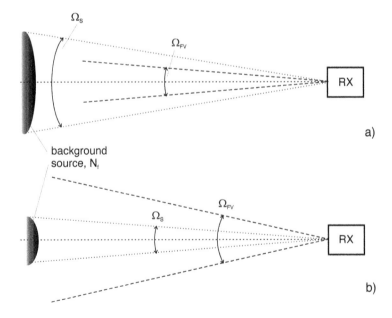

FIGURE 7.5   Geometry relating background source and optical receiver. Ω_{FV} is the receiver's field of view, Ω_S is the solid angle under which the receiver sees the extended background source. (a) $\Omega_{FV} < \Omega_S$ and (b) $\Omega_{FV} > \Omega_S$.

This relationship can also be deduced from Figure 7.5a, following purely geometric considerations. The factor 2 in Eq. (7.18) takes into account the fact that background radiation is depolarized. In case of a polarizer placed before the photodetector, one has $M_{pol} = \Omega_{FV}/\Omega_{DL}$.

The total received background power is obtained by multiplying N_{RX} with the bandwidth B_O of the optical filter. The spectral density per mode, N_{back}, follows as:

$$N_{back} = N_f \lambda^2. \tag{7.19}$$

If, on the other hand, the receiver's field of view Ω_{FV} is larger than Ω_S (see Figure 7.5b), we have:

$$N_{RX} = N_f A_{RX} \Omega_S, \tag{7.20}$$

and

$$M_{pol} = 2 \cdot \Omega_S / \Omega_{DL}. \quad (\text{if } \Omega_{FV} > \Omega_S) \tag{7.21}$$

However, combining Eqs. (7.16), (7.17), and (7.18), we arrive again at $N_{back} = N_f \lambda^2$. Thus, while the number of spatial and polarization modes received is different for the cases $\Omega_{FV} < \Omega_S$ and $\Omega_{FV} > \Omega_S$, Eq. (7.19) gives the spectral density per mode, N_{back}, applied equally, as expected.

Next, we show how to find the spectral radiance function, N_f. In case of dominating self-emission, a background radiator may sufficiently well be modeled by a blackbody radiator, characterized by its absolute temperature, T_B. The spectral radiance N_f of a blackbody is given by Planck's law:

$$N_f = \frac{hf^3}{c^2 \left[exp\left(\dfrac{hf}{kT_B} \right) - 1 \right]} \tag{7.22}$$

With Eq. (7.19), we obtain, for the power spectral density per mode:

$$N_{back} = \frac{hf}{\left[exp\left(\dfrac{hf}{kT_B} \right) - 1 \right]} \tag{7.23}$$

This result is identical to the expression for the average energy per mode of a blackbody in thermal equilibrium. Equation (7.23) is quite well suited to estimate the radiation from the Sun (with $T_B = 5700$ K). For Sun-illuminated bodies like the Moon ($T_B = 400$ K) or the Earth ($T_B = 300$ K) or Venus (with $T_B = 330$ K), Eq. (7.23) yields good results only for the infrared region above 3 μm. At shorter wavelengths, the Sun reflectance has to be taken into account and in fact is the dominating contribution if $\lambda < 2$ μm [23]. To determine N_{back} in these instances, one may resort to the values of

TABLE 7.1

Power Spectral Density per Mode, N_{back} (in [W/Hz]), Produced by Various Celestial Bodies at Selected Wavelengths

	N_{back} (W/Hz)		
λ (μm)	0.85	1.06	1.55
Sun	1.3×10^{-20}	1.9×10^{-20}	4.1×10^{-20}
Moon	1.8×10^{-26}	3.6×10^{-26}	8.6×10^{-26}
Venus	2.1×10^{-25}	3.1×10^{-25}	3.5×10^{-26}
Earth	2.6×10^{-26}	4.0×10^{-26}	4.6×10^{-25}

spectral irradiance, H_λ. This quantity gives the wavelength-dependent radiant power incident on a surface (e.g., the receiver) per unit area and unit wavelength increment (dimension [W/m²m]). The relation to N_{back} is [23]:

$$N_{back} = \frac{H_\lambda \lambda^4}{c\Omega_S}, \quad (7.24)$$

with c being the velocity of light in vacuum. For radiation caused by reflected sunlight, the spectral irradiance, H_λ, also follows Planck's λ dependence, but its maximum is not given by the body's temperature. In such cases, H_λ is determined by an effective temperature and a case-specific maximum value of H_λ. More details and numerical values can be found in the literature [23–25]. Table 7.1 cites a few examples.

If an optical preamplifier is employed in the receiver, its ASE acts as background radiation too. The spectral density per mode caused by this process is then given by Eq. (7.25):

$$N_{back,ASE} = hf \frac{F_{RX} G_{RX}}{2}, \quad (7.25)$$

where F_{RX} and G_{RX} are the preamplifier's noise figure and gain. Note that for typical EDFA preamplifier parameters (λ =1.55 μm, G_{RX} = 40 dB, F_{TX} = 4 dB), the ASE-induced background radiation at the detector amounts to about 1.6×10^{-15} W/Hz, while the corresponding background radiation per mode at the detector for a receiver looking directly into the Sun is on the order of $G_{RX}= 4 \times 10^{-20}$ W/Hz ~ 4×10^{-16} W/Hz. Hence, in this example, the preamplifier ASE contributes about four times as much noise as the Sun.

If an optical booster amplifier is implemented at the transmitter, its ASE may also contribute to the background radiation received [26]. The booster amplifier will—in general—constitute a wide band, point-like, i.e., spatially coherent, background source. The associated power spectral density per mode at the receiver is:

$$N_{back,ASE,booster} = \frac{hfG_{TX}F_{TX}}{2} g_{TX} g_{RX} \frac{\lambda^2}{(4\pi L)^2}, \quad (7.26)$$

where G_{TX} and F_{TX} are the booster amplifier's gain and noise figure, g_{TX} and g_{RX} are the transmit antenna gain and the receive antenna gain, and L is the link distance. For a high-gain booster EDFA ($\lambda = 1.55$ μm and, e.g., $G_{TX} = 35$ dB, $F_{TX} = 6$dB), $N_{back,ASE,booster}$ may take on disturbingly high values if the link distance L is small. Then, the beating between the signal and the transmit booster ASE may dominate all other noise terms, leading to a SNR independent of the link distance L [26].

7.2.6 CROSS TALK

Finally, we want to mention yet another source of degradation similar to background radiation: In a transceiver, non-negligible optical cross talk may occur from the (high-power) transmitter into the (highly sensitive) receiver. The causes for such cross talk are reflections and stray light, especially if one and the same antenna is employed for outgoing and receiving signals. The resulting beat noise terms are similar to the signal-background and background-background beating and their influence on receiver performance can be treated the same way as *multipath interference* in fiber optic wavelength division multiplexed systems [27]. To first order, an interfering signal produces the same amount of beat noise as if background radiation or ASE were added at the power spectral density of the interfering signal. Since beat noise only occurs for co-polarized interferers, the polarization dimension can be used to improve transmit-receive isolation, e.g., by transmitting left circular polarized beams and receiving only right circular beams. (The use of circular instead of linear polarization conveniently makes the system independent of axial rotations in free-space optical links.)

7.3 DEVICES

7.3.1 PHOTODIODES

For free-space laser communication receivers, photodiodes are considered the most appropriate devices to convert the incoming optical radiation into an electrical signal [28]. These semiconductor elements provide high conversion efficiency, fast response (i.e., high bandwidths), and low inherent noise. They can be produced as small, rugged, and reliable elements in the form of either pin-diodes[5] or APDs and are available for the visible and near-infrared wavelength regime.

The devices discussed below are operated as reverse-biased diodes, where photons to be associated with the incident optical power p are absorbed to generate electron-hole pairs, leading to a current i in the electrical circuit connected to the diode's electrodes. To this end, the energy of the photon, hf (h, Planck's constant; $f = c/\lambda$, optical frequency; c, velocity of light in vacuum; λ, wavelength in vacuum), has to exceed the bandgap energy, E_g, separating the valence band and the conduction

---

[5] The letters "pin" indicate the basic sequence of layers in the semiconductor structure: p-doped, intrinsic, and n-doped.

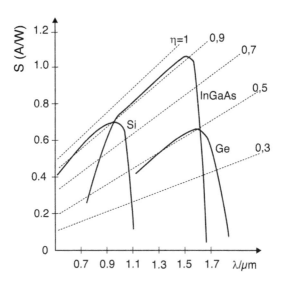

FIGURE 7.6   Spectral responsivity of pin photodiodes made from Si, Ge, and InGaAs.

band of the semiconductor material. Therefore, the longest wavelength that can be detected, also called the cut-off wavelength, is given by:

$$\lambda_g = \frac{hc}{E_g}. \tag{7.27}$$

The light-to-current conversion is characterized by the quantum efficiency, η, defined as the ratio of the number of carrier pairs generated per time interval within the diode to the number of photons incident on the detector. The detector responsivity, S, defined as the ratio of electrical current to optical power then reads as:

$$S = \frac{i}{p} = \frac{\eta e}{hf}, \tag{7.28}$$

with e denoting the elementary charge. Figure 7.6 gives typical responsivities vs. wavelength for three common diode materials (Si, Ge, and InGaAs) together with lines of constant quantum efficiency η (dashed). The figure mirrors the proportionality of S with λ, expressed by Eq. (7.14) for $\lambda \leq \lambda_g$. It also shows that the response at cut-off is not abrupt in a real-world device. This is a consequence of the wavelength dependence of the number of absorbed photons through the thermally broadened uncertainty of the bandgap energy at temperatures above 0 K. Table 7.2 gives the cut-off wavelengths and the operating ranges for photodiodes made of various semiconductor materials.

An exemplary cross-section of a *pin-diode* is shown in Figure 7.7. An antireflection (AR) coating reduces the Fresnel loss of the incident radiation. Absorption takes place preferentially in the intrinsic layer, governed by an exponential decay of the optical power along the direction z. Under the influence of the internal electric field

TABLE 7.2

Cut-Off Wavelength and Operating Ranges of Some Photodiodes

Material	Cut-Off Wavelength (μm)	Typical Operating Range (μm)
Si	1.1	0.5–0.9
Ge	1.85	1.0–1.7
InGaAs	1.65	1.0–1.6
$Hg_xCd_{1-x}Te$	Up to 18	4–11

For HgCdTe diodes to operate in the mid-infrared, cooling to 77 K is usually employed.

in the depletion zone, the resulting electrons and holes travel to the n$^+$ and the p$^+$ regions, respectively, thus creating a current i in the external circuit. To achieve a high quantum efficiency, most of the photons have to be absorbed before entering the n$^+$ region. This calls for an i-zone with an extension of several photon absorption lengths.[6] A long i-zone advantageously reduces the capacitance of the diode and thus the RC time constant of the external circuit. However, it increases the carrier transit time, which in turn results in reduced internal speed. Thus, a trade-off between overall bandwidth and quantum efficiency has to be accepted. Note also that a large bandwidth can only be obtained if the sensitive area is small to keep the diode capacitance low. Figures 7.8 and 7.9 show the current-voltage (i-u) characteristic and the equivalent circuit of a photodiode. For no illumination ($p = 0$), the i-u dependence is that of an ordinary semiconductor diode, with a dark current I_D caused by thermal carrier excitation. For $p > 0$, the characteristic shifts and the operating point moves along the line given by the bias voltage, U_0, and the effective load resistance, R. The equivalent circuit is modeled by the two current sources representing the photo-induced current, $i = Sp$, and the dark current, I_D, by the junction capacitance

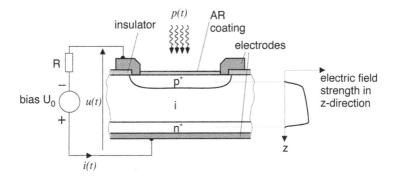

FIGURE 7.7   Cross-section and bias supply of a pin photodiode. Also shown is the electric field causing the transport of the photo-generated carriers.

[6] The photon absorption length characterizes the degree of exponential decrease of optical power caused by the absorption process.

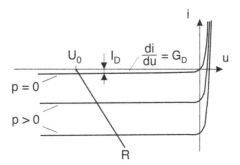

FIGURE 7.8 Current-voltage characteristic of a pin photodiode.

C_D in parallel with a conductance G_D and a series resistance R_S. A diode mount may contribute a further capacitance, C_M, and a series inductance, L_M.

In order to decouple the direction of charge transport and photon absorption, *side-illuminated* (or *edge-illuminated*) diode structures as well as *waveguide-coupled* diode structures have been implemented for high-speed operation [29–31]. Further speed improvements have been shown using *uni-traveling carrier* (UTC) diodes [32].

As in every diode, and as explained along with Eq. (7.27), the current through the pin-structure is associated with shot noise whose variance is given by:

$$\sigma_{shot}^2 = 2e(I_D + iS)B_e, \tag{7.29}$$

where B_e is the effective (electrical) bandwidth. The dark current portion $2eI_D B_e$ of the shot noise variance is signal-independent and is therefore often considered part of the electronics noise (cf. Section 7.2.3)

The dynamic range of the pin-diode typically spans several orders of magnitude, extending from the noise level up to diode currents of a several milliamperes (optical input power of several milliwatts). When entering saturation, two major effects are observed: First, the dependence between optical power and electrical current starts to become sublinear. Second, and often more importantly, the diode gets slower, which can severely impact digital reception quality.

Table 7.3 presents typical parameters of pin photodiodes developed for fiber optic communication systems at data rates of several Gbit/s. As with most devices

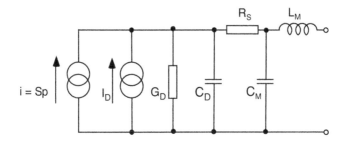

FIGURE 7.9 Equivalent circuit of a pin photodiode.

TABLE 7.3

Typical Parameters of Pin Photodiodes Made from Indium-Gallium-Arsenide

	InGaAs	InGaAs	InGaAs
Operating range (μm)	1.0–1.6	1.0–1.6	1.48–1.62
Responsivity, S (A/W)	0.95 @ 1550 nm	0.9 @ 1550 nm	0.6 @ 1550 nm
Bandwidth @ 50 Ω (GHz)	3.5	14	75
Capacitance (pF)	0.6	0.17	n/a
Dark current I_D (nA) @ 290 K	0.2	0.01	5
Bias voltage (V)	−5	−5	−2.8
Diameter of active area (μm)	50	28	Waveguide coupled

developed for fiber-optic communications, these diodes are equally well suited for optical free-space links. Note that the dark current I_D strongly depends on temperature, rising by a factor of approximately three for a temperature increase of 10°C. As to be expected, capacitance and response time increase with increasing active area.

To achieve reasonable sensitivity in free-space optical communication receivers, pin photodiodes are used nowadays only in combination with optical pre-amplification, or with coherent reception. As a consequence, pin photodiodes made from silicon are of no practical relevance.

Higher-bandwidth InGaAs diodes of up to 100 GHz are commercially available, but typically have a narrower spectral operating range and lower responsivity. This is usually not a problem for optically pre-amplified receivers. On the other hand, responsivity/quantum efficiency has to be kept in mind as an important parameter for coherent receivers *without* optical pre-amplification.

7.3.2 TRANS-IMPEDANCE AMPLIFIERS

The very low currents delivered by a pin-diode—and also the low currents output by an APD—ask for an electronic preamplifier to follow the detector. This has to be a broadband device with a low cutoff frequency at or near DC and an upper cutoff frequency somewhere close to 0.7 times the data rate. To achieve a sufficiently short rise time of the diode-amplifier combination, a low input impedance is asked for, i.e., a low input capacitance and a low input resistance (typically 50 Ω). However, this concept suffers from large thermal noise. To overcome this problem, one often implements the trans-impedance amplifier (TIA) concept. Here, feedback of the amplified signal to the amplifier input results in an enlarged input impedance without reducing the upper cutoff frequency. One useful characterization of the electronic noise due to the preamplifier is done by specifying the equivalent noise current density i_n at the photodiode-amplifier interface (see Section 7.2.3, Eq. (7.29)). This number depends on parameters of the photodiode, like the junction capacitance, on the input impedance of the amplifier, on the noise generated within the transistors, and on the unavoidable stray capacitance and inductance between diode and amplifier. In general, lower values of i_n can be achieved for lower bandwidths B_e, i.e., with receivers

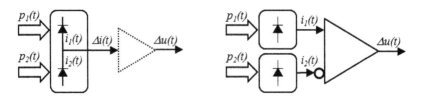

FIGURE 7.10   Balanced photodetector using current subtraction before entering a single-ended amplifiers (a) and using individual signals entering a differential amplifier (b).

for low data rates, R. Typical values are $i_n = 30$ fA/$\sqrt{Hz}$ at $B_e = 50$ kHz, $i_n = 5$ pA/$\sqrt{Hz}$ at $B_e = 10$ MHz, $i_n = 10$ pA/$\sqrt{Hz}$ at $B_e = 10$ GHz, and $i_n = 40$ pA/$\sqrt{Hz}$ at $B_e = 30$ GHz.

7.3.3   BALANCED DETECTORS AND BALANCED RECEIVERS

As discussed in Sections 7.4.2 and 7.4.4, certain optical receiver structures optimally use a pair of *balanced detectors* as shown in Figure 7.10, optionally integrated with an electrical amplifier [33]. The two diodes are typically set up to receive a pair of optical signals whose powers $p_1(t)$ and $p_2(t)$ carry both an (undesired) common-mode component and a (signal-bearing) differential-mode component, i.e.:

$$i_1(t) = S_1 p_1(t) = S_1 p_{common}(t) + S_1 p_{signal}(t) \tag{7.30}$$

$$i_2(t) = S_2 p_2(t) = S_2 p_{common}(t) - S_2 p_{signal}(t), \tag{7.31}$$

with $S_1 \approx S_2$. The difference in photocurrent then yields about *twice* the desired signal component plus a small residual common-mode determined by the balanced detectors' common-mode rejection ratio (CMRR):

$$CMRR = (S_1 - S_2)/(S_1 + S_2), \tag{7.32}$$

resulting in:

$$\Delta i(t) = i_1(t) - i_2(t) = \frac{S_1 + S_2}{2}\left[2\,CMRR\,p_{common}(t) + 2 p_{signal}(t)\right]. \tag{7.33}$$

In practice, the CMRR of high-speed integrated balanced detectors can be frequency dependent, with typical integrated values better than −20 dB [34–36]. Balanced detectors with bandwidths of 25–30 GHz are widely available today [33, 36], with some balanced detectors even ranging up to 100-GHz bandwidths.

7.3.4   AVALANCHE PHOTODIODES

In an *APD*, primary electrons and holes generated by photons experience internal amplification by an effect known as impact ionization. By proper doping (and strong reverse-biasing), a zone with very high field strength is established, and the primary carriers are accelerated to such high velocities that they can effectuate secondary electron-hole pairs. These, in turn, will produce further pairs, and an avalanche builds up in the gain region. Similar to the case of a pin-diode (Figure 7.10), a single photon generated carrier pair results in an external current impulse. However, since m electrons are generated per photon in an APD, each impulse now carries a charge of $m \times e$. Apart from the generation of primary carriers, which is a statistical process due to quantum mechanics, the avalanche multiplication process, too, is random. Hence, m is slightly different for each individual primary charge carrier. On average, we observe a multiplication factor:

$$M_{APD} = \bar{m} = \frac{i}{i_{prim}}, \tag{7.34}$$

where the overbar denotes averaging. The average avalanche multiplication is thus defined as the ratio of average multiplied photocurrent to average primary photocurrent, where i is the average current provided by the APD to the external circuit and i_{prim} is the average current only due to the primary carriers. The photon-induced external current can thus be written as:

$$i = pM_{APD} \frac{\eta e}{hf}. \tag{7.35}$$

Cross-section, internal electric field ,and external circuit of an APD are sketched in Figure 7.11.

Similar to pin-diodes, the incident optical power is mostly absorbed in a zone with only light doping (p⁻). The carriers that drifted into the high field zone then each experience a multiplication by a random factor m (on average, M_{APD}). The statistical fluctuations of m can be quantified by the excess noise factor, F_{APD}:

$$F_{APD} = \frac{\overline{m^2}}{M_{APD}^2} = 1 + \frac{\sigma_m^2}{M_{APD}^2}. \tag{7.36}$$

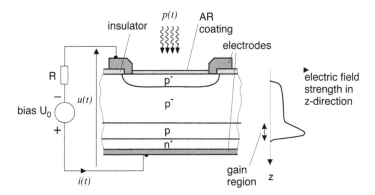

FIGURE 7.11 Cross-section and bias supply of an avalanche photodiode. Also shown is the electric field causing the transport and multiplication of the photo-generated carriers.

In case of deterministic amplification (i.e., when no variance is associated with the multiplication process, $\sigma_m^2 = 0$), one would have $F_{APD} = 1$. In practice, $F_{APD} > 1$ and an excess noise has to be associated with the avalanche amplification. This turns up in the expression for the shot noise variance of the APD current. Equation (7.37), repeated here, is in a more device-oriented form:

$$\sigma_{shot}^2 = 2e(I_{DM} + iS)B_e M_{APD}^2 F_{APD} + 2eI_{DO}B_e; \qquad (7.37)$$

I_{DM} is a dark current contribution that experiences multiplication, and I_{DO} is one that does not (e.g., because it flows along the surface of the device). Accordingly, the dark current of an APD may be written as:

$$I_D = I_{DM}M + I_{DO}. \qquad (7.38)$$

Some APD vendors characterize their devices by including dark current shot noise into a signal power independent NEP or an equivalent electronics noise current density, which leads to the multiplied shot noise Eq. (7.38) that only represents the signal power dependent term of Eq. (7.7). Equation (7.7) can be considered as an extension of the shot noise equations for the pin-diode, i.e., Eqs. (7.4) and (7.37), where the standard deviation of the primary noise current is now enlarged by the factor $M_{APD}\sqrt{F_{APD}}$ and where there exists also a non-multiplied dark current I_{DO}. As with the pin-diode, the response time of the APD is influenced by the carrier transit time in the absorption zone and by the external RC time constant, both making up for the impulse response $h(t)$ of detection circuitry, which was introduced in Section 7.2.3. However, in many cases the response time is dominated by an effect specific to the APD, namely the time it takes to develop the avalanche within the high field zone. Thus, the overall APD rise time τ becomes a function of the gain, M_{APD}. A very simple model predicts that when varying M_{APD} in a given device, the product $M_{APD} \times \tau$ is roughly constant. Short rise times are achieved if the probability of ionization by one carrier type (e.g. the holes) is much smaller than that by the other (e.g., the electrons). In this case of a small ionization ratio k_{ion}, the avalanche build-up is completed faster compared to the case that both carriers contribute equally ($k_{ion} = 1$). A small value of k_{ion} turns out to be also beneficial to achieve a low noise factor, F_{APD}. To give some examples: For silicon, we typically have $k_{ion} \approx 0.03$, while for InGaAs, $k_{ion} \approx 0.5$, with the consequence that APDs made from Si are less noisy and allow for shorter rise times than those made from InGaAs. A simple model of the avalanche noise process leads to [14]:

$$F = M_{APD}k_{ion} + (1 - k_{ion})\left(2 - \frac{1}{M_{APD}}\right). \qquad (7.39)$$

The dynamic range of the APD is in general smaller than that of a pin-diode. Because of the current multiplication and the rather large reverse voltage needed to achieve the avalanche effect, the maximum rating for the dissipated electric power may be reached at comparatively low optical input power. This effect has to be observed

TABLE 7.4

Typical Parameters of Commercially Available APDs Made from Silicon and Indium-Gallium-Arsenide

	Si	InGaAs	
Operating range (μm)	0.4–1.0	0.9–1.6	
Quantum efficiency η	0.75	0.8	
Bandwidth @ 50Ω (MHz)	1000	2500	1500
Capacitance (pF)	1.0	0.35	1.5
Total dark current at room temp. (nA)	0.5	10	50
Bias voltage (V)	200–250	60	
Ionization ratio k_{ion}	0.04	0.2	
Multiplication factor M	100	10	
Noise figure F_{APD}	5	3.5	
Overall responsivity SM (A/W) @ peak	50 @ 800 nm	9 @ 1550 nm	
diameter of active area (μm)	200	80	200

Diameters on the order of 200 μm are often used in ground-based receivers, where the focal spot size is determined by atmospheric turbulence. If this is not the limiting factor, one usually strives for smaller detector size and the associated benefits (lower dark current, lower capacitance, resulting in lower receiver noise).

especially in the case of large background radiation, where background light may drive the diode into saturation and may eventually even damage it. On the other hand, the dependence of the gain, M_{APD}, on the applied reverse voltage allows a rather easy implementation of an automatic gain control (AGC) within an optical receiver. Table 7.4 presents typical parameters of APDs. The dark current increases significantly with increasing reverse voltage. Further, dark current (I_{DO}, I_{DM}) and gain M_{APD} are strongly temperature dependent.

For the relatively popular wavelength of Nd:YAG lasers, 1064 nm, near infrared (NIR)-enhanced silicon APDs are available from several manufacturers, often targeting light-detection and ranging (LIDAR) applications. Quantum efficiencies of ~0.3 are achieved at that wavelength.

Recently, InGaAs APDs with multiple gain stages [37] have been brought to the market, with a significant improvement in terms of effective ionization factor for this material, on the order of 0.02–0.03. These devices are available with similar active areas as shown in Table 7.4, but can be operated at multiplication factors up to 40, with a noise figure below 3. Furthermore, HgCdTe APDs are subject of research and have been proposed for free-space optical communications [38]. To our knowledge, no commercial product is available yet.

InGaAs APDs with smaller diameters (30 μm) are commercially available as well and can be used in single-mode 10 Gbit/s receivers. For even higher data rates, Si/Ge APDs supporting 25 Gbit/s operation are becoming commercially available for fiber-optic access applications. A good review of APD technologies and performance is given in Ref. [39].

When comparing pin, diodes and APDs, pin-diodes have the advantage of:

- low bias voltage,
- very fast response,
- simple driving circuit, and
- large sensitive areas available (with reduced bandwidth B_e),

while in favor of the APD one would stress:

- internal amplification, thus
- easy implementation of electronic AGC via variation of the APD bias voltage, and
- reduced low-noise requirements on a following electronic preamplifier.

One can also identify situations where an APD performs worse than a pin photodiode. One example is the case when the background contribution dominates the overall SNR, another is a well-designed receiver with optical pre-amplification, i.e., one where the signal-ASE beat noise dominates, making avalanche multiplication unnecessary, and a third one is a coherent receiver with dominating LO shot noise, which constitutes the desired operating condition and again makes avalanche multiplication redundant. In general, an APD yields a better overall SNR only in cases where the noise of the electrical preamplifier dominates the overall receiver noise.

7.3.5 PHOTON-COUNTING DETECTORS

For low data rate applications requiring extreme sensitivity, which is less an issue for LEO applications but more for optical links for deep-space probes, Geiger-Mode avalanche photodiodes (GM-APDs) or superconducting nanowires may be considered as detectors.

GM-APDs operating at 1.06 μm and 1.55 μm have been developed with detection efficiencies greater than 50%, timing resolution less than 1 ns, and dark count rates less than 100 kHz [40]. One limitation of such devices is the relatively long reset time, which requires the detector to be quenched for a period of time before it can be reactivated after a detection event. GM-APDs operating at infrared wavelengths typically ask for reset times of tens of microseconds. Using conventional pulsed modulation formats, a single GM-APD with a 30-μs reset time would limit the achievable data rate to ~33 kbps. However, using higher-order modulation formats, such as PPM, multiple bits can be received for each detection event. In a specific experiment, a commercially available fiber-coupled GM-APD was biased at 47 V and operated at −40°C [41]. It showed a dark count rate of 428 kHz and a detection efficiency of 28%. In connection with 64-ary pulse position modulation and FEC, the scheme allowed to detect a signal at a source data rate of 100 kbit/s with a sensitivity as high as 1.5 photons/bit. Superconducting nanowires have been studied as detectors for ultra-high sensitivity deep-space links as well, achieving sensitivities of 0.5 photons/bit [42, 43].

7.3.6   OPTICAL FILTERS

As discussed in Section 7.2.3, in a free-space optical communication link, background radiation may severely degrade the sensitivity of a receiver. To reduce the detrimental effect of background radiation, one may place an optical bandpass filter in front of the photodiode. For this device, the system designer usually asks for:

- narrow bandwidth,
- low insertion loss,
- a wide blocking range with high rejection outside the filter's passband,

and, depending on the position of the filter within the receiver, for

- a wide field of view.

In the following, we review several concepts of optical filters and give their main characteristics.

Thin film interference filters rely on the wavelength dependence of the transmission of a stack of thin films of different dielectric materials deposited on a transparent substrate. By proper choice of film materials and film thickness (e.g., alternating quarter wavelength and half wavelength), such filters can be centered at any wavelength. Filters with bandwidths down to 0.1 nm are available today and come with transmissions of above 90%. Because these multilayer filters are based on the interference effect (similar to a Fabry-Perot resonator), the center wavelength shifts with the angle of incidence. This shift may amount to 1 or 2 nm per degree of tilt and thus limits the filter's field of view. In some applications this effect may by counteracted by employing a curved substrate. A very compact realization of filter action consists in directly applying the multilayer structure onto the photodiode's surface instead of the AR coating shown in Figure 7.11.

Fiber Bragg filters have been developed primarily for fiber systems. They rely on the distributed feedback action caused by periodic changes of the refractive index along the fiber axis. They offer band-stop filter characteristics, but the combination with a (fiber) circulator yields the desired bandpass filter (see Figure 7.12). Filter bandwidths below 0.1 nm may be achieved in connection with low insertion loss. On the other hand, such filters intrinsically operate as single-mode device, which means that their field of view is restricted to the narrow cone to be associated with the fundamental (Gaussian-like) fiber mode. Thus, their application in free-space systems is restricted to those cases where the received radiation has already been coupled to a single-mode fiber.[7]

Absorption filters may be considered in special cases where narrow bandwidth is not required and where, by accident, a favorable closeness of the received wavelength, λ; the cut-off wavelength of the photodiode; and the band edge of a semiconductor material exists. Consider, e.g., $\lambda = 980$ nm and a silicon photodiode

---

[7] An example is a receiver with an EDFA acting as optical preamplifier [51], or a coherent receiver based on single-mode fiber components.

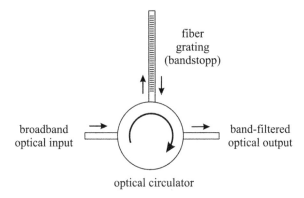

FIGURE 7.12   Optical bandpass filter consisting of a fiber grating and an optical circulator.

(cut-off near 1050 nm). If we arrange a window of undoped GaAs (cutoff at 920 nm, antireflection coated) in front of the photodiode, the overall setup has a bandpass characteristic with a width of ~130 nm. While this bandwidth is very large, the field of view is essentially unrestricted.

Tunable Fabry-Perot filters may be of interest for communication systems experiencing large Doppler shifts. In one implementation, two properly coated fiber ends face each other, and the gap in between can be varied piezoelectrically. The entire device usually comes with single-mode fiber interfaces. The inherent periodicity of the transmission is a clear disadvantage for the purpose of background reduction. To achieve sufficient off-band rejection across a wide bandwidth, we should ask for further wide-band filtering.

In case that an optical free-space transmission system employs wavelength division multiplexing (WDM), the optical demultiplexer in the receiver could be implemented in the form of *a waveguide grating router* (WGR).[8] These planar optical devices do not only serve as demultiplexer, but, at the same time, provide narrow bandpass filtering for each of the optical channels [44]. In contrast to the multiple interference filters mentioned above, WGRs can be designed without phase distortions accompanying the filter's amplitude response at the band edges.

7.3.7   OPTICAL PREAMPLIFIERS

As demonstrated widely in fiber systems, optical pre-amplification may considerably improve the sensitivity of optical receivers. This is especially true for systems where APDs cannot provide enough bandwidth and/or enough internal low-noise amplification, i.e., in high data rate receivers with pin-diodes where the noise is primarily due to the electronic amplifier. The prime specifications of an optical preamplifier are its noise figure, F, and its small signal gain, G. The physical effect responsible for producing excess noise during amplification is ASE. One can show the noise

---

[8] These devices are also known under the name arrayed waveguide grating (AWG) and Phased Array (PHASAR).

power density N_{ASE} (measured in W/Hz) into a single mode at the output of an optical amplifier is [45, 46].

$$N_{ASE} = (G-1)hfn_{SP}. \tag{7.40}$$

Here, $hf = hc/\lambda$ is the energy of a photon and n_{SP} is the amplifier's inversion factor, which characterizes the degree of material inversion achieved by pumping electrons from the lower to the upper (excited) energy states. The ASE occupies the entire spectral region within which optical amplification is possible, which is much wider than any reasonable optical signal bandwidth. Therefore, amplifier noise can be modeled as white noise. A perfect amplifier is fully inverted, corresponding to the lowest possible value of $n_{SP} = 1$. Values only slightly larger than $n_{SP} = 1$ can be obtained with commercially available EDFAs. The optical noise power P_N (in W) at the output of a "single-mode" amplifier is then given by:

$$P_N = 2N_{ASE}B_O, \tag{7.41}$$

where B_O is the optical bandwidth in Hertz. The factor 2 in Eq. (7.41) takes into account that even a "single-mode" device allows for two polarization modes to propagate. This factor of 2 has to be omitted if a polarization filter is implemented at the amplifier's output. If we follow RF practice and define the noise figure F as the quotient of SNR at the amplifier's input to that at its output, one finds[9]:

$$F = 2\left(1 - \frac{1}{G}\right)n_{SP}. \tag{7.42}$$

In general, $G \gg 1$. This implies that the lowest possible value of the noise figure of an optical amplifier, achievable for complete inversion ($n_{SP} = 1$), is $F = 2$, corresponding to a noise figure of 3 dB. Figure 7.13 gives an equivalent circuit for an optical amplifier: Noise-less amplification with gain G is followed by the addition of white

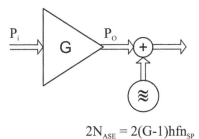

$$2N_{ASE} = 2(G-1)hfn_{SP}$$

FIGURE 7.13 Equivalent circuit of an optical amplifier. Noiseless amplification with gain, $G = P_O/P_i$, is followed by the addition of white noise with power density N_{ASE} per polarization.

---

[9] A more exact analysis reveals an additional term $1/G$ in Eq. (7.42) (see, e.g., Ref. [22]), but for the typical application in mind this is negligible.

noise with power density N_{ASE} per polarization. In an equivalent circuit with the noise source at the amplifier input, for $G \gg 1$ and $n_{SP} = 1$, this noise source would have a power density of hf per polarization. Combining Eqs. (7.40) and (7.41) yields:

$$N_{ASE} = \frac{hfGF}{2}. \qquad (7.43)$$

an expression that is very useful for a systems design engineer.

The optical amplifier most widely employed in optical communication systems is the EDFA, which operates close to the theoretical noise limit in that it routinely offers a noise figure of $F \le 4$ dB. Also, a gain of $G > 30$ dB is easily achieved. The energy levels of Erbium ions allow operation covering both the C-band (1530–1565 nm) and the L-band (1565–1625 nm). A strong advantage of EDFAs is that they can be easily connected or spliced to the transmission fiber with small coupling loss. Note in this context that any coupling loss at the input of an optical amplifier increases its noise figure by the same amount. Most EDFAs commercially available today operate on only the fiber's fundamental mode, which means that the received radiation has to be delivered to the amplifier in a single-mode fiber. In recent years, however, considerable research effort has been put also into the development of optical amplifiers operating on multiple spatial modes (see, e.g., [47] and also Section 7.1), and we expect to see some commercially available few-mode and multimode optical amplifiers in the near future.

Another optical amplifier that has been attracting considerable attention in recent years for its compactness and small mass is the SOA. It can be designed for any wavelength where sufficient inversion can be achieved in the semiconductor material and thus can amplify signals over a very broad spectral range from 1200 nm to 1650 nm [48]. While its typical gain is only somewhat lower than that of an EDFA, it is considerably worse concerning its noise figure: The noise figure of SOAs is typically 3–4 dB higher than that of EDFAs mainly due to a higher input coupling loss. As a consequence, it becomes difficult to build receivers with SOAs that exhibit sensitivities below some 10 dB above the quantum limit, making the EDFA currently the preferable choice for pre-amplification in optical space communication systems. Furthermore, SOAs have a much shorter relaxation time and interaction length compared to EDFAs. This makes the SOA more prone to pattern-dependent amplification and to cross talk in systems with wavelength multiplexing, especially if the SOA has to be operated with high power.

7.3.8 LOCAL LASER OSCILLATORS

A coherent optical receiver, be it a heterodyne or a homodyne receiver, requires a local laser oscillator. The most important requirements of such a device are

- transverse and longitudinal single-mode operation in order to achieve an intermediate frequency (IF) signal centered at a single carrier,
- narrow linewidth, i.e., little laser phase noise to avoid sensitivity degradations; the closer a receiver operates to the quantum limit, the more pronounced is the influence of non-zero linewidth,

- sufficient output power so that LO shot noise dominates all other noise contributions, and
- electronic frequency tuneability to achieve constant IF despite variations in transmitter laser frequency and varying Doppler shift; in case of analog homodyne detection, tuneability must be sufficiently fast to control the optical phase between received carrier and local laser to small fraction of 2π. This requirement is much relaxed for heterodyne or intradyne detection.

Following the almost exclusive use of intradyne coherent receivers in fiber-optic communications at 100 Gbit/s and above [49, 50], fully C-band tunable lasers with ~100 kHz linewidth and output powers of ~17 dBm are commercially available in miniature packages today.

7.3.9 DELAY INTERFEROMETERS

In order to detect phase modulation based on a photodetector that is only sensitive to the optical field's intensity, optical preprocessing must be performed, as discussed in detail in Section 7.4. A *delay interferometer* (DI; also known as *delay-line interferometer*, DLI) is a key component in this context. A DI is typically implemented as a Mach-Zehnder interferometer (cf. Figure 7.13), whose two interference branches are designed to exhibit a differential delay corresponding to the symbol period of the employed modulation. Operation based on larger delays (covering integer multiples of symbol periods) has also been discussed for data rate flexible receivers [51], but achieving rate adaptability through multi-symbol delays usually comes with performance penalties. A more promising approach for rate adaptability, especially for transmitter architectures that use an EDFA booster-amplifier in free-space applications, is based on a single fixed one-symbol DI in combination with continuously adaptable burst-mode operation [52]. By cleverly utilizing the fact that booster EDFAs are average-power limited [53], the optical power of the data burst is increased at constant average output power (and hence at constant electrical supply power) proportional to the burst duty cycle, which in turn increases transmit power and hence improves the link budget at a reduced burst-averaged symbol rate.

In the spectral domain, DIs exhibit a sinusoidally periodic transmission characteristic, whose periodicity (called *free-spectral range*, FSR) is given by the inverse of the differential time delay between the two interferometer arms. The *extinction ratio* of the periodic DI spectrum is an important metric for DI performance, as is the *polarization-dependent frequency shift* that some DI structures induce, owing to slight differences in the effective refractive indices between the two polarizations in the DI. The latter plays less of a role in free-space communications, as polarization multiplexing is not usually employed and the channel shows no random polarization rotations, hence polarization filtering at the receiver can therefore be implemented. In addition to the macroscopic differential delay between the two interference arms of the DI, a microscopic delay (on the order of a wavelength) is induced by a phase shifter to control the interference properties at the two interferometer output ports and establish perfect constructive and perfect destructive interference at the two DI output ports, respectively. As these conditions depend on the exact frequency of the incoming signal light, adaptive DI phase control is required, especially for fast Doppler shifts in space-borne applications. Phase shifters typically operate based on the thermo-optic or electro-optic effect. Key characteristics of DIs are reviewed in detail in Ref. [54].

DIs are commercially available as free-space optical devices [54], all-fiber devices [55], and integrated optics devices [57], with interferometer delays of typically between 20 and 400 ps, corresponding to symbol rates between 2.5 and 50 Gbaud. All-fiber devices offer the lowest insertion losses (<0.5 dB), going up to ~3 dB for integrated-optics devices.

7.3.10 Optical Hybrids

A key component of intradyne coherent receivers is the 90° optical hybrid whose functionality is discussed in Section 7.4.5 (cf. Figure 7.17). Its role is to combine the information-bearing signal with two 90° phase-shifted copies of a LO laser. Owing to the strong demand for intradyne receivers for virtually all fiber-optic communication systems exceeding 100 Gbit/s per wavelength [50], 90° hybrids are commercially available from many vendors and at various stages of optoelectronic integration. Since fiber-optic receivers inherently require polarization diversity due to the random polarization coupling occurring along the transmission fiber, com-mercial 90° hybrids typically include two separate single-polarization hybrids in a polarization-diversity arrangement [57]. Devices based on free-space optical com-ponents [57], all-fiber components [58], planar lightwave integrated circuits [59], as well as devices that integrate the four pairs of high speed balanced photodetectors [60, 61] are available.

As an alternative to the more traditional 90° hybrid structure of a coherent receiver, a 3×3 fused optical coupler may be used as a 120° hybrid. Its three output ports contain full information on the optical field. This type of coherent optical front end avoids the need for balanced photodetectors and is based on all-fiber technology, which can be advantageous when coherent optical front-end loss plays an important role, as is the case for shot noise limited detection. This intradyne receiver structure is described in detail in Ref. [62].

A simple 2×2 fused optical coupler may be used as well as a 180° hybrid feeding a balanced photodetector, for use in analog coherent receivers.

7.4 OPTICAL RECEIVERS: STRUCTURES, PERFORMANCE, AND OPTIMIZATION

The choice of receiver best suited for a particular free-space laser link depends on a variety of fundamental as well as hardware parameters, the most important of which are as follows:

Receiver sensitivity: This parameter specifies the required optical receive power to achieve a target receiver output performance, such as a target BER. A 3-dB increase in receiver sensitivity can be traded for a 3-dB reduction in optical transmit power, a 41% increase in free-space communication distance, 16% smaller telescopes, or an increased tolerance to beam mis-alignment or beam profile distortions brought by atmospheric turbulence.

Modulation format: Not every reception technique is suited for every modulation format. For example, direct-detection receivers are insensitive to phase and polarization information unless this information is converted into intensity modulation by means of external optical components (e.g., delay demodulation of DPSK) [7, 54]. Coherent receivers, on the other hand, detect the optical field directly, and therefore allow for any modulation format without additional optical preprocessing.

Hardware availability, reliability, space qualification, and cost: Different types of receivers require different hardware building blocks, which are not always available at reasonable cost, at the desired wavelength, or readily space-qualified. For example, efficient and low-noise optical amplification is predominantly available within the fiber-telecom wavelength range around 1.55 μm; this is also where compact, low-cost, and high-speed optical transmitter modules are widely available from the fiber-optic communication industry. As another example, high-gain APDs are based on silicon technology, and therefore only work at wavelengths below 1.1 μm (see Section 7.4.3).

7.4.1 OPTICAL RECEIVER PERFORMANCE MEASURES

Before comparing different optical receiver concepts and discussing the most relevant receiver design trade-offs, we introduce some important receiver performance measures.

The parameter of ultimate interest in any digital communication system is the *BER*, defined as the average ratio of wrong bit decisions to the total number of detected bits. The BER is an important measure for transmission quality: Different digital applications ask for widely different BERs to operate properly. If a system cannot meet a certain BER target, the "raw BER" (i.e., the BER measured right after the receiver's decision circuit) can be improved by means of *FEC codes*. The benefits of FEC, however, come at the expense of a bit rate overhead of typically 7% for legacy terrestrial multi-Gbit/s fiber-optic systems, and up to 25% for stronger codes used in submarine and advanced terrestrial coherent fiber communications [63]. Deep-space systems use even higher-overhead codes, described in more detail in Chapter 6 The FEC overhead is required to generate redundancy (such as parity bits), which is exploited for error correction within the receiver. When using FEC, the "decoded BER," i.e., the BER measured *after* the FEC decoder, is significantly lower than the raw BER. Typical FECs working in the multi-Gbit/s range need an input BER in the range of 10^{-2} or 10^{-3} to achieve an output BER of better than 10^{-15}.

Apart from the BER, the temporal distribution of errors plays an important role in transmission quality. While occasional, randomly distributed bit errors may not significantly affect a digital communication system, the occurrence of extended *error bursts* can lead to short periods of complete link failure, also called a *system outage*. Note that the occurrence of error bursts is not captured by the BER, since the BER is a long-term average quantity. It may happen that the BER as such is within specifications, but that the occurrence of error bursts jeopardizes system performance. To counteract outage, one may use system margins that capture the worst-case system operating conditions. However, if these worst-case conditions happen only rarely, the system would be over-engineered for the average case and this has insufficient performance. Trading *system margin* for *outage probability* is an important system engineering task. Alternatively, *bit interleaving* can be employed to break up reasonably short error bursts, such that prior to decoding one FEC block does not see an overly large amount of errors while other FEC blocks are completely error free, but that errors are evenly distributed among many FEC blocks and are hence block-by-block correctable. However, bit interleaving comes at the expense of latency, as received bits need to be brought into their correct order across the entire interleaving depth after FEC decoding. For free-space applications, channel interleavers with latencies between 0.1 s and 1 s are being discussed, making links robust against fading events due to atmospheric turbulence. Decoding and descrambling latencies should be seen in context of the system's propagation delay, hence are more of an issue in LEO scenarios than in deep-space missions.

Another important parameter for characterizing optical receiver performance is Personick's Q-factor, defined for binary modulation as [13]:

$$Q = \frac{i_1 - i_0}{\sigma_1 + \sigma_0}, \tag{7.44}$$

where i_1 and i_0, respectively, represent the noise-free electrical signal current for a "1"-bit and a "0"-bit at the receiver's decision gate, and σ_1 and σ_0 are the noise standard deviations associated with these signal levels, as discussed in Section 7.4.4. Recall that some important noise sources encountered in optical communications are *signal-dependent*, i.e., their noise variance is a function of the optical signal power. This necessitates the distinction between "1"-bit and "0"-bit noise in Eq. (7.45). Assuming the Gaussian noise, the Q-factor relates to the BER via:

$$BER = \frac{1}{2} erfc\left\{\frac{Q}{\sqrt{2}}\right\}, \tag{7.45}$$

where $erfc\{x\} = (2/\sqrt{\pi})\int_x^\infty \exp\{-\xi^2\}d\xi$ denotes the complementary error function. For purely signal-independent (e.g., thermal) noise, we find $\sigma_1 = \sigma_0 = \sigma$, and Eq. (7.45) reduces to [65, 66]:

$$BER = \frac{1}{2} erfc\left\{\frac{|i_1 - i_0|}{2\sqrt{2}\sigma}\right\}. \tag{7.46}$$

Equation (7.46) is a powerful tool that allows for intuitive interpretations of receiver performance and that can give deep insight in certain receiver design trade-offs. However, any receiver analysis based on the Q-factor automatically comprises several important assumptions that may not be met in reality.[10] Thus, some care has to be taken whenever quantitative predictions based on the Q-factor as used in Eq. (7.45) are being made.

In the fiber-optics literature, researchers frequently measure the BER by counting errors, which remains the most accurate receiver characterization, and then conveniently use the inverse of Eq. (7.45) to calculate an "equivalent Q-factor" (sometimes also called Q^2-factor), frequently given in dB:

$$Q = \sqrt{2}\, erfcinv\{2\, BER\}, \tag{7.47}$$

$$Q_{dB} = 20 \log_{10} Q = 10 \log_{10} Q^2.$$

Note that this alternative use of the Q-factor is *independent* of Eq. (7.47) and all its assumptions. It is merely a convenient way to express a measured BER in logarithmic units, and is hence universally valid, irrespective of assumptions on noise statistics, etc.

While the BER specifies an external requirement that makes an optical receiver useful for a given digital transmission task, the *receiver sensitivity* specifies the average optical signal power P_{av} that is required at the receiver input to achieve a target BER.[11] Obtaining the highest possible receiver sensitivity (i.e., meeting the target BER with the lowest possible input signal power) is one of the top priorities in free-space optical receiver design. The receiver sensitivity is often given in terms of the average number of *photons per bit*:

$$n_{av} = \frac{P_{av}}{hfR}, \tag{7.48}$$

here hf denotes the photon energy at the transmit wavelength and R is the bit rate. This specification eliminates the wavelength dependence and data rate dependence of P_{av}, and therefore allows for an easier performance comparison of different receivers as well as for a straightforward benchmarking to a receiver's ultimate performance limit, the quantum limit.

The *quantum limit* represents a strict lower bound on the performance of any receiver in optical communications. It is usually specified in terms of n_{av}, and

---

[10] The most important assumptions for the validity of any Q-factor-based analysis are [67] the following: (i) The statistics of the electrical decision variable are Gaussian, which is not the case for beat noise or APD shot noise. However, it turns out that analyses based on Gaussian statistics work well for OOK, while they fail for balanced detection of DPSK. (ii) Receiver performance is dominated by noise rather than by inter-symbol interference (ISI), since the Q-factor only uses the two discrete signal levels i_1 and i_0; in the case of severe ISI, there are many different "1"-bit and "0"-bit amplitudes, which requires further averaging. (iii) The receiver's decision threshold is dynamically optimized, which is an inherent and important assumption for Gaussian detection statistics to work, but is not always done in practical receivers.

[11] Note that the receiver sensitivity not only characterizes the receiver itself, but also to some extent the properties of the transmitter and its interplay with the receiver, such as extinction ratio or ISI generated by transmitter opto-electronics. Thus, knowledge of the receiver sensitivity alone does not allow trustworthy predictions on how the receiver will perform with different transmitters or with other modulation formats.

depends on the target BER, the modulation format, and the receiver type. It is calculated by assuming an ideal transmitter (perfect extinction, no ISI, etc.) as well as an ideal receiver (no thermal noise, optimum receiver filtering, perfect sampling, and decision, etc.). Obviously, and unlike some other frequently cited limits which will be encountered below, the quantum limit can only be *approached* by careful engineering, but it can never be reached in practice.

Speaking about receiver sensitivities, an important remark related to coded systems is in order: Since FEC always introduces some amount of bit rate overhead, a certain portion of the optical power at the receiver input is not used for information transport, but rather for redundancy. For a fair comparison, receiver performance for coded systems therefore has to be given in terms of the average number of *photons per information bit* at the target BER after decoding. This number is always larger than the number of photons per bit on the channel, which inherently reduces the FEC's coding gain, and leads to the notion of a *net coding gain* [68].

Another important performance measure of digital optical receivers is the *OSNR* required to achieve a certain BER. This quantity is predominantly used to characterize receivers for fiber-optic communication systems, but is uniquely related to receiver sensitivity for the class of beat noise limited receivers (e.g., the optically pre-amplified receiver discussed below). We briefly introduce this parameter here in order to allow the reader to leverage from the vast literature based on fiber-optic communications. The *required OSNR* specifies the minimum ratio of average optical signal power to the power of all background radiation sources (including an optical preamplifier) that is needed to guarantee a target BER. Background radiation is typically assumed depolarized and its power is measured within an optical reference bandwidth of 12.5 GHz. Note that the required OSNR is defined directly *at the photodetector*, in contrast to the receiver sensitivity, which is defined at the receiver input.[12]

7.4.2 DIRECT-DETECTION RECEIVER

The *pin*-receiver depicted in Figure 7.14 is the simplest optical receiver structure. It consists of a *pin* photodiode (see Sect. 7.5.3), some electronic amplification, some

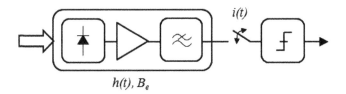

$h(t), B_e$

FIGURE 7.14 Setup of a direct-detection pin-receiver.

---

[12] The reason for specifying the OSNR at the detector is the huge amount of ASE generated by in-line optical amplification in fiber-optic systems. This source of background radiation is much stronger than what is added by the optical preamplifier within the receiver, so that the OSNR is essentially left unchanged upon optical pre-amplification. If the optical preamplifier contributes a significant amount of noise, though, as is the case in space-based systems that cannot have any in-line optical amplification between transmitter and receiver, the OSNR is only a reasonable quantity when defined at the detector.

inherent or intentionally introduced post-detection electrical filtering, and a sampling-and-decision device that restores the digital data. As discussed along with Eq. (7.46), the photocurrent $i(t)$ is linearly proportional to the incident optical power. The relevant noise terms in a *pin*-receiver are *electronics noise*, *shot noise*, and *beat noise*, the latter only in the presence of background radiation, either due to celestial bodies or due to optical amplifiers.

Although electronics noise usually dominates shot noise, and can in principle be engineered to insignificance by cooling the receiver, this is hardly ever done in practice. Shot noise, on the other hand, is fundamentally present in any optical receiver. The limit when shot noise dominates all other noise terms ($\sigma^2_{shot} \gg \sigma^2_{elec}$) is referred to as the *shot noise limit*. Note that the shot noise limit is a *condition* rather than a *quantity*. A shot noise limited *pin*-receiver approaches quantum-limited performance if all other engineering optimization measures are fully exploited (photon-counting operation, elimination of ISI, etc.).

The quantum limit for the *pin*-receiver using OOK is obtained by ignoring electronics noise, by assuming a perfect photodiode (quantum efficiency $\eta = 1$), and by evaluating the BER for the Poissonian photon statistics of perfect laser light [14]. Leaving the derivation to more detailed texts on optical receivers [69, 70], we merely cite the result:

$$BER = 0.5 \cdot \exp\{-2n_{av}\}. \tag{7.49}$$

For BER $= 10^{-9}$, the quantum-limited receiver sensitivity of the *pin*-receiver is $n_{av} \approx$ 10 photons/bit. This intriguingly low receiver sensitivity, however, does not apply to receivers that can be implemented in practice, since in reality electronics noise by far dominates shot noise ($\sigma^2_{shot} \ll \sigma^2_{elec}$). As a consequence, receiver sensitivities achieved by *pin*-receivers are typically 15–30 dB off the quantum limit, depending on the data rate: Assuming an equivalent noise current density of 10 pA/$\sqrt{Hz}$ and a 10-Gbit/s receiver with ~7 GHz electrical bandwidth operating at a wavelength of 1550 nm, the electronics noise variance amounts to $7 \times 10^{-13} A^2$ (cf. Eq. (7.49)), while the "1"-bit shot noise variance going with the detection of 10 photons/bit comes to about $4 \times 10^{-17} A^2$ (cf. Eq. (7.49)), four orders of magnitude below the electronics noise. For a realistic BER, the Q-factor is thus entirely determined by electronics noise, and for $Q = 6$, we arrive at a receiver sensitivity of $n_{av} \approx 5000$ photons/bit, 27 dB above the quantum limit.

To achieve higher receiver performance, more advanced receiver types have to be employed. There are basically three ways to proceed: *Avalanche photodetection*, *coherent detection*, and *optically pre-amplified detection*. These rather diverse techniques, which will be discussed in the following paragraphs, have one common attribute: They all amplify the received optical signal before or at the stage of photodetection, while at the same time unavoidably introducing additional noise. As soon as the newly introduced noise terms dominate electronics noise, any further increase of the employed gain mechanism does *not* affect receiver performance any more. In contrast to *pin*-receivers, the respective quantum limits can be closely approached with these receiver types in experimental reality.

7.4.3 AVALANCHE PHOTODIODE RECEIVER

As discussed in Section 7.4.2, an APD multiplies the generated primary photo-electrons, on average, by its avalanche gain M_{APD}. This multiplication comes at the expense of enhanced shot noise, captured in the APD's excess noise figure, F_{APD} (see Eq. (7.36)). In the desired limit when multiplied shot noise dominates electronics noise ($\sigma^2_{elec} \ll \sigma^2_{mult.shot}$), the Q-factor for high signal extinction ratios ($i_0 \ll i_1$) approaches[13]:

$$Q = \frac{S M_{APD} P_1}{\sigma_{elec} + \sqrt{\sigma^2_{mult.shot} + \sigma^2_{elec}}} \rightarrow \sqrt{\frac{\eta n_{av} R}{F_{APD} B_e}} \approx \sqrt{2 n_{av} / F_{APD}} \qquad (7.50)$$

with $P_1 = 2 P_{av}$ equal to the "1"-bit optical signal power for NRZ-OOK for an equally probable number of ones and zeros. Thus, the excess noise factor F_{APD} takes the role of a noise figure in degrading detection performance. Note that optimum performance of an APD receiver is in general *not* attained at the highest possible multiplication, M_{APD}, since F_{APD} is a complicated and highly technology-dependent function of M_{APD} (see Sect. 7.5.3), necessitating joint optimization of M_{APD}, F_{APD}, and σ_{elec} [13, 14]. Good APDs ($M_{APD} \approx 100$, $F_{APD} \approx 5$) for operation up to 1 Gbit/s are available in silicon technology, which limits their operating range to wavelengths below ~1100 nm. Receiver sensitivities of 200 photons/bit have been achieved at 50 Mbit/s [71]. InGaAs or InAlAs-based APDs for use in the 1550-nm wavelength region, on the other hand, exhibit fairly low multiplication ($M_{APD} \approx 10$) for 10-Gbit/s detection. Receiver sensitivities of ~1000 photons/bit have been demonstrated at 10 Gbit/s [72].

7.4.4 ANALOG COHERENT RECEIVERS

Coherent detection provides another way of amplifying the signal while at the same time boosting detection noise above the electronics noise floor. Before the advent of EDFAs, coherent receivers were widely studied for optical fiber communications, reflected in many books and review articles [73, 79], and saw a revival around 2005 [49, 50] due to the need to increase per-wavelength bit rates and spectral efficiencies, with today's fiber-optic systems at 100 Gbit/s and above being largely based on coherent detection. We restrict ourselves to the most basic operation principles of coherent receivers in this chapter.

The basic structure of a coherent receiver is depicted in Figure 7.15: The optical input signal is combined with light from a LO laser by means of an optical beam combiner. Upon detection, the two optical fields beat against each other, resulting in a photocurrent according to Eqs. (7.14) and (7.15). Neglecting the filtering impact of post-detection electrical filtering, the electrical signal reads:

$$i(t) = S\left\{ \varepsilon p_s(t) + (1-\varepsilon) P_{LO} + 2\sqrt{\mu} \sqrt{\varepsilon(1-\varepsilon)} \sqrt{p_s(t) P_{LO}} \cos\left(2\pi\Delta ft + \phi(t)\right) \right\}, \qquad (7.51)$$

---

[13] Note that Eq. (7.50) only reveals general trends, since in reality the decision variable is Poissonian rather than Gaussian, which violates the assumptions of Q-factor analyses (see footnote [11]): Specializing Eq. (7.50) for pin-reception ($F_{APD} = 1$), we arrive at a quantum limit of $n_{av} = 18$ photons/bit, which is off its correct value by 2.6 dB.

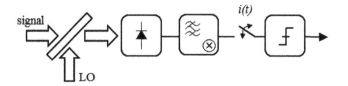

FIGURE 7.15 Setup of an analog coherent receiver. Signal and LO are optically combined and detected by a pin photodiode. Before making bit decisions, the intermediate-frequency signal is electronically post-processed, including filtering and mixing to baseband.

where $p_s(t)$ and P_{LO} denote the signal and LO power before combination, and $0 < \varepsilon < 1$ is the transmission factor of the (lossless) beam combiner. In the desired mode of operation, the LO power at detection is chosen much stronger than the signal power, $(1 - \varepsilon)P_{LO} \gg \varepsilon p_s(t)$, such that the first term in Eq. (7.51) can be neglected compared to the second and third one. Filtering out (through DC coupling) the temporally constant second term, we are left with the beat term, revealing an exact replica of the received optical field's *amplitude* $\sqrt{p_s(t)}$ and *phase* $\varphi(t)$. Since, in contrast to direct-detection receivers, both amplitude *and* phase of the optical field are translated into an electrical signal, any amplitude or phase modulation scheme can directly be used in combination with coherent receivers. Note that the beat term in general oscillates at the difference frequency between LO and signal light Δf, also called the *IF*. With reference to Figure 7.16, if the frequency of the LO differs from the signal frequency, we speak of a *heterodyne* receiver. In a coherent system with a standard transmitter, Δf is typically chosen to be two to three times the symbol rate, necessitating an electronic front-end bandwidth of approximately five times the symbol rate. Using spectral shaping at the transmitter, either through a digital-to-analog converter (DAC) or through optical pulse shaping [50], Δf can be reduced to as little as half the symbol rate and the optoelectronic front-end bandwidth to the symbol rate. In a heterodyne receiver, the IF signal is then transposed to baseband after photodetection using standard (analog or digital) microwave mixing techniques. If LO and signal have the exact same frequency, such that $\Delta f = 0$, we speak of a *homodyne* receiver. Homodyne detection requires strict analog optical phase locking between LO and signal optical fields in addition to keeping the IF zero. Since the stable operation of an optical phase locked loop (PLL) implies significant technological effort [73, 79], homodyning is

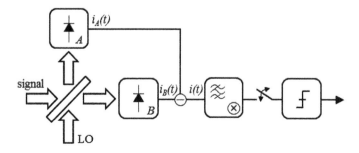

FIGURE 7.16 Setup of a balanced coherent receiver.

TABLE 7.5

Different Operation Modes of a Coherent Receiver, Depending on the Choice of the IF

	Heterodyne	Homodyne	Intradyne
Front-end bandwidth	~ 5* Symbol rate	Symbol rate	~Symbol rate
Phase/frequency locking	Frequency locking	Analog optical PLL	Digital electronic PLL (free-running LO)
Spectral sketch			

only taken into consideration if utmost receiver sensitivities are desired, as detailed below. As an alternative to strict homodyne detection, a *phase diversity* [74], *or intradyne* receiver [75–77] can be constructed by detecting both quadratures of the IF signal oscillating at $\Delta f \approx 0$. This approach offers the advantage of reduced bandwidth requirements for the detection optoelectronics and does not require analog phase locking, as this task can then be performed in the *digital domain*, provided that sufficiently fast digital signal processing (DSP) electronics are available. This forms the basis for digital coherent detection as discussed in Section 7.4.5.

In the single-ended setup of coherent receivers shown in Figure 7.16, the transmission factor ε of the optical beam combiner has to be chosen as high as possible such as not to waste too much signal power, and as low as acceptable to let sufficient LO power reach the detector in order to achieve LO shot noise limited performance, as will be discussed below. The heterodyne efficiency, μ, accounts for the degree of spatial overlap as well as for any polarization mismatch between the LO field and the signal field. When using free-space beam combination, μ directly reflects a mismatch in the transversal mode structure of the two interfering beams. Such a mismatch becomes particularly important if the signal beam has acquired random phase distortions across its transversal beam profile, as may be the case in the presence of atmospheric turbulence. If both signal and LO are provided, co-polarized in single-mode optical fibers, the mode structure of signal and LO are inherently equal at the point of beam combination. However, coupling the free space signal beam into a single-mode fiber is associated with a certain, mode-dependent coupling efficiency that exactly equals μ, and then takes the role of the heterodyne efficiency [78].

Due to the high LO power reaching the detector, the main noise contribution in a coherent receiver, in the absence of background radiation, is the *shot noise* produced by the LO (cf. Eq. (7.52)).

$$\sigma^2_{LO,shot} = 2eS\,(1-\varepsilon)\,P_{LO}B_e. \tag{7.52}$$

If this noise term dominates electronics noise ($\sigma^2_{LO,shot} \gg \sigma^2_{elec}$), the Q-factor becomes independent of electronics noise. The limit of dominating LO shot noise is known as the *shot noise limit* in the context of coherent receivers. In the shot noise limit, receiver performance is independent of the LO power, since both the information-bearing beat term in Eq. (7.53) and the noise standard deviation, i.e., the square-root of Eq. (7.54), scale with $\sqrt{P_{LO}}$, which lets the LO power cancel in the Q-factor. This independence of receiver performance on the amplification mechanism is common to all types of high-performance receivers: In the case of the APD receiver, increasing the multiplication factor does not help once multiplied shot noise dominates electronics noise, and in the case of the optically pre-amplified receiver discussed below, it is of no use to increase the amplifier gain beyond its value necessary to let beat noise dominate all other noise terms.

Unlike direct-detection *pin*-receivers, where electronics noise can typically not be engineered away in practice, a coherent receiver can very well be made shot noise limited, and by careful design can closely approach quantum-limited performance. The highest receiver sensitivity with the potential of practical implementation that is known today can be achieved using homodyne detection of PSK, where the data bits are directly mapped onto the phase of the optical signal $\{0,1\} \rightarrow \{0, \pi\}$. To derive this important quantum limit, we use the definition (7.51) of the Q-factor, and insert the beat term of Eq. (7.2) for i_1 and i_0, setting $\Delta f = 0$ (homodyning), $p_s(t) = P_{av}$, and $\varphi_0 = 0$ or $\varphi_1 = \pi$ (PSK). We also make use of the shot noise expression (7.52) as well as of the fact that LO shot noise affects both the detection of a "1"-bit and of a "0"-bit, i.e., $\sigma_1 + \sigma_0 = 2\sigma_{LO,shot}$. Assuming further ISI-free detection with a matched electrical filter[14] of noise bandwidth, $B_e = R/2$, we arrive at:

$$Q = 2\sqrt{\eta\mu\varepsilon}\sqrt{n_{av}}. \tag{7.53}$$

We see that both reduced heterodyne efficiency μ and reduced quantum efficiency, η, of the photodetector linearly affect receiver sensitivity. Assuming an ideal receiver ($\mu = 1$, $\varepsilon \rightarrow 1$, $\eta = 1$), we immediately arrive at the quantum-limited Q-factor for homodyne PSK:

$$Q = 2\sqrt{n_{av}}. \tag{7.54}$$

For BER = 10^{-9} ($Q \approx 6$), we thus require $n_{av} \approx 9$ photons/bit at the receiver input. Note that this number is almost identical to the quantum limit of a direct detection *pin*-receiver, Eq. (7.54), however, with the big difference that the quantum limit can be closely approached in a coherent receiver in practice, while in a *pin*-receiver it cannot due to its dominating electronics noise. The best reported receiver sensitivity for homodyne PSK to date is 20 photons/bit at 565 Mbit/s [80]. Using OOK instead of PSK, the sensitivity degrades by 3 dB. Going to heterodyne detection results in an additional loss of 3 dB in terms of receiver sensitivity due to the doubled noise bandwidth of a heterodyne receiver. Table 7.6 summarizes the quantum limit of a selection of modulation formats in combination with various receivers, together with some experimentally achieved results. A detailed discussion on the derivation of various quantum limits can be found in, e.g., Ref. [69].

---

[14] A matched filter is the theoretically optimum filter for the detection of a pulse in white Gaussian noise, provided that no ISI is present [73]. The matched filter's impulse response equals the temporally inverted, complex conjugated pulse shape to be detected.

TABLE 7.6
Overview of Detection Techniques and Their Sensitivities for Various Modulation Formats at BER = 10⁻⁹

	Coherent								Optically Pre-Amplified			
	Homodyne				Heterodyne							
	QL	Off by	R	Ref.	QL	Off by	R	Ref.	QL	Off by	R	Ref.
PSK	9 ppb	3.5 dB 9.0 dB 10.5 dB	565 Mb/s 4 Gb/s 10 Gb/s	[80] [81] [82]	18 ppb							
DPSK	10 ppb				20 ppb	8.2 dB 10.2 dB 13 dB	565 Mb/s 4 Gb/s 10 Gb/s	[83] [84] [85]	20 ppb	1.8 dB 2.8 dB	10 Gb/s 42.7 Gb/s	[89] [90]
OOK	18 ppb				36 ppb	6.9 dB 12 dB	4 Gb/s 10 Gb/s	[84] [85]	38 ppb	0.5 dB 1.4 dB 3.1 dB	5 Gb/s 10 Gb/s 40 Gb/s	[91] [92] [93]
FSK					40 ppb	2.1 dB 2.2 dB 6.8 dB	1 Gb/s 2.5 Gb/s 4 Gb/s	[86] [87], [88] [84]	40 ppb			

Note that not all combinations of modulation formats and detection techniques have been actively pursued. (QL, quantum limit; ppb, photons per bit [BER = 10⁻⁹, uncoded]; R, data rate).

It became evident during our above analysis of coherent receivers that the single-detector receiver structure of Figure 7.16 has the fundamental drawback of wasting a significant fraction ($\varepsilon \to 1$) of the available signal and LO power, with the receiver sensitivity depending linearly on the beam combiner's splitting ratio, ε, cf. Eq. (7.52). A frequently employed alternative implementation of coherent receivers that overcomes this problem is shown in Figure 7.16. It makes use of *balanced detection* with $\varepsilon \approx 1/2$. In this case, *both* output ports of the beam combiner, which are inherently out of phase by 180° [14] are detected individually, yielding:

$$i_A(t) = S_A \left\{ (1-\varepsilon)p_s(t) + \varepsilon P_{LO} - 2\sqrt{\mu}\sqrt{\varepsilon(1-\varepsilon)}\sqrt{p_s(t)P_{LO}} \cos\left(2\pi\Delta f + \phi(t)\right) \right\}, \quad (7.55)$$

and

$$i_B(t) = S_B \left\{ \varepsilon p_s(t) + (1-\varepsilon)P_{LO} + 2\sqrt{\mu}\sqrt{\varepsilon(1-\varepsilon)}\sqrt{p_s(t)P_{LO}} \cos\left(2\pi\Delta f + \phi(t)\right) \right\}. \quad (7.56)$$

The difference of the two photocurrents then reads:

$$i(t) = i_A(t) - i_B(t) = \left\{ (1-\varepsilon)S_A - \varepsilon S_B \right\} p_s(t) + \left\{ \varepsilon S_A - (1-\varepsilon)S_B \right\} P_{LO} - \\ -2(S_A + S_B)\sqrt{\mu}\sqrt{\varepsilon(1-\varepsilon)}\sqrt{p_s(t)P_{LO}} \cos\left(2\pi\Delta f + \phi(t)\right) \quad , \quad (7.57)$$

where S_A and S_B denote the responsivities of the two photodetectors. Ideally ($S_A = S_B$, $\varepsilon = 1/2$), the first two terms vanish, and we are left with the desired beat term only. Note that a balanced coherent receiver has *exactly* the same quantum-limited sensitivity as its single-detector equivalent. This is easily verified by evaluating the Q-factor and taking note of the fact that the shot noise variances of both detectors have to be added to obtain the noise variance of the difference signal, since the two shot noise processes are statistically independent. Apart from making more efficient use of signal and LO power, a balanced receiver is more robust to *relative intensity noise (RIN)* of the LO, i.e., to random fluctuations of the LO power P_{LO}: Depending on the type of LO laser and the IF of the system, the RIN spectrum of the strong LO power may extend into the signal band, where it acts as an additional noise source. By carefully balancing the receiver, the second term in Eq. (7.57), which includes possible LO power fluctuations translating to the difference signal can be greatly suppressed. In addition, the first term in Eq. (7.57), representing the direct-detection component of the signal, is cancelled in a balanced receiver and no longer acts as cross talk to the desired beat term.

Another important effect with the potential of seriously affecting coherent receiver performance is laser phase noise [73, 79]. In coherent receivers, which always operate on the beating of two optical fields, it is the *sum* of the LO's linewidth and the transmit laser's linewidth that is responsible for sensitivity penalties. Both lasers therefore have to meet high spectral requirements. Note, however, that this hardware requirement depends only on the *ratio* of laser linewidth and bit rate, which makes it easier to fulfill as data rates increase: For example, homodyne PSK needs a laser linewidth to data rate ratio of $\leq 5 \times 10^{-4}$, which translates to the technologically easily realizable value of ≤ 5 MHz at 10 Gbit/s.

7.4.5 DIGITAL COHERENT RECEIVERS

The rapid evolution of CMOS DSP capabilities has enabled a wide variety of sophisticated DSP techniques to enter high-speed fiber-optic optical communication solutions, starting with the implementation of FEC in optical systems in the early 1990s [94–96], followed by maximum likelihood sequence estimation (MLSE) at 10 Gbit/s starting in 2004 [97, 98], and the commercialization of digital intradyne receivers at 40 Gbit/s [99] and 100 Gbit/s [100] in 2008 and 2010. Today, integrated coherent DSP ASICs are able to digitally process up to 500 Gbit/s and ASICs with Tb/s processing capabilities are in development. Originally proposed in 1992 [75], an intradyne receiver uses a 90° hybrid as shown in Figure 7.17 (see also Section 7.3.10) to convert both in-phase and quadrature components of the optical signal to electronic baseband:

$$i_{in-phase}(t) = 4S\sqrt{\mu}\sqrt{\varepsilon(1-\varepsilon)}\sqrt{p_s(t)P_{LO}}\cos(2\pi\Delta f + \phi(t)), \tag{7.58}$$

$$i_{quadrature}(t) = 4S\sqrt{\mu}\sqrt{\varepsilon(1-\varepsilon)}\sqrt{p_s(t)P_{LO}}\sin(2\pi\Delta f + \phi(t)). \tag{7.59}$$

The LO laser is typically only coarsely frequency stabilized, making it fall within a few GHz of the signal's center frequency, i.e., Δf drifts slowly over time. The outputs of the photodiodes are sampled using high-speed analog-to-digital converters (ADCs), typically with a 2× oversampling factor. In order to avoid power-hungry interfacing between the ADCs and the CMOS DSP, ADCs are typically built in CMOS, integrated with the DSP on the same chip. Such devices exist today with sampling rates of ~90 GSamples/s and 8-bit nominal resolution.

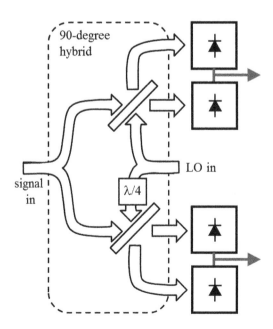

FIGURE 7.17   Setup of an intradyne coherent receiver.

Once in the digital domain, the sampled optical field can be processed using DSP algorithms for frequency locking, phase tracking, and digital equalization. Reviews of digital intradyne receivers as used in optical fiber communications can be found in, e.g., [49, 50, 99, 101]. The use of intradyne detection in free-space laser communications has also recently been reported [102]. However, one must be careful when porting intradyne technology from fiber-optics to space-borne applications, as the associated ASICs are typically power hungry, owing to their distortion compensation capabilities specific to the fiber-optic channel (such as large amounts of chromatic dispersion or fiber nonlinearities) as well as their ability to rapidly respond to random polarization variations occurring on a fiber-optic link. Neither of these effects is present in a typical free-space optical link. Intradyne transponders for free-space optical communications most conveniently leverage ASICs developed for the increasingly important class of pluggable coherent transponders for short-reach fiber-optic applications.

In terms of receiver sensitivity, intradyne receivers show the same 3-dB sensitivity penalty compared to homodyne receivers as heterodyne receivers, albeit for different reasons: While the 3-dB sensitivity penalty originates from a doubled noise bandwidth for heterodyne receivers, it stems from the fact that the signal needs to be split into two paths in an intradyne receiver for I/Q detection.

7.4.6 OPTICALLY PRE-AMPLIFIED RECEIVER

The historically youngest class of highly sensitive optical receivers uses *optical pre-amplification* to boost the weak received signal to appreciable optical power levels prior to detection, as shown in Figure 7.18.

At the same time, and fundamentally unavoidable, ASE is introduced by the amplification process, as discussed in Section 7.4.2. The intriguing performance characteristics of EDFAs have made optically pre-amplified receivers an attractive detection technique. Since the gain spectrum (and thus also the ASE spectrum) is much broader than the signal spectrum (typ. 30 nm in the 1550-nm wavelength band), an optical bandpass filter is employed to suppress out-of-band ASE. ASE enters the detection process in complete analogy to any other (external) source of incoherent background radiation, and leads to signal-ASE beat noise and to ASE-ASE beat noise, as discussed in Section 7.2.3. If the beat noise variance, given by Eqs. (7.55) through (7.57), dominates over all other noise sources in the system (in particular, electronics noise), we speak of *beat noise limited* detection. Since both the photocurrent (i_1 and i_0) and the beat noise standard deviations (σ_1 and σ_2) scale linearly with the amplifier gain G, the Q-factor becomes independent of the amplifier gain

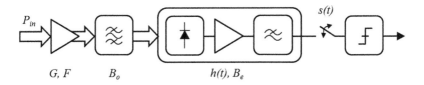

FIGURE 7.18   Setup of an optically pre-amplified receiver.

in the beat noise limit, and any further increase of G does not improve the receiver's performance (cf. Eq. [7.50]). Using the definition of the Q-factor, Eq. (7.53), together with Eq. (7.54) and the simplified signal-background beat noise expression, Eq. (7.52), and under the assumptions of (i) no background radiation at the receiver input, such that the entire background radiation at the detector is given by the optical preamplifier's ASE, (ii) reasonably narrow-band optical filtering such that $\sigma_{back\text{-}back}$ $\ll \sigma_{sig\text{-}back}$, (iii) beat-noise limited reception of OOK modulation ($P_1 = 2P_{av}$), and (iv) an electrical receiver bandwidth of $B_e = R/2$, we get a good estimate for the optically pre-amplified receiver's performance:

$$Q = \frac{2SGP_{av}}{\sqrt{4S^2 2GP_{av}N_{ASE}B_e}} = \sqrt{\frac{P_{av}}{hfFB_e}} = \sqrt{\frac{2n_{av}}{F}}, \qquad (7.60)$$

where $N_{ASE} = hfGF/2$ is the power spectral density per (spatial and polarization) mode of the ASE generated by the optical preamplifier, Eq. (7.26). As expected, receiver performance is *independent*[15] of the amplifier gain G. Note that, unlike in the case of coherent reception, receiver performance is *independent* of the detector's quantum efficiency, η, and is only degraded by an imperfect optical amplifier noise figure. By assuming an ideal optical preamplifier ($F = 3$ dB), we finally arrive at the approximation for the quantum-limited optically pre-amplified receiver performance:

$$Q = \sqrt{n_{av}}. \qquad (7.61)$$

At BER $= 10^{-9}$ ($Q \approx 6$), we therefore require approximately 36 photons/bit for optically pre-amplified direct detection of OOK.[16] Experimentally achieved sensitivities are listed in Table 7.6; note that the quantum limit has been approached to within some remarkable few tenths of a decibel with this type of receiver. This has been achieved by tailoring the optical filter prior to detection (cf. Figure 7.18) to approximate a *matched optical filter*, i.e., an optical filter whose impulse response equals the data pulse shape; such an optical filter can be shown to yield optimum performance for an optically pre-amplified receiver in the absence of significant electrical post-detection filtering and ISI [79]. However, since the latter two effects can usually not be ignored at multi-Gbit/s data rates, a joint optimization of optical and electrical filters, trading off the effects of ISI, signal distortion, and beat noise accumulation is typically required [3, 103].

---

[15] From a noise perspective, the beat noise limited receiver is solely characterized by the *OSNR*, measured at the photodetector: Using the definition $OSNR = GP_{av}/(2N_{ASE}B_{ref})$, Eq. (7.60) can be written as $Q = \sqrt{OSNR} \sqrt{B_{ref}/B_e}$, where B_{ref} is the optical reference bandwidth for defining the *OSNR*, typically 12.5 GHz. This relation between the Q-factor and the *OSNR* can be used to link the large amount of results reported for fiber-optic communications to problems of free-space laser communications.

[16] Including ASE-ASE beat noise from a single (spatial and polarization) mode, and also taking note of the fact that the Q-factor, based on Gaussian detection statistics, does not accurately represent the non-Gaussian detection noise found in this receiver, the exact value for optically pre-amplified receiver's quantum limit turns out to be 38 photons/bit [63, 64].

7.4.7 The Role of Modulation Formats

Coherent receivers are the only class of receivers that can be directly used to detect any optical modulation format, since they translate the full optical field information (amplitude *and* phase) to the electronic domain. Any receiver based on direct detection can *per se* only be used to detect *intensity modulation (IM)*, either in the form of *OOK* or in the form of *PPM*. Pulse position modulation has significantly lower quantum limits than OOK, the exact values depend on the dominant noise sources and the dimensionality of the PPM alphabet [104, 105], however, at the expense of requiring short, high-intensity pulses that are timed with sub-bit accuracy, as well as the need for photon-counting detectors [40–43]. Therefore, PPM is mainly considered for deep-space applications, which require utmost receiver sensitivities at comparatively low data rates [106, 107]. By combining PPM with coherent detection, record receiver sensitivities have been achieved without photon-counting detectors [108]. If direct detection is used, optical modulation formats that encode information on the phase, frequency, or polarization of the optical signal have to be optically converted to intensity modulation prior to detection (see, e.g., [7, 69] for a comprehensive overview). The most prominent non-intensity modulated format is *DPSK*. Here, the phase of the optical carrier is changed by π for each "1"-bit to be transmitted, and is left unchanged for each transmitted "0"-bit. At the receiver, a DI is used to let each bit interfere with its preceding bit. If the interferometer is tuned for destructive interference in the absence of phase modulation, a π-phase jump between successive bits will produce constructive interference at the DI output, and will thus be decoded as a "1"-bit. Every DI has two output ports (see Figure 7.19): If destructive interference is observed on one output port, the other port will show constructive interference. Therefore, the decoded signals at the two ports are logically inverted copies, and balanced detection can be used in order not to waste signal power, similar to the balanced coherent receiver of Figure 7.20. In contrast to balanced coherent receivers, which exhibit fundamentally the same performance as their single-ended equivalents, the performance of balanced DPSK receivers is *better by about 3 dB* than the performance of a receiver that only makes use of a single DI output port [54]. In the balanced reception case, optically pre-amplified detection of DPSK yields a quantum limit of 20 photons/bit, 3 dB better than for OOK (cf. Table 7.6). At multi-Gbit/s data rates, the absolute record receiver sensitivities have not been achieved by homodyne PSK, but rather by optically pre-amplified balanced DPSK (cf. Table 7.6). Critical aspects of practical DPSK systems are discussed in [54].

In the case of *Pol-SK*, conversion to intensity modulation is done by means of a polarization filter or a polarization beam splitter prior to (balanced or single-ended)

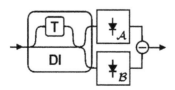

FIGURE 7.19 Delay demodulation for balanced direct detection of DPSK.

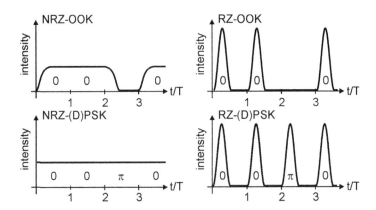

FIGURE 7.20 Optical intensity waveforms of OOK and (D)PSK modulation using NRZ and RZ coding; T is the bit duration, 0 and π indicate the relative optical phase.

detection. If received single-ended, *Pol-SK* suffers a 3-dB penalty compared to OOK, which is recovered upon balanced detection. Nevertheless, *Pol-SK* cannot outperform OOK. Decoding of *frequency shift keying (FSK)* is done by means of narrow-band optical filters used as optical frequency discriminators [69]; FSK yields about the same performance as OOK (cf. Table 7.6). Apart from binary digital formats, multilevel formats are also considered in situations where the available electronics bandwidth limits the communication bit rate; the interested reader is referred to [7] and the references cited therein.

Essentially independent of the choice of a modulation format is the choice of whether to encode information on *continuous* (cw) laser light, or whether to encode information on optical *pulses*. The former class of modulation formats is generally referred to as *NRZ*, while the latter is termed *return-to-zero (RZ)* modulation,[17] *pulsed* modulation, or *impulsive coding*. Figure 7.20 illustrates this definition using OOK and (D)PSK modulation. The primary advantage of *RZ* over *NRZ* formats in free-space laser communications is the higher peak power together with the stronger confinement of pulses to the bit slot, which results in an increased robustness to ISI generated within the transmitter or within the receiver. Typically, a gain of 1–3 dB in direct-detection receiver sensitivity is observed using RZ coding [17] at the expense of a more complex transmitter.

7.4.8 COHERENT VS. OPTICALLY PRE-AMPLIFIED DIRECT DETECTION

A frequent dispute between designers of free-space optical laser links pertains to the question of whether *coherent detection* or *optically pre-amplified direct detection* would be the most appropriate reception technique (see, e.g., [109, 110]), especially in view of their comparable performance (cf. Table 7.6 or [69, 111]).

---

[17] The names NRZ and RZ are derived from OOK modulation: For NRZ, the optical intensity does *not* return to zero between successive "1" bits, while the pulsed nature of RZ *always* lets the intensity return to zero between 2 bits.

Disregarding PPM, which can in principle achieve arbitrarily good sensitivities at the expense of electronics bandwidth, *homodyne PSK* is the optical communication technique with the best theoretical performance. All other coherent techniques (homodyne OOK, heterodyne DPSK) suffer at least a 3-dB degradation, which makes them *equivalent* in quantum-limited performance to optically pre-amplified direct detection. However, it has to be kept in mind that quantum limits only reveal the *potential* of a certain communication technique, which may be far off what can actually be realized in a stable, space-qualified manner. The best receiver sensitivities reported to date at multi-Gigabit/s data rates are all based on *optically pre-amplified* receivers (see Table 7.6). The reason for the experimental superiority of direct detection is undoubtedly the vast effort that has gone into developing and maturing this type of receiver by the fiber-optic communications industry over the last decade, which points at another important aspect: Wherever possible, a free-space laser communication receiver should *leverage from terrestrial fiber-optic communications* in order to provide high-performance and cost-effective solutions.

Since free-space communication terminals ask for utmost reliability criteria, the question of *receiver complexity* is of high importance. Coherent receivers require a signal laser and an LO laser with a linewidth requirement on the order of several hundred kilohertz to a few megahertz at 10 Gbit/s (depending also on the underlying modulation format), and correspondingly less at lower data rates. Such lasers are available today. However, setting up a stable phase-locked loop required for homodyne receivers is a nontrivial engineering task. Heterodyne receivers require higher optoelectronic front-end bandwidths, and intradyne receivers require fast CMOS ASICs to be developed, which is a costly value proposition for the low-volume FSO market. Reusing intradyne ASICs from terrestrial fiber communication systems can be problematic due to their significant power consumption, partially spent on DSP functions that are not needed in FSO links, such as chromatic dispersion compensation or fast polarization tracking. Implementing intradyne DSP on an FPGA platform is even more costly in terms of electrical supply power. An optically pre-amplified receiver based on EDFA technology requires an Erbium-doped fiber, together with a pump laser source that neither has to meet stringent linewidth requirements, nor has to be stabilized in frequency or phase. Stabilization is required only for the narrow-band optical filter used to suppress out-of-band ASE. This filter also has to follow potential Doppler shifts of the input signal, a task that can be done by tuning the LO laser in the case of any coherent reception scheme. Optimum optical filter bandwidths for optically pre-amplified receivers are on the order of the data rate, with only moderate penalties for bandwidths that are as much as 10 times the data rate [103]. Filters with bandwidths of 10 GHz are commercially available today (see Section 7.2.3). In the case of DPSK detection, an optical DI in combination with balanced detection has to be employed. DIs are commercially available in stable integrated optics, fiber-based, and packaged free-space implementations (cf. Section 7.3.9). Nevertheless, the interferometer has to be carefully tuned to the incoming wavelength, or severe penalties have to be expected. (For a frequency offset of 4% of the data rate, we find a 1-dB sensitivity penalty [54].) Another interesting difference between coherent and optically pre-amplified reception that is related to system complexity is optical loss: While optical loss after the optical preamplifier has no impact on receiver performance in beat noise limited operation, optical losses linearly degrade coherent receiver performance.

7.4.9 MULTI-APERTURE AND MULTIMODE RECEIVERS

Increasing the antenna diameter is one obvious way to increase signal power at the photodetector but such an approach becomes quickly prohibitive as large optical telescopes with apertures beyond a few tens of centimeters are not only heavy but also expensive to build. Furthermore, higher-order wavefront distortions caused by the Earth's atmosphere have a direct impact on the coupling efficiency to the active area of a photodiode or core of an optical fiber, and the compensation of such distortions becomes more difficult, the larger the area of the wavefront sample collected by the antenna. Adaptive optics (AO) [112] is a common approach to combat negative effects of atmospheric turbulence, e.g., on astronomical observations, but in terms of free-space laser communications the presence of AO could significantly impact the receiver's cost, weight, and power consumption.

One possible solution to increase the collecting area on the receiver side without resorting to AO is to split up the incoming beam into many sub-beams by an array of small apertures and coherently combine their output. The size of the aperture may be chosen small enough that only the lowest order of distortion, i.e., beam tilt, needs to be compensated for at each sub-aperture to efficiently couple the light into an optical fiber. Originally, these sub-beams were coherently combined in the optical domain by optical PLLs [113, 114] that significantly added to the receiver's complexity. However, owing to recent advances in the field of digital coherent receivers and DSP, the need for controlling the phase on the beams in the optical domain could be made obsolete as the DSP could now take over the task of coherently combining all sub-beams in the electrical domain as illustrated in Figure 7.21 [115].

An alternative way to increase signal power at the receiver in presence of wavefront distortions including simple beam tilt is to use multimode optics in the focal plane of the telescope and thereby enlarge the effective (i.e., light-collecting) area of either the photodetector or the fiber. Coupling first to a fiber instead of a photodetector allows for optical pre-amplification and therefore constitutes a promising means to approach quantum-limited receiver performance. In addition, the strong interest in space division multiplexing in recent years, mostly with regard to optical fiber transmission systems, has brought forth new multimode amplification techniques and all-fiber-based mode converters such as photonic lanterns [116, 117] that can efficiently couple the light from multimode fibers to an array of single-mode fibers. By using a multimode pre-amplified receiver in combination with a multimode photodetector, a 6-dB improvement in the power budget of free-space optical links has been demonstrated [118]. On the other hand, when optical pre-amplification with single-mode optical amplifiers is preferred, the light in the multimode fiber may be efficiently split up by a photonic lantern into an array of single-mode fibers before amplification, as depicted in Figure 7.22. In the latter case, the SMF array would then be connected to an array of coherent receivers for coherently combining all signals [119].

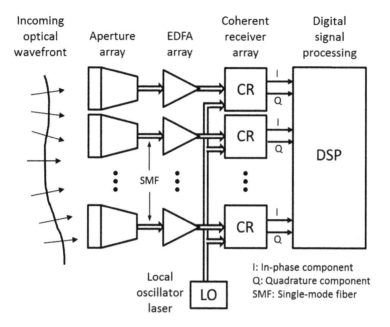

FIGURE 7.21 Digital coherent multi-aperture receiver [115].

7.5 RECEIVER ARCHITECTURES IN (OR CLOSE TO) OPERATION

After having discussed many different receiver architectures, it is also interesting to see which concepts have been, or are being implemented in various near-Earth laser communication systems.

With the launch of the European ARTEMIS satellite in 2001, the European "Semiconductor Laser Intersatellite Experiment" (SILEX) employed both NRZ-OOK and 2-PPM modulation, using direct detection with avalanche photo diodes [120]. Links were primarily established to the French SPOT-4 satellite (on a daily basis), but there were also experimental links demonstrated at least with the ESA optical ground station, with the Japanese OICETS satellite, and with an aircraft.

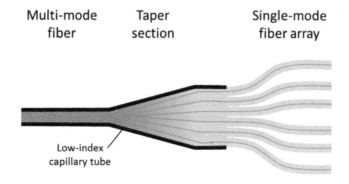

FIGURE 7.22 Schematics of a photonic lantern.

Also in 2001, a classified U.S. satellite called "Geosynchronous Lightweight Technology Experiment" (GeoLITE) was launched, carrying a radiometer and a laser terminal as payloads. The primary application was geosynchronous orbit-to-ground laser communication [121]. Implementation details seem to remain classified.

At the time of this writing, the European Data Relay System (EDRS) had two laser communication terminals in geo-stationary orbit, and four in low Earth orbit—and had started routine operation [122]. It employs binary phase-shift keying (BPSK) transmission at 1.8 Gbit/s, combined with homodyne reception. The receiver is based on a precursor development [123], and comprises an optical 90° hybrid, two balanced photo receivers, and a Costas-type optical phase-locked loop, controlling a Nd:YAG solid state laser as LO. Satellite-to-ground communications, using AO to compensate for the effects of atmospheric turbulence, has been demonstrated as well [124].

NASA's Laser Communications Relay Demonstration (LCRD) mission was still in development and planned for launch in 2019 [125]. The LCRD system relies on DPSK with a RZ waveform [126]. Furthermore, thanks to burst-mode transmission, average data rates from 72 Mbit/s to 2.88 Gbit/s can be achieved at almost identical receiver sensitivity (in terms of average photons per bit). The receiver comprises several preamplifier stages and optical filters (the whole chain implementing a matched optical filter for the RZ waveform), a DLI, and a balanced photo receiver.

Numerous optical LEO-direct-to-Earth communication experiments were in operation or development as well—to our knowledge, all using receiver front ends comprising an APD of suitable diameter (again, to overcome the effect of atmospheric turbulence), and a TIA. Due to the relatively short link distances, the quest for utmost receiver sensitivity is less pronounced in this field.

7.6 SUMMARY AND OUTLOOK

In the preceding sections, we have discussed optical receivers with a view on their application in free-space systems. We described the process of photodetection and reviewed the properties of the main building blocks (photodiodes, optical filters, optical preamplifiers, etc.). We also accounted for sources of background radiation, explained the various receiver concepts (direct detection with and without optical pre-amplification, coherent detection), and developed expressions for the receivers' sensitivities.

Today, reception of high-data rate optical signals is well understood, concerning both the physical limits as well as how to design devices with specified properties. Many aspects of optical receiver technology have reached high maturity and sufficient reliability. This is mostly a result of the huge efforts expended on the development of fiber optic transmission systems. Despite the existence of many commonalities among the requirements of receivers for fiber systems, on the one hand, and those for space systems, on the other, a number of differences exist which may be the cause for different implementations—also for differing status of maturity. In this respect one has to point out that the free-space receiver

- does not have to cope with pulse envelopes distorted by chromatic dispersion or nonlinearities along the channel, thus the received signal is just a replica of the transmitted one, with known shape,

- will usually strive for utmost sensitivity because no in-line amplification is available in space and because transmit power is extremely valuable,
- will have to face wavefront tilt or distortion, e.g., caused by antenna vibrations, or atmospheric turbulence, not completely compensated for by an automatic tracking system—potentially resulting in input power fluctuations, and
- will have to cope with Doppler shift of the received signal.

These effects ask for an additional effort in the receiver development. While a coherent receiver in combination with a homodyne PSK modulation format is known to have the highest receiver sensitivity, it is likely that many advances in terms of intradyne detection, DSP, and space division multiplexing owing to the rapid development of fiber optic transmission systems in recent years will also find practical applications in future space systems. Furthermore, signal routing inside the receiver terminal through optical fibers allows for a flexible and low-weight receiver design, enables optical pre-amplification, and thus ensures close to quantum-limited detection. As such, multimode fiber optics in combination with digital coherent combining seems to be a very promising means to increase receiver sensitivity for Earth-to-Space communications, while keeping down both cost and complexity.

Last, but not least, the application in space asks for space qualification of all devices and functionalities. With fiber optics, somewhat similar requirements exist for undersea communication links. However, it is still space which poses the biggest demands in this respect, and despite the excellent status of development of optical receivers for terrestrial applications, time and costs of space qualifying a specific receiver must not be underestimated.

REFERENCES

1. J. Renaudier et al., "First 100-nm continuous-band WDM transmission system with 115Tb/s transport over 100 km using novel ultra-wideband semiconductor optical amplifiers," Proc. ECOC, Th.PDP.A.3 (2017).
2. B. Huang et al., "Turbulence-resistant free-space optical communication using few-mode preamplified receivers," Proc. ECOC, Tu.2.E (2017).
3. M. Pfennigbauer, M. M. Strasser, M. Pauer, and P. J. Winzer, "Dependence of optically preamplified receiver sensitivity on optical and electrical filter bandwidths—measurement and simulation," Photon. Technol. Lett. 14, 831–833 (2002).
4. G. M. Williams and E. J. Friebele, "Space radiation effects on erbium-doped fiber devices: sources, amplifiers, and passive measurements," IEEE Trans. Nucl. Sci. 45, 1531–1536 (1998).
5. T. S. Rose, D. Gunn, G. C. Valley, "Gamma and proton radiation effects in Erbium-doped fiber amplifiers: active and passive measurements," J. Lightwave Technol. 19, 1918–1923 (2001).
6. E. Haddad et al., "Comparison of Gamma Radiation Effect on Erbium Doped Fiber Amplifiers," Proceedings of International Conference on Space Optics, Biaritz, October 2016.
7. P. J. Winzer and R.-J. Essiambre, "Advanced optical modulation formats," Proc. IEEE 94(5), 952–985 (2006).
8. R. H. Kingston, Detection of optical and infrared radiation, Springer-Verlag, Berlin, 1978.

9. H. Kogelnik, R. M. Jopson, L. E. Nelson, "Polarization-Mode Dispersion," in Optical Fiber Telecommunications IV-B, I. Kaminow, T. Li., eds., Academic Press, Waltham, MA, 2002.

10. R. G. Frehlich and M. J. Kavaya, "Coherent laser radar performance for general atmospheric refractive turbulence," Appl. Opt. 30, 5325–5352 (1991).

11. L. Mandel and E. Wolf, Optical Coherence and Quantum Optics, Cambridge University Press, Cambridge, 1995.

12. A. Papoulis, Probability, Random Variables, and Stochastic Processes, McGraw-Hill, 1991.

13. S. D. Personick, "Receiver design for digital fiber optic communication systems, I," Bell Syst. Tech. J., 52: 843–874, 1973. B. L. Kasper, O. Mizuhara, Y.-K. Chen, "High bit-rate receivers, transmitters, and electronics," in Optical fiber telecommunications IVA, I. Kaminow and T. Li, eds., Academic Press, 2002; T. V. Muoi, Receiver design for high-speed optical-fiber systems, J. Lightwave Technol. 2, 243–267 (1984); S. B. Alexander, "Optical communication receiver design, SPIE tutorial texts in Optical Engineering," TT22, 1997. E. Saeckinger, Broadband Circuits for Optical Fiber Communication, John Wiley & Sons, Hoboken, NJ, 2004.

14. B. E. A. Saleh and M. C. Teich, Fundametals of Photonics, John Wiley, Hoboken, NJ, 1991.

15. B. E. A. Saleh, Photoelectron Statistics, Springer-Verlag, Berlin, Heidelberg, New York, 1978.

16. J. W. Goodman, Statistical Optics, John Wiley & Sons, Hoboken, NJ, 1985.

17. P. J. Winzer and A. Kalmár, "Sensitivity enhancement of optical receivers by impulsive Coding," J. Lightwave Technol. 17(2), 171–177 (1999).

18. N. A. Olsson, "Lightwave Systems with Optical Amplifiers," J. Lightwave Technol. 7, 1071–1082 (1989).

19. P. J. Winzer, "Receiver noise modeling in the presence of optical amplification," Proc. OAA'01, OTuE16 (2001).

20. R.-J. Essianbre et al., "Capacity limits of optical fiber networks," J. Lightwave Technol. 28(4), 662–701 (2010).

21. W. R. Leeb, "Degradation of signal to noise ratio in optical free space data links due to background illumination," Appl. Opt., 28, 3443–3449 (1989).

22. A. E. Siegman, "The antenna properties of optical heterodyne receivers," Proc. IEEE 54, 1350–1356 (1966).

23. R. C. Ramsey, "Spectral Irradiance from Stars and Planets, above the atmosphere, from 0.1 to 100.0 Microns," Appl. Opt. 1, 465–471 (1962).

24. W. K. Pratt, Laser Communication Systems, Wiley, New York, 1969, Chap. 6.

25. G. Thuillier, et. al., "The solar spectral irradiance from 200 to 2400 nm as measured by the SOLSPEC spectrometer from the ATLAS and EURECA missions," Sol. Phys. 214, 1–22 (2003).

26. P. J. Winzer, A. Kalmar, W. R. Leeb, "The role of amplified spontaneous emission in optical free-space communication links with optical amplification—impact on isolation and data transmission; utilization for pointing, acquisition, and tracking," Proc. SPIE 3615, 104–114 (1999).

27. J. Bromage, P. J. Winzer, R.-J. Essiambre, "Multiple-path interference and its impact on system design," in Raman Amplifiers and Oscillators in Telecommunications, M. N. Islam ed., Springer-Verlag, Berlin, 2003.

28. S. B. Alexander, Optical Communication Receiver Design, SPIE Optical Engineering Press, 1997.

29. K. Kato, "Ultrawide-Band/High-Frequency photodetectors," IEEE Trans. Microw. Theory Tech. 47(7), 1265–1281 (1999).

30. H.G. Bach et al., "InP-Based waveguide-integrated photodetector with 100-GHz bandwidth," IEEE J. Sel. Top. Quantum Electron 10(4), 668–672 (2004).

31. A Beling et al., "Miniaturized waveguide-Integrated p-i-n photodetector with 120-GHz bandwidth and high responsivity," Phot. Technol. Lett. 17(10), 2152–2154 (2005).

32. N. Shimitsu et al., "InP–InGaAs Uni-Traveling-Carrier photodiode with improved 3-dB Bandwidth of over 150 GHz," Phot. Technol. Lett. 10(3), 412–414 (1998).

33. A. Umbach et al., "Integrated limiting balanced photoreceiver for 43 GbiUs DPSK transmission," Proc. ECOC, We3.6.4 (2005).

34. Y. Painchaud et al., "Performance of balanced detection in a coherent receiver," Opt. Express 17(5), 3659–3672 (2009).

35. H.G. Bach et al., "Monolithic 90°Hybrid with Balanced PIN Photodiodes for 100 Gbit/s PM-QPSK Receiver Applications," Proc. OFC, OMK5 (2009)

36. A. Beling et al., "High-speed balanced photodetector module with 20 dB broadband common-mode rejection ratio," Proc. OFC'03, WF4 (2003).

37. G.M. Williams et al., "Multi-gain-stage InGaAs avalanche photodiode with enhanced gain and reduced excess noise," IEEE J. Electron. Dev. Soc. 1(2), 54–65 (2013).

38. J. Rothman et al., "HgCdTe APDs for free space optical communications," Proc SPIE 9647, 96470N (2015).

39. J. Campbell, "Recent advances in avalanche photodiodes," J. Lightwave Technol. 34(2), 278–285 (2016).

40. K. A. McIntosh, et al., "Arrays of III-V semiconductor Geiger-mode avalanche photo-diodes," LEOS Annual Meeting, WZ1, 2003.

41. B. S. Robinson, D. O. Caplan, M. L. Stevens, R. J. Barron, E. A. Dauler, S. A. Hamilton, "1.5-photons/bit photon-counting optical communications using Geiger-mode avalanche photodiodes," IEE LEOS Newsletter 19(5), October 2005.

42. M. E. Grein et al., "Demonstration of a 1550-nm photon-counting receiver with <0.5 Detected photon-per-bit sensitivity at 187.5Mb/s," Proc. CLEO CWN5 (2008).

43. B. S. Robinson et al., "781 Mbit/s photon-counting optical communications using a superconducting nanowire detector," Opt. Lett. 31(4), 444–446 (2006).

44. C. R. Doerr, Optical Fiber Telecommunications IVA, I. Kaminow and T. Li., eds., Academic Press, Waltham MA, 2002.

45. A. Yariv, Quantum Electronics, John Wiley & Sons, New York, 1989.

46. E. Desurvire, Erbium-doped Fiber Amplifiers, John Wiley & Sons, New York, 1994.

47. Z. S. Eznaveh, N. K. Fontaine, H. Chen, J. E. Antonio Lopez, J. C. Alvarado Zacarias, B. Huang, A. Amezcua Correa, C. Gonnet, P. Sillard, G. Li, A. Schülzgen, R. Ryf, and R. Amezcua Correa, "Ultra-low DMG multimode EDFA," in Optical Fiber Communication Conference (OFC), 2017, paper Th4A.4.

48. D. R. Zimmerman and L. H. Spiekman, "Amplifiers for the masses: EDFA, EDWA, and SOA amplets for metro and access applications," J. Lightwave Technol. 22(1), 63–70 (2004).

49. K. Kikuchi, "Fundamentals of coherent optical fiber communications," J. Lightwave Technol. 34, 157–179 (2016).

50. P.J. Winzer, "High-spectral-efficiency optical modulation formats," J. Lightwave Technol. 20(26), 3824–3835 (2012).

51. Y. K. Lize et al., "Tolerances and receiver sensitivity penalties of multibit delay dif-ferential-phase shift-keying demodulation," Photon. Technol. Lett. 19(23), 1874–1876 (2007).

52. D. O. Caplan et al., "Multi-rate DPSK optical transceivers for free-space applica-tions," in Free-Space Laser Communication and Atmospheric Propagation XXVI, H. Hemmati and D. M. Boroson, eds., Proc. SPIE Vol. 8971, 89710K (2014), doi: 10.1117/12.2057570.

53. M. Pauer, P. J. Winzer, and W. R. Leeb, "Booster EDFAs in RZ coded links: are they average-power limited?" Proceedings of SPIE 4272, Free-Space Laser Communication Technologies XIII, San José, CA, 118–127, 2001.

54. A. H. Gnauck and P. J. Winzer, "Optical phase-shift-keyed transmission," J. Lightwave Technol. 23(1), 115–130 (2005).

55. X. Liu et al., "Athermal optical demodulator for OC-768 DPSK and RZ-DPSK signals," Phot. Technol. Lett. 17(12), 2610–2612 (2005).

56. F. Seguin and F. Gonthier, "Tuneable all-fiber delay-line interferometer for DPSK demodulation," Proc. OFC, OFL5 (2005).

57. J. Gamet and G. Pandraud, "C- and L-Band planar delay interferometer for DPSK decoders," Photon. Technol. Lett. 17(6), 1217–1219 (2005).

58. S. Chandrasekhar and X. Liu, "Enabling components for future high-speed coherent communication systems," Proc. OFC, OMU5 (2011).

59. S. Savory et al., "Digital equalisation of 40Gbit/s per wavelength transmission over 2480 km of standard fibre without optical dispersion compensation," Proc. ECOC (2006), doi: 10.1109/ECOC.2006.4800978.

60. T. Inoue and K. Nara, "Ultrasmall PBS-integrated coherent mixer using 1.8%-delta silica-based planar lightwave circuit," Proc. ECOC, Mo.2.F.4 (2010).

61. R. Kunkel et al., "First monolithic InP-based 90°-hybrid OEIC comprising balanced detectors for 100GE coherent frontends," Proc. IPRM (2009), doi: 10.1109/ICIPRM.2009.5012469.

62. P. Dong et al., "Monolithic silicon photonic integrated circuits for compact 100+Gb/s coherent optical receivers and transmitters," IEEE J. Sel. Top. Quantum Electron. 20(4), 6100108 (2014).

63. C. Xie et al., "Colorless coherent receiver using 3x3 coupler hybrids and single-ended detection," Opt. Express, 20(2), 1164–1171 (2012).

64. A. Leven and L. Schmalen, "Status and recent advances on forward error correction technologies for Lightwave systems," J. Lightwave Technol. 32, 2735–2750 (2014).

65. R. D. Gitlin, J. F. Hayes, and S. B. Weinstein, Data Communications Principles, Plenum Press, New York, 1992.

66. E. A. Lee and D. G. Messerschmitt, Digital Communication, 2nd ed., Kluwer Academic Press, 1994.

67. P. J. Winzer, "Optical transmitters, receivers, and noise," Wiley Encyclopedia of Telecommunications, J. G. Proakis, ed., John Wiley & Sons, New York, 1824–1840, 2002. Available at: http://www.mrw.interscience.wiley.com/eot/eot_sample_fs.html.

68. T. Mizuochi, K. Kubo, H. Yoshida, H. Fujita, H. Tagami, M. Akita, and K. Motoshima, "Next generation FEC for optical transmission systems," in Proc. OFC, 2003, Paper ThN1.

69. G. Jacobsen, Noise in Digital Optical Transmission Systems. Artech House, Norwood, MA, 1994.

70. G. Einarsson, Principles of Lightwave Communications, John Wiley & Sons, New York, 1996.

71. G. Planche et al., "SILEX final ground testing and in-flight performance assessment," Proc. SPIE 3615, 64–77 (1999).

72. K. Sato et al., "Record highest sensitivity of -28 dBm at 10 Gb/s achieved by newly developed extremely compact superlattice-APD module with TIA-IC," Proc. OFC'02, 2002, paper FB11.

73. S. Betti, G. De Marchis, E. Iannone, Coherent optical communication systems, Wiley-Interscience, 1995; S. Ryu, Coherent Lightwave Communication Systems, Artech House, Norwood MA, 1995; R. A. Linke et al., "High-capacity coherent lightwave systems," J. Lightwave Technol. 6(11), 1750–1769 (1988); Y. Yamamoto et al., "Coherent optical fiber transmission systems," J. Quantum Electron. QE-17(6), 919–934 (1981); L. Kazovsky, S. Benedetto, and A. Willner, Optical Fiber Communication Systems, Artech House, Norwood, MA, 1996.

74. L. G. Kazovsky, P. Meissner, and E. Patzak, "ASK multiport optical homodyne receivers," J. Lightwave Technol. 5, 770–790 (1987).

75. F. Derr, "Coherent optical QPSK intradyne system: concept and digital receiver realization," J. Lightwave Technol. 10(9), 1290–1296 (1992).

76. R. Noe, "Phase noise-tolerant synchronous QPSK/BPSK baseband-type intradyne receiver concept with feedforward carrier recovery," J. Lightwave Technol. 23(2), 802–808 (2005).

77. S. Tsukamoto et al., "Coherent demodulation of optical multilevel phase-shift-keying signals using homodyne detection and digital signal processing," Photon. Technol. Lett. 18(10), 1131–1133 (2006).

78. P. J. Winzer and W. R. Leeb, "Fiber coupling efficiency for stochastic light and its applications to LIDAR," Opt. Lett. 23(13), 986–988 (1998).

79. L. Kazovsky, S. Benedetto, and A. Willner, Optical Fiber Communication Systems, Artech House, Norwood, MA, 1996.

80. B. Wandernoth, "20 photon/bit 565 Mbit/s PSK homodyne receiver using synchronisation bits," Electron. Lett. 28, 387–388 (1992).

81. J. M. Kahn et al., "4-Gb/s PSK homodyne transmission system using phase-locked semiconductor lasers," Photon. Technol. Lett. 2(4), 285–287 (1990).

82. S. Norimatsu et al., "10-Gbit/s optical BPSK homodyne detection experiment with solitary DFB laser diodes," Electron. Lett. 31(2), 125–127 (1995).

83. B. Clesca et al., "Highly sensitive 565 Mb/s DPSK heterodyne transmission experiment using direct current modulation of a DFB laser transmitter," Photon. Technol. Lett. 3(9), 838 (1991).

84. A. H. Gnauck et al., "4-Gb/s heterodyne transmission experiments using ASK, FSK, and DPSK modulation," Photon. Technol. Lett. 2(12), 908–910 (1990).

85. C. Wree et al., "Measured noise performance for heterodyne detection of 10-Gb/s OOK and DPSK," Photon. Technol. Lett. 19(1), 15–17 (2007).

86. S. B. Alexander et al., "1 Gbit/s coherent optical communication system using a 1 W optical power amplifier," Electron. Lett. 29(1), 114–115 (1993).

87. T. Imai et al., "Over 300km CPFSK transmission experiment using 67 photon/bit sensitivity receiver at 2.5Gbit/s," El. Lett. 26, 357–358 (1990).

88. N. Ohkawa et al, "A highly sensitive balanced receiver for 2.5 Gb/s heterodyne detection systems," Photon. Technol. Lett. 3(4), 375–377 (1991).

89. W. A. Atia and R. S. Bondurant, "Demonstration of return-to-zero signaling in both OOK and DPSK formats to improve receiver sensitivity in an optically preamplified receiver," in Proceedings of LEOS 12th Annual Meeting, vol. 1, San Francisco, CA, 1999, pp. 226227.

90. J. H. Sinsky et al., "RZ-DPSK transmission using a 42.7-Gb/s integrated balanced optical front end with record sensitivity," Photon. Technol. Lett. 22(1), 180–182 (2004).

91. D. O. Caplan and W. A. Atia, "A quantum-limited optically-matched communication link," Proc. OFC'01, 2001, paper MM2.

92. M. M. Strasser et al., "Experimental verification of optimum filter bandwidths in direct-detection (N)RZ receivers limited by optical noise," Proc. LEOS'01, 2001, paper WK5.

93. P. J. Winzer et al., "40-Gb/s return-to-zero alternate-mark-inversion (RZ-AMI) transmission over 2000 km," Photon. Technol. Lett. 15(5), 766–768 (2003).

94. P. Moro and D. Candiani, "565 Mb/s optical transmission system for repeaterless sections up to 200 km," Proc. ICC (1991), doi: 10.1109/ICC.1991.162546.

95. A. Robinson, "Experiences with forward error correction for optically amplified submarine systems," IEE Colloquium on International Transmission Systems (1994).

96. J. L. Pamart et al., "Forward error correction in a 5 Gbit/s 6400 km EDFA based system," Electron. Lett. 30(4), 342–343 (1994).

97. N. Alic, G. C. Papen, and Y. Fainman, "Performance of maximum likelihood sequence estimation with different modulation formats," IEEE/LEOS Workshop on Advanced Modulation Formats (2004).

98. J.P. Elbers et al., "Measurement of the dispersion tolerance of optical duobinary with an MLSE-receiver at 10.7 Gb/s," Proc. ECOC, OThJ4 (2005).

99. H. Sun, T. T. Wu, and K. Roberts, "Real-time measurements of a 40 Gb/s coherent system," Opt. Express 16(2), 873–879 (2008).

100. P. J. Winzer, "Beyond 100G Ethernet," IEEE Commun. Mag. 48(7), 26–30 (2010).

101. S. J. Savory, "Digital coherent optical receivers: algorithms and subsystems," IEEE J. Sel. Top. Quantum Electron. 16(5), 1164–1179 (2010).

102. J. Surof, J. Poliak, and R. M. Calvo, "Demonstration of intradyne BPSK optical free-space transmission in representative atmospheric turbulence conditions for geostationary uplink channel," Opt. Lett. 42, 2173–2176 (2017).

103. P. J. Winzer, M. Pfennigbauer, M. M. Strasser, and W. R. Leeb, "Optimum filter bandwidths for optically preamplified RZ and NRZ receivers," J. Lightwave Technol. 19(9), 1263–1273 (2001).

104. A. J. Phillips et al., "Novel laser inter-satellite communication system employing pre-amplified PPM receivers," IEE Proc. Commun. 142(1), 15–20 (1995).

105. A. Kalmar, P. J. Winzer, K. H. Kudielka, and W. R. Leeb, "Multifunctional optical terminals for microsatellite clusters—design tradeoffs," AEÜ Int. J. Electron. Commun. 56(4, Special Issue on Optical Communications), 279–288 (2002).

106. D. O. Caplan et al, "Demonstration of 2.5-Gslot/s optically-preamplified M-PPM with 4 photons/bit receiver sensitivity," Proc. OFC'05, PDP32 (2005).

107. N. W. Spellmeyer et al., "Design for a 5-Watt PPM transmitter for the Mars laser communications demonstration," LEOS Summer Topical Meetings, San Diego, CA, 51–52 (2005).

108. X. Liu et al., "Demonstration of Record Sensitivities in Optically Preamplified Receivers by Combining PDM-QPSK and M-Ary Pulse-Position Modulation," J. Lightwave Technol. 30(4), 406–413 (2012).

109. G. S. Mecherle, "Comparison of coherent detection and optically preamplified receivers," Proc. SPIE 3266, 111–119 (1998).

110. S. Yamakawa et al., Trade-off between IM/DD and coherent system in high-data rate optical inter-orbit links, Proc. SPIE 3615, 80–89 (1999).

111. O. K. Tonguz and R. E. Wagner, "Equivalence between preamplifed direct detection and heterodyne receivers," IEEE Photon. Technol. Lett. 3, 835–837 (1991).

112. R. K. Tyson, Principles of Adaptive Optics, 4th ed., CRC Press, San Diego CA, 2016.

113. J. Xu, A. Delaval, R. G. Sellar, A. Al-Habash, P. Reardon, R. L. Phillips, L. C. Andrews, "Experimental comparison of coherent array detection and conventional coherent detection for laser radar and communications," Proc. SPIE, Free-Space Laser Communication Technologies XI, 3615, 54–63 (1999).

114. K. Kudielka, W. Neubert, A. Scholtz, W. Leeb, "Adaptive optical multiaperture receive antenna for coherent intersatellite communications," Proc. SPIE 2210, 61–70 (1994).

115. D. J. Geisler, T. M. Yarnall, M. L. Stevens, C. M. Schieler, B. S. Robinson, S. A. Hamilton, "Multi-aperture digital coherent combining for free-space optical communication receivers," Opt. Express 24(12), 12661–12671 (2016).

116. D. Noordegraaf, P. M. W. Skovgaard, M. D. Nielsen, J. Bland-Hawthorn, "Efficient multi-mode to single-mode coupling in a photonic lantern," Opt. Express 17(3), 1988–1994 (2009).

117. S. G. Leon-Saval, N. K. Fontaine, J. R. Salazar-Gil, B. Ercan, R. Ryf, J. Bland-Hawthorn, "Mode-selective photonic lanterns for space division multiplexing," Opt. Express 22(1), 1–9 (2014).

118. B. Huang, C. Carboni, H. Liu, J. C. Alvarada-Zacarias, F. Peng, Y. Lee, H. Chen, N. K. Fontaine, R. Ryf, J. E. Antonio-Lopez, R. Amezcua-Correa, G. Li, "Turbulence-resistant free-space optical communication using few-mode preamplified receivers," in European Conference of Optical Communications (ECOC), 2017, paper Tu.2.E.4.

119. I. Ozdur, P. Toliver, and T.K. Woodward, "Photonic-lantern-based coherent LIDAR system," Opt. Express 23(4), 5312–5316 (2015).
120. T. T. Nielsen and G. Oppenhaeuser, "In-orbit test result of an operational intersatellite link between ARTEMIS and SPOT4, SILEX," Proc. SPIE 4635, 1–15 (2002).
121. Intellipedia information on GeoLITE, declassified 8 March 2016, NRO FOIA Case F-2016-00021.
122. F. Heine et al., "Progressing towards an operational optical data relay service," Proc. SPIE 10096, 100960X (2017).
123. R. Lange and B. Smutny, "Optical inter-satellite links based on homodyne BPSK modulation: heritage, status and outlook," Proc. SPIE 5712, 1–12 (2005).
124. E. Fischer et al., "Use of adaptive optics in ground stations for high data rate satellite-to-ground links," International Conference on Space Optics, Biarritz, France (2016).
125. E. Luzhansky et al., "Overview and status of the laser communication relay demonstration," Proc. SPIE 9739, 97390C (2016).
126. D. O. Caplan et al., "Multi-rate DPSK optical transceivers for free-space applications," Proc. SPIE 8971, 89710K (2014).

8 Atmospheric Channel

Sabino Piazzolla

CONTENTS

8.1 INTRODUCTION

When an optical beam propagates in the atmosphere, it can experience attenuation loss of its irradiance and random degradation of the beam quality itself. The first effect is caused by absorption and scattering, operated by molecular constituents and particulates present in the atmosphere. The second effect is related to clear air turbulence (or optical turbulence) that induces (among other things) phase fluctuations of the laser signal, focusing or defocusing effects, local deviations in the direction of electromagnetic propagation, and signal intensity fluctuations at the receiver (also known as signal scintillation). The purpose of this chapter, therefore, is to introduce a theoretical basis of the beam propagation characteristics concerning atmospheric absorption loss and clear air turbulence. Particularly, we first introduce a description of the cause of clear air turbulence, a description of scintillation of the optical signal, deterioration of the receiver performance resulting from phase degradation of the signal (due to turbulence), and other beam effects. Finally, we describe the mechanisms that induce atmospheric loss and sky background noise radiance.

8.1.1 STATISTICAL DESCRIPTION OF ATMOSPHERIC TURBULENCE

Clear air turbulence phenomena affect the propagation of an optical beam because the refractive index randomly varies in space and time. Mainly, random variation of the refractive index of air depends on the air mixing due to temperature variation in the atmosphere. In fact, sunlight incident upon the earth's surface causes heating of the earth's surface and the air in its proximity. This sheet of warmed air becomes less dense and rises to combine with the cooler air of the above layers, which causes air temperature to vary randomly (from point to point). Because the atmospheric refractive index depends on air temperature and density, it varies in a random fashion in space and time, and this variation is the origin of clear air turbulence.

To describe clear air turbulence, one should consider the atmosphere as a fluid that is in continuous flow. A fluid flow at small velocity is first characterized by a smooth laminar phase. In fluid dynamics, a figure of merit of the fluid flow is the Reynolds number (Re), which is the ratio between fluid inertial and viscous forces [1]:

$$Re = V_c l / v_k, \tag{8.1}$$

where, V_c and l are the characteristic velocity scale and length given in m/s and m, respectively, and v_k is the kinematics viscosity given in m²/s.

The laminar flow of the fluid is stable only when the Reynolds number does not exceed a certain critical value ($Re \sim 2300$). When the Reynolds number exceeds the critical value (e.g., by increasing flow velocity), motion becomes unstable and the flow changes from laminar to a more chaotic, turbulent state. To describe this turbulent state, Kolmogorov developed a theory based on the hypothesis that kinetic energy associated with larger eddies is redistributed without loss to eddies of decreasing size, until they are finally dissipated by viscosity [1]. The structure of the turbulence according to this theory is depicted in Figure 8.1.

The scale of the turbulence can be divided into three ranges: input range, dissipation range, and inertial subrange. The input range, where the energy is injected in the turbulence, is characterized by eddies of size greater than the outer scale of turbulence (L_0). Since the turbulence in this range greatly depends on local conditions, there is no mathematical approach capable to describe it. The dissipation range is characterized by eddies of size smaller than the inner scale of turbulence ($l_0 \ll L_0$). In this case, turbulent eddies disappear, the remaining energy is dissipated as heat, and energy loss from eddies (due to viscosity) dominates. The inertial subrange is at the core to the Kolmogorov's theory, here the turbulence energy is transferred from eddies size L_0 down to eddies of size l_0, Figure 8.1 [2].

Mathematical description of the turbulent regime is quite complex and is approached with the help of statistics. In this regard, the study by Kolmogorov [1] showed that in a turbulent regime for a statistically homogenous medium, the longitudinal structure function of wind velocity between two observation points, $\mathbf{R}_1$ and $\mathbf{R}_2$, is given in the inertial subrange by:

$$D_v(R) \equiv D_v(\mathbf{R}_1, \mathbf{R}_2) = \left\langle [v(\mathbf{R}_1) - v(\mathbf{R}_2)]^2 \right\rangle$$
$$= \left\langle [v(\mathbf{R}_1) - v(\mathbf{R}_1 + \mathbf{R})]^2 \right\rangle = C_v^2 R^{2/3}, l_0 \ll R \ll L_0, \tag{8.2}$$

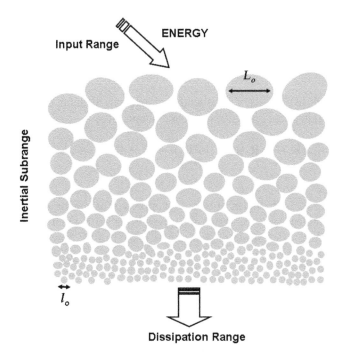

FIGURE 8.1 Depiction of turbulence regime. The energy is exchanged from larger eddies to smaller ones. In the inertial subrange, the largest eddies have size L_0 and the smallest have size l_0.

where

> $\langle \ \rangle$ is the mean operator.
> C_v^2 is defined as the structure constant of the wind velocity, which indicates the strength level of the turbulence.
> Exponent 2/3 of the distance R expresses a power law that is true in the inertial subrange.

Kolmogorov also derived the three-dimensional power spectrum of wind velocity that is related to C_v^2 and the spatial frequency κ as:

$$\Phi_v(\kappa) = 0.033 C_v^2 \kappa^{-11/3}, \quad 1/L_0 \ll \kappa \ll 1/l_0, \tag{8.3}$$

noting that the above spectrum is valid in the inertial subrange.

To extend the Kolmogorov theory to describe the variation of the refractive index in clear air turbulence, one must first describe the refractive index, n, in terms of observable atmospheric variables. In the optical domain, the refractive index can be described by the following relationship [3]:

$$n - 1 \approx 79 \times 10^{-6} \frac{p}{T}, \tag{8.4}$$

where

> p is the atmospheric pressure in mbar.
> T is the temperature in K.

Again, under the hypothesis of statistically homogenous medium, one can derive the structure function of the refractive index and obtain [3]:

$$D_n\left(n(\mathbf{R})\right)=\left\langle[n(\mathbf{R}_1)-n(\mathbf{R}_2)]^2\right\rangle=\left\langle[n(\mathbf{R}_1)-n(\mathbf{R}_1+\mathbf{R})]^2\right\rangle=C_n^2 R^{2/3};\quad l_0<R<L_0,$$

(8.5)

where the term C_n^2 is the index of refraction structure parameter that indicates the strength of turbulence. At the same time, given Equation 8.4, one can demonstrate that the refractive index structure constant is related to the temperature structure constant, C_T^2, as:

$$C_n^2=\left(\frac{78p}{T^2}\times10^{-6}\right)^2 C_T^2.$$

(8.6)

Equation 8.6 indicates that a way to determine the refraction index structure parameter is to measure temperature, pressure, and the temperature spatial fluctuations (to derive C_T^2) and finally through Equation 8.6, we obtain C_n^2 [4].

8.1.2 MODELING OF THE REFRACTIVE INDEX STRUCTURE PARAMETER

As previously mentioned, the refractive index structure parameter is the most significant parameter that determines the turbulence strength. Clearly, C_n^2 depends (among different variables) on the geographical location, altitude, and time of day. Different locations can have different characteristics of temperature distribution (e.g., tropical location vs. temperate location) that are reflected on the values assumed by C_n^2. Close to the ground, there is the largest gradient of temperature associated with the largest values of atmospheric pressure (and air density); therefore, one should expect larger values of C_n^2 at the sea level. As the altitude increases, the temperature gradient decreases and so the air density (and atmospheric pressure) with the result of smaller values of C_n^2.

Considering the temperature dynamics during the day, one should expect turbulence to be stronger around noon. Conversely, at sunset and dawn, due to a form of thermal equilibrium along the atmosphere vertical profile, one should expect C_n^2 to have lower values.

In applications that envision a horizontal path even over a reasonably long distance, one can assume C_n^2 to be practically constant. Typical value of C_n^2 for a weak turbulence at the ground level can be as little as 10^{-17} m$^{-3/2}$, while for a strong turbulence it can be up to 10^{-13} m$^{-3/2}$ or larger. If we instead consider application from ground to space, or more generally, along a slant path, C_n^2 must vary due to the different temperature gradient, air pressure, and density along the altitude. Modeling and determining the profile of C_n^2 with altitude is not a simple task. In fact, it is not easy to capture properly the variations of C_n^2 profile measured usually with thermosondes. Generally, experimental data of C_n^2 are not readily available. However, a number of parametric models have been formulated to describe the $C_n^2(h)$ profile and among those, one of the more used models is the Hufnagel-Valley [5] given by:

$$C_n^2(h) = 0.00594(v/27)^2 \left(10^{-5}h\right)^{10} \exp(-h/1000)$$
$$+ 2.7 \times 10^{-16} \exp(-h/1500) + A_0 \exp(-h/100), \tag{8.7}$$

where

  h is the altitude in m.
  v and A_0 are parameters to be set by the user.

Particularly, A_0 (dimensionally $m^{-2/3}$) defines the turbulence strength at the ground level, while v is the rms (root mean square) wind speed at high altitude (dimensionally m/s). The Hufnagel-Valley model, therefore, allows an easy modification of the $C_n^2(h)$ profile by changing values of the parameters A_0 and v. Particularly, for an $A_0 = 1.7 \times 10{-14}$ $m^{-2/3}$ and $v = 21$ m/s, the Hufnagel-Valley is commonly termed as HV5/7 because it generates the conditions for an atmospheric coherence length of 5 cm and an isoplanatic angle of 7 μrad (definitions of these last two terms will be introduced later in this chapter) at a wavelength of 500 nm.

  Beside the Hufnagel-Valley, a number of other empirical/experimental models of the $C_n^2(h)$ are available. Again, great care must be applied when one uses these models because (beside their daytime variability) they are most accurate when applied to describe $C_n^2(h)$ in the geographical location where the measurements were made. For instance, the CLEAR 1 model [3] is pertinent to the New Mexico desert, and describes the nighttime $C_n^2(h)$ profile for altitude as $1.23 < h < 30$ km as:

$C_n^2(h) = 10^{-17.025-4.3507h+0.8141h^2}$ for $1.23 < h \leq 2.13$ km

$C_n^2(h) = 10^{-16.2897+0.0335h-0.01341h^2}$ for $2.13 < h \leq 10.34$ km

$C_n^2(h) = 10^{-17.0577-0.0449h-0.00051h^2+0.6181\exp(-0.5(h-15.5617)/12.0173)}$ for $10.34 < h \leq 30$ km,

$$\tag{8.8}$$

where altitude h is expressed in km, while above 30 km, the refractive index structure constant is supposed to be zero. The CLEAR 1 model was obtained by averaging and statistically interpolating a number of thermosonde observation measurements obtained over a large number of meteorological conditions. The lower limit of 1.23 km of ground elevation is pertinent to the altitude of the New Mexico region where such measurements took place [3]. Another $C_n^2(h)$ model is Submarine Laser Communication (SLC) that was developed for the AMOS observatory in Maui, Hawaii. The daytime profile of $C_n^2(h)$ according to SLC model is given by [2]:

$$C_n^2(h) = 8.40 \times 10^{-15} \qquad \text{for } h \leq 18.5\text{m}$$
$$C_n^2(h) = \left(3.13 \times 10^{-13}\right)/h \qquad \text{for } 18.5 < h \leq 240\text{m}$$
$$C_n^2(h) = 1.3 \times 10^{-15} \qquad \text{for } 240 < h \leq 880\text{m} \qquad . \qquad (8.9)$$
$$C_n^2(h) = 8.87 \times 10^{-7}/h^3 \qquad \text{for } 880 < h \leq 7,200\text{m}$$
$$C_n^2(h) = 2.0 \times 10^{-16}/h^{1/2} \qquad \text{for } 7,200 < h \leq 20,000\text{m}$$

A graphic comparison of the three different parametric models of $C_n^2(h)$ (HV5/7, CLEAR 1, and SLC Day) are plotted in Figure 8.2. Again, a careful choice of $C_n^2(h)$ model that must be taken when used to calculate the different figures of merit describing propagation of a beam in a turbulent atmosphere. For instance, variation of the rms wind velocity in the Hufnagel-Valley will lead to a different profile of $C_n^2(h)$ at higher levels of the atmosphere, while variation of daytime and nighttime of $C_n^2(h)$ can be described by a multiplier of the parametric model selected. In any case, great care is suggested when adopting a model or another (including experimental measurements)

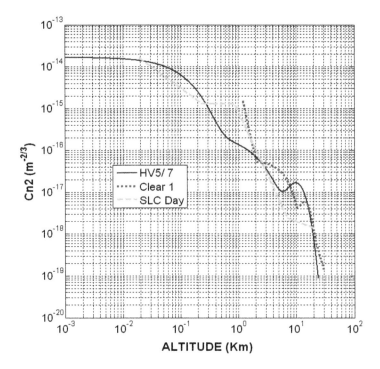

FIGURE 8.2 Comparison of three models of atmospheric profile of the structure constant of the refractive index. HV5/7, CLEAR 1, and SLC Day are plotted. Notice that, due to geographical specificity of each model, not all models extend to the sea level.

because the large dynamics of atmospheric conditions can lead to a great variation of the $C_n^2(h)$ profile.

As analogously expressed for the wind velocity in Equation 8.2, the Kolmogorov three-dimensional spectrum of the refractive index spectrum is:

$$\Phi_n(\kappa, h) = 0.033 C_n^2(h)\kappa^{-11/3}, \quad 1/L_0 \ll \kappa \ll 1/l_0. \tag{8.10}$$

Here, the spectrum depends on the altitude, h, and (of course) has the limitation to be applicable only in the inertial subrange [6]. A modification to the Kolmogorov spectrum is the so-called modified von Karman spectrum [4] that is given by:

$$\Phi_n(\kappa, h) = 0.033 C_n^2(h) \frac{\exp(-\kappa^2 / \kappa_m^2)}{(\kappa^2 + \kappa_0^2)^{11/6}}, 0 \leq \kappa < \infty, \tag{8.11}$$

where

$$\kappa_m = 5.92/l_0.$$

$$\kappa_0 = 1/L_0.$$

The modified von Karman spectrum (as other spectrum available in literature of the refractive index) has the same behavior of the Kolmogorov spectrum in the inertial subrange of spectrum, but it is analytically easier to handle outside the inertial subrange having the singularity of the Kolmogorov spectrum at $\kappa = 0$ removed.

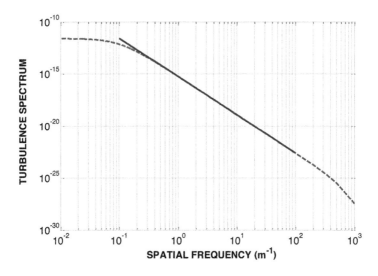

FIGURE 8.3 Comparison between the Kolmogorov's turbulence spectrum (solid line) and the von Karman's turbulence spectrum (dashed line). Notice that in the figure, $C_n^2 = 10^{-13} (m^{-2/3})$, while the actual subinertial range is between 10 cm and 10 m.

8.1.3 PROPAGATION IN TURBULENT ATMOSPHERE

To describe the propagation of a monochromatic electromagnetic wave $\mathbf{E}$ in clear air turbulence, one should start (as usual) from Maxwell's equations:

$$\nabla^2\mathbf{E}+k^2n^2\,\mathbf{E}+2\nabla[\mathbf{E}\cdot\nabla\log n]=0, \tag{8.12}$$

where

   $\mathbf{E}$ is the vector electric field.
   n is the (random) refractive index.
   $k = 2\pi/\lambda$ is the propagation constant.
   Laplacian operator is described by $\nabla^2 = \partial^2/\partial x^2 + \partial^2/\partial y^2 + \partial^2/\partial z^2$.

Neglecting depolarization of the wave, and under the condition that the wavelength λ is much smaller than the scale of turbulence, one obtains that the vector wave Equation 8.12 can be rewritten in the scalar form:

$$\nabla^2 E + k^2 n^2\,E = 0, \tag{8.13}$$

where E is a scalar component of the electric field.

A number of approaches have been considered to solve the above stochastic differential equation (notice, again, that n is here a random term). Among these, one of the best known is the Rytov method [7], whose solution holds interesting and practical consequences for the characterization of the wave propagation in random media.

According to the Rytov method, one first describes the scalar field as the complex exponential form:

$$E = \exp(\psi), \tag{8.14}$$

with $\psi = \log(E)$. Then, rewriting the scalar Maxwell equation in terms of the exponential term ψ, one gets the Riccati equation:

$$\nabla^2\psi + \nabla\psi.\nabla\psi + k^2 n^2 = 0. \tag{8.15}$$

A closed solution of the above equation is not obtainable. However, using the perturbation expansion of the electric field one can write:

$$E = \exp\left(\psi_0 + \psi_1\right), \tag{8.16}$$

where $E_0 = \exp(\psi_0)$ is the solution of Maxwell's equation in absence of perturbation. The term ψ_1 the field perturbation related to the first-order scattering of the field. As a consequence of Equations 8.14 and 8.16, one can write the exponential turbulence term as:

$$\psi_1 = \psi - \psi_0 = \log\left(\frac{|E|}{|E_0|}\right) + j(\phi - \phi_0) = \chi + jS \tag{8.17}$$

where

χ is the field log amplitude.
S is the field phase fluctuations.

Demonstration of the solution of the Riccati stochastic differential equation using Rytov method is not shown here because it goes beyond the scope of this chapter [3]. However, understanding the main results of the Rytov method has great implications in the definition of the most meaningful figure of merits of the problem. From the solution of the Riccati equation, one finds that the lognormal amplitude χ and the phase term S depends on the variation of the random refraction index along propagation path. Because of the integration of the random effects along the propagation path, one can consider that overall random perturbation implies a normal distribution of χ and S, with the consequence that both wave amplitude [proportional to $\exp(\chi)$] and wave intensity [proportional to $\exp(2\chi)$] of the wave have lognormal distribution, a conclusion that has also been proved experimentally correct for weak turbulence [3].

8.1.4 SCINTILLATION INDEX

One of the effects of clear air turbulence is the fluctuation of the signal irradiance. In nature such irradiance fluctuation is also manifested in the twinkling of a star seen with the naked eye. The scintillation index, σ_I^2, describes such intensity fluctuation (otherwise known as scintillation) as the normalized variance of the intensity fluctuations given by:

$$\sigma_I^2 = \frac{<(I-<I>)^2>}{<I>^2} = \frac{<I^2>}{<I>^2} - 1 \tag{8.18}$$

where $I = |E|^2$ is the signal irradiance (or intensity).

Using results from the Rytov method, one can express the scintillation index in terms of the variance of the field log-amplitude or Rytov's number, σ_χ^2, as [3]:

$$\sigma_I^2 = \exp(4\sigma_\chi^2) - 1 \approx 4\sigma_\chi^2 \tag{8.19}$$

where the right-hand side approximation is valid for the condition of weak turbulence mathematically corresponding to $4\sigma_\chi^2 < 1$. Expressions of lognormal field amplitude variance depend on (among other things) the nature of the electromagnetic wave traveling in the turbulence and on the link geometry. Concerning the geometry of the optical link, one can generalize that the case studies are divided into downlink, uplink, and horizontal path. According to the Rytov method, one can write the downlink lognormal variance for a plane wave as [3]:

$$\sigma_\chi^2 = 0.56 k^{7/6} \sec(\theta)^{11/6} \int_{h_0}^{h_0+L} C_n^2(h)(h-h_0)^{5/6} \, dh, \tag{8.20}$$

where Equation 8.20 describes a slant path between the two altitudes at $h_0 + L$ (source location) and h_0 (receiver location) with an angle from the zenith θ for total path length, $L/\cos(\theta)$.

For a spherical wave, the variance of the lognormal field amplitude is instead:

$$\sigma_x^2 = 0.56k^{7/6}\sec(\theta)^{11/6}\int_{h_0}^{h_0+L} C_n^2(h)(h-h_0)^{5/6}(1-(h-h_0)/L)^{5/6}\,dh. \qquad (8.21)$$

Concerning the uplink, the expression of the variance of the lognormal amplitude for a plane wave is:

$$\sigma_x^2 = 0.56k^{7/6}\sec(\theta)^{11/6}\int_{h_0}^{h_0+L} C_n^2(h)(L+h_0-h)^{5/6}\,dh, \qquad (8.22)$$

for a spherical wave instead, one can get:

$$\sigma_x^2 = 0.56k^{7/6}\sec(\theta)^{11/6}\int_{h_0}^{h_0+L} C_n^2(h)(L+h_0-h)^{5/6}(1-(L+h_0-h)/L)^{5/6}\,dh. \qquad (8.23)$$

The horizontal path is characterized by a constant, C_n^2, the variance of the log-amplitude of the field for a plane wave over a path L will then be:

$$\sigma_\chi^2 = 0.307k^{7/6}C_n^2 L^{11/6}, \qquad (8.24)$$

and for a spherical wave over a path L:

$$\sigma_\chi^2 = 0.124k^{7/6}C_n^2 L^{11/6}. \qquad (8.25)$$

Other closed-form expressions of σ_χ^2 can be obtained for the propagation of a Gaussian beam [2]. Such expressions, however, can have much more complicated mathematical descriptions than those presented in this section.

8.1.5 Scintillation Statistics

As indicated in Section 8.13, the lognormal amplitude can be statistically described by a Gaussian distribution. As a result of the Rytov method, for weak turbulence of the irradiance, fluctuation of a propagating beam can be described by a lognormal distribution. Therefore, one can derive that the irradiance probability density function (PDF) is given by:

$$p_I(I) = \frac{1}{\sqrt{2\pi}I\sigma_I}\exp\left\{-\frac{\left[\ln\left(\frac{I}{<I>}\right)+\frac{1}{2}\sigma_I^2\right]^2}{2\sigma_I^2}\right\}, \qquad (8.26)$$

where $\langle I\rangle$ is the average irradiance, Figure 8.4.

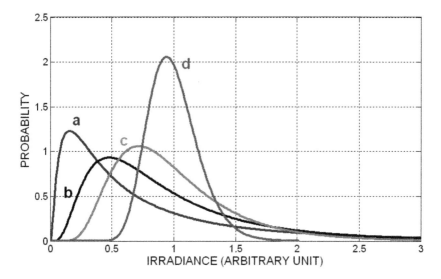

FIGURE 8.4 Irradiance PDF for a weak turbulence for different values of variance of the lognormal field amplitude: (a) $\sigma_\chi^2 = 0.2$, (b) $\sigma_\chi^2 = 0.1$, (c) $\sigma_\chi^2 = 0.05$, and (d) $\sigma_\chi^2 = 0.025$. The average irradiance here is $\langle I \rangle = 1$ in arbitrary unit.

Once defined the PDFs for weak turbulence, one can describe the statistics of the signal fluctuations above and below the nominal average irradiance $\langle I \rangle$. During a link design, those fluctuations below the average irradiance manifest themselves as dynamical signal loss (or signal fade), and it is of concern for system engineers to quantify fade probability or (in other terms) the amount of time that the received signal is below a determined fade limit. If one defines a lower limit of the irradiance at the receiver as I_F, the signal fade threshold F_T, in dB is:

$$F_T = 10 \log_{10} \left(\frac{\langle I \rangle}{I_F} \right), \tag{8.27}$$

where the fade threshold, F_T, can be considered as an additional loss that can be tolerated by the system. Using Equation 8.27, one can calculate the probability that the signal will be below I_F (or to have an additional loss larger than F_T) as:

$$P_F = \int_0^{I_F} p_I(I) dI = \int_0^{I_F} \frac{1}{\sqrt{2\pi} I \sigma_I} \exp \left\{ -\frac{\left[\ln\left(\frac{I}{\langle I \rangle} \right) + \frac{1}{2}\sigma_I^2 \right]^2}{2\sigma_I^2} \right\} dI = \frac{1}{2} \left\{ 1 + erf\left(\frac{\sigma_I^2 - 0.46 F_T}{2\sqrt{2}\sigma_I^2} \right) \right\}.$$

$$\tag{8.28}$$

Scintillation of the irradiance affects the overall system performance as well as degrades the bit error rate (BER) of the receiver. As an example, one can consider the direct detection of a signal modulated on-off keying. In Chapter 2, it was shown that the BER in the absence of turbulence is:

$$BER(Q) = \frac{1}{2} \text{erfc}\left(\frac{Q}{\sqrt{2}} \right). \tag{8.29}$$

The above term Q is related to the detector current as:

$$Q(I) = \frac{R_s A_R I L_T}{\sqrt{\sigma_1^2} + \sqrt{\sigma_0^2}}, \tag{8.30}$$

where

$R_s A_R I L_T$ is the current generated at the receiver during the high state (usually) corresponding to the digit "1."

R_s is the detector responsivity.

I is the signal irradiance when the digit "1" is transmitted.

A_R is the entrance pupil area.

L_T is the optical system loss.

σ_1^2 is the current noise power during the high signal state.

σ_0^2 is the current noise power during the low state.

In the presence of atmospheric turbulence, the BER is calculated by averaging it over the intensity PDF, which is given by:

$$\langle BER \rangle = \int_0^\infty p_I(I) BER[Q(I)] dI. \tag{8.31}$$

8.1.6 Aperture Averaging Factor

So far we have described the irradiance fluctuation and scintillation index experienced by a point collecting aperture. However, as the collecting aperture of a receiver system increases, the receiver experiences a reduction of the fluctuation of total received power. This fluctuation reduction is related to the averaging of the nonuniform irradiance integrated by the collecting aperture, referred to as aperture averaging [8]. To quantify the aperture averaging, it is usually introduced by the concept of averaging aperture factor, A_v, defined as the ratio between $\sigma_1^2(D)$ (the normalized flux variance of the collection aperture of diameter D) over the scintillation index:

$$A_v = \frac{\sigma_1^2(D)}{\sigma_1^2} < 1. \tag{8.32}$$

In Ref. [9], the aperture averaging factor is written in terms of the covariance of the irradiance ($C_1(x)$) and the optical transfer function of the circular aperture ($M(x)$) as:

$$A_v = \frac{1}{8} \int_0^1 \frac{C_1(xD)}{C_1(0)} M(x)x \, dx. \tag{8.33}$$

Finding an exact analytical solution to the above is a very challenging problem. However, for the case of a link from space to Earth (downlink), an approximated and very practical solution to Equation 8.33 has been presented in Ref. [10] for weak turbulence for both plane and spherical waves, where the aperture averaging factor is expressed by the formula:

$$A_v = \left[1 + 1.1\left(\frac{D^2}{\lambda h_s \cos(\theta)}\right)^{7/6}\right]^{-1}, \tag{8.34}$$

where

θ is the angle from the zenith for slant path.
h_s is a scale height of the turbulence defined as:

$$h_s = \left[\frac{\displaystyle\int_{h_0}^{h_0+L} C_n^2(h)(h-h_0)^2(1-(h-h_0)/L)^{-1/3} \, dh}{\displaystyle\int_{h_0}^{h_0+L} C_n^2(h)(h-h_0)^2(1-(h-h_0)/L)^{5/6} \, dh}\right]^{6/7} \tag{8.35}$$

where the above equation is valid for a spherical wave, and L is the difference in altitude between the source (at $h_0 + L$) and the receiver (at h_0). At the same time, one can derive the scale height of the turbulence for a plane wave as:

$$h_s = \left[\frac{\displaystyle\int_{h_0}^{h_0+L} C_n^2(h)(h-h_0)^2 \, dh}{\displaystyle\int_{h_0}^{h_0+L} C_n^2(h)(h-h_0)^{5/6} \, dh}\right]^{6/7}. \tag{8.36}$$

In the instance of a horizontal path with constant C_n^2 over a link distance L, the aperture averaging factor for a plane wave is [2]:

$$A_v = \left[1 + 1.06\left(\frac{kD^2}{4L}\right)\right]^{-7/6}, \tag{8.37}$$

while for spherical waves it holds that [2]:

$$A_v = \left[1 + 0.33\left(\frac{kD^2}{4L}\right)\right]^{-7/6}.$$ (8.38)

8.1.7 MODELING OF SCINTILLATIONS IN STRONG TURBULENCE

Scintillation statistics and theory so far described apply to weak turbulence. As the strength of the turbulence increases, reaching the condition of strong turbulence ($4\sigma_\chi^2 > 1$), the lognormal statistical characterization of the irradiance PDF fails to adequately represent the process [11]. However, a number of other statistical models have been suggested in the literature to describe scintillation statistics in a regime of strong turbulence. For instance, one of the first models used for this purpose was the K distribution [12, 13]. Later, statistical models were presented that were based on the lognormal Rician distribution, which presented the computational and numerical problem of not being available in closed form [14, 15]. However, it must be pointed out that both K distribution and lognormal Rician distribution cannot easily relate their mathematical parameters with the observables of the atmospheric turbulence, which limits their applicability and utilization. Alternatively, the gamma-gamma distribution is quite successfully used to describe scintillation statistics for weak and strong turbulence conditions [16]. While still mathematically complex, the gamma-gamma distribution can be expressed in closed form (unlike the lognormal Rician) and can relate its main parameters to the scintillation conditions (scintillation index, large-scale scintillation, etc.). Particularly, for a downlink signal the PDF of the fluctuation of the normalized irradiance $\left(\hat{I} = \frac{I}{<I>}\right)$ is expressed by the gamma-gamma distribution as [2]:

$$p(\hat{I}) = \frac{2(\alpha\beta)^{(\alpha+\beta)/2}(\hat{I})^{(\alpha+\beta)/2-1}K_{\alpha-\beta}(2\sqrt{\alpha\beta\hat{I}})}{\Gamma(\alpha)\Gamma(\alpha)},$$ (8.39)

where $K(\cdot)$ is a modified Bessel function of second kind, α and β are related to the scintillation index as:

$$\sigma_I^2 = \frac{1}{\alpha} + \frac{1}{\beta} + \frac{1}{\beta\alpha}.$$ (8.40)

The scintillation index in Equation 8.40 is for downlink beam experiencing weak to strong turbulence conditions, and it calculated as [2]:

$$\sigma_I^2 = \exp\left(\frac{0.49\sigma_R^2}{\left(1+1.11\sigma_R^{12/5}\right)^{7/6}} + \frac{0.51\sigma_R^2}{\left(1+0.69\sigma_R^{12/5}\right)^{5/6}}\right) - 1,$$ (8.41)

where $\sigma_R^2 = 4\sigma_\chi^2$, with the Rytov's number as $0 \le \sigma_\chi^2 \le \infty$.

In terms, the parameters α and β are expressed as:

$$\alpha = \left\{ \exp\left(\frac{0.49\sigma_R^2}{\left(1+1.11\sigma_R^{12/5}\right)^{7/6}} \right) - 1 \right\}^{-1}, \qquad (8.42)$$

$$\beta = \left\{ \exp\left(\frac{0.51\sigma_R^2}{\left(1+1.11\sigma_R^{12/5}\right)^{5/6}} \right) - 1 \right\}^{-1}. \qquad (8.43)$$

As an alternative to analytical models, wave optics simulation has largely been used with success to simulate and study the effects of clear air turbulence on a propagating optical beam [17]. Compared to analytical methods, wave optics numerically simulates the propagation in the time domain of a two-dimensional wave front of an electric field sampled in a spatial grid described by a mesh of **N × N** elements over an area of interest [18]. In a wave optics simulation, the atmosphere along the direction of propagation is divided into a cascade of **M** layers [17]. In the case of laser beam propagation (e.g., from transmitter to receiver), the propagation of the two-dimensional wave front of the electric field from the beginning of one layer to the beginning of the next layer takes place in two steps. In the first step, the phase of the spatially sampled electric field is modulated by a random phase screen [19] whose strength is determined by the C_n^2 in the layer. This phase screen describes the phase aberrations experienced by the propagating wave that are induced by the atmospheric turbulence within the layer (described in Section 8.1.5). In the next step, the electrical field is propagated to the beginning of the next layer using fast Fourier transform techniques [17]. The propagation process is repeated through the cascade of atmospheric layers until the wave reaches the collecting aperture of the receiver where the wave intensity is integrated over the aperture surface. This procedure describes the propagation of the wave in one single realization at the initial time (t_1), while wave optics allows a simulation over an extended time period. Therefore, to simulate the propagation over the next time interval, the process is updated to the next time step $t_1 + \Delta t$, in which the phase screen of each layer experiences a spatial shift as $v_T\Delta t$, where v_T is the transversal component of the wind speed at the corresponding layer. Once the phase screens are updated, the wave propagation through the **M** layers is repeated. Then, the wave intensity is (again) collected at the receiver at the time $t_1 + \Delta t$. The algorithm (phase screen update and propagation) is repeated n times until the simulation reaches its final target time $(t_1 + n\Delta t)$ and the simulation output time evolution of process is provided.

Wave optics simulations are extensively used, not only in simulating intensity scintillation dynamics, but also in a number of applications of adaptive optics, optical tracking, and laser communications. Finally, one must also consider the not trivial advantage over analytical approaches that wave optics simulation techniques can be used regardless of the turbulence strength (weak or strong), nature, and profile.

8.1.8   PHASE STATISTICS

In clear air turbulence, the random variation of the atmospheric refractive index induces a random degradation of the phase of wave front of a propagating electromagnetic wave. Again, to describe the random variation of the phase (ϕ) of an electromagnetic wave, one can use the phase structure function between two points ($\mathbf{R}_1$ and $\mathbf{R}_2 = \mathbf{R}_1 + \mathbf{R}$) given by [3]:

$$D_\phi(\mathbf{R}) = \left\langle [\phi(\mathbf{R}_1 - \phi(\mathbf{R}_1 + \mathbf{R})]^2 \right\rangle = 2.914 k^2 \sec(\theta) \mathbf{R}^{5/3} \int\limits_{h_0}^{h_0+L} C_n^2(h) dh, \quad (8.44)$$

where the above equation relates to a generic plane wave traveling between the altitudes $h_0 + L$ and h_0 with an inclination angle from the zenith of θ.

Clearly, due to the random spatial variation of the refractive index, the wave experiences distortion and aberration. The magnitude of the wave front aberration is related to the refractive index structure constant and the path length.

Particularly, the phase aberration manifests itself when an imaging system is used. Consider a generic imaging system (e.g., a telescope) with an entrance pupil aperture of diameter D and focal length F. According to the diffraction theory, when a plane wave is projected on the focal plane, the spatial distribution of the intensity is described by the Airy pattern as:

$$I(r) = I_o \left(\frac{2 J_1\left(\frac{\pi D r}{\lambda F} \right)}{\left(\frac{\pi D r}{\lambda F} \right)} \right)^2 \quad (8.45)$$

where r is the distance from the center of the focal plane and I_o is the peak intensity at the center. The full width at half maximum of the Airy pattern is the diffraction-limited spot size at the focal plane (d_{dl}) is:

$$d_{dl} = \frac{\lambda}{D} F, \quad (8.46)$$

which indicates the larger is the aperture diameter D, the better is the system resolution. The resolving power (or the resolution) of a diffraction telescope [20] can be defined as the integral of the optical transfer function of the optical system, that for a diffraction limited system is:

$$R_{dl} = \frac{\pi}{4} \left(\frac{D}{\lambda F} \right)^2 \text{ (cycle/m)}^2 \quad (8.47)$$

In the presence of clear air turbulence, the telescope resolving power will be degraded because of the wave front phase aberration will affect the system resolving power. In other words, one should expect that in the presence of clear air turbulence, the telescope resolution (and the minimum spot size at the focal plane) would be bounded not only by the telescope aperture, but also by the strength of the turbulence. Particularly in Ref. [20], it is described how in the presence of clear air turbulence the telescope resolving power degrades, this degradation can be approximately expressed in closed form as [2]:

$$R = \frac{\pi}{4} \left\{ \frac{1}{\lambda F} \frac{D}{[1+(D/r_o)^{5/3}]^{3/5}} \right\}^2$$

(8.48)

where r_0 is the Fried parameter or atmospheric coherence length (to be defined later), which is related to the strength of the optical turbulence. Equation 8.48 indicates that large aperture telescope resolution is lower, bounded by the diffraction-limited value when $D \ll r_0$ (weak or no optical turbulence) and to a saturation value when $D \gg r_0$ (strong turbulence):

$$R_{MAX} = \frac{\pi}{4} \left(\frac{r_o}{\lambda F} \right)^2$$

(8.49)

Explanation of the physical meaning of the atmospheric coherence length can be visualized when the telescope resolution as R/R_{MAX} is plotted as function of the ratio D/r_0 as seen in Figure 8.5. For $D/r_0 < 1$, the telescope resolution is that one of a diffraction limited aperture, Equation 8.47. However, at $D/r_0 \sim 1$, there is the knee of the telescope resolution curve, and for $D/r_0 > 1$, the telescope resolution starts deviating

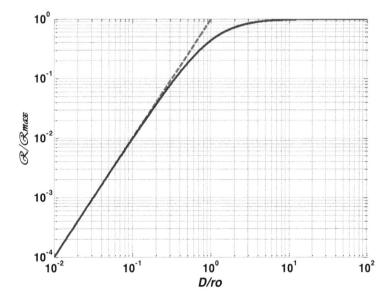

FIGURE 8.5 Telescope resolution in the presence of clear air turbulence. For $D/r_0 > 1$, the telescope resolution converge to R_{MAX}. At $D/r_0 \sim 1$, there is the knee of curve.

from the diffraction limit of R. In fact, for $D/r_0 \sim 2$–3, the telescope resolution rapidly converges to a resolution close to R_{MAX} as in Equation 8.49.

As already done for other figure of merits of the turbulence, r_0 can be described as a function of the refractive index structure parameter, wavelength, and wave front. Particularly, in the case of a plane wave propagating from the altitude $h_0 + L$ to h_0 (downlink), one has [3]:

$$r_0 = \left[0.423 \sec(\theta) k^2 \int_{h_0}^{h_0+L} C_n^2(h) dh \right]^{-3/5}, \qquad (8.50)$$

where θ is the angle from the zenith in the slant path; for the horizontal path of length L at constant C_n^2:

$$r_0 = (0.423 k^2 C_n^2 L)^{-3/5}. \qquad (8.51)$$

For a spherical wave instead, we have (downlink):

$$r_0 = \left[0.423 \sec(\theta) k^2 \int_{h_0}^{h_0+L} C_n^2(h) \{ (L + h_0 - h) / L \}^{5/3} dh \right]^{-3/5}, \qquad (8.52)$$

while for a horizontal path:

$$r_0 = (0.158 k^2 C_n^2 L)^{-3/5}. \qquad (8.53)$$

Notice that, with the help of the Fried parameter one can conveniently rewrite the phase structure function in Equation 8.44 as:

$$D_\phi(\mathbf{R}) = 6.88 \left(\frac{\mathbf{R}}{r_0} \right)^{5/3}. \qquad (8.54)$$

The atmospheric coherence length, r_0, is a crucial parameter describing the optical turbulence and its effect on the quality of the wave front that is propagating in the atmosphere, particularly it follows that:

1. r_0 is the diameter of an equivalent aperture where the phase variation in rms is approximately 1 rad;
2. r_0 is the hypothetical diameter of a telescope whose resolution is approximately the same of that one in absence of turbulence: the larger the r_0 the smaller is the effects of turbulence on the propagating wave;
3. r_0 varies with the wavelength as $\lambda^{6/5}$; therefore, for the same $C_n^2(h)$ profile, at larger wavelengths of operation the Fried parameter has larger values that imply a less severe turbulence effects on the propagating wave front;

4. r_0 values vary depending on location, and time of the day; usually the Fried parameter is indicate at zenith and for 500nm: for an site of an astronomical telescope, r_0 can vary from few centimeters to tens of centimeter in a good location;

5. according to Equation 8.52, the turbulence in the proximity of the receiver (pupil plane) predominatly affects the atmospheric coherence length. Hence, in a space-to-Earth satellite link one should expect large atmospheric coherence length at the satellite (uplink), and smaller r_0 with more severe phase distortion at the ground station receiver (downlink) Figure 8.6;

6. for $D/r_0 \gg 1$, the angular resolution of a telescope is limited by the astronomical seeing, defined as [21]:

$$\text{seeing} = \frac{\lambda}{r_0}. \text{ rad} \tag{8.55}$$

The seeing is also related to the size of the full-width-half-maximum or spot size of the intensity pattern at the focal plane. In fact, for a receiving optical system with focal length F and aperture diameter $D \gg r_0$, the related spot size at the focal plane will be:

$$d_{\text{seeing}} = \frac{\lambda}{r_0} F, \tag{8.56}$$

with $d_{\text{seeing}} \gg d_{\text{dl}}$.

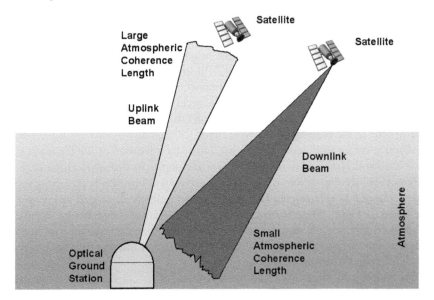

FIGURE 8.6 Atmospheric coherence length is affected in different fashions for uplink and downlink. For downlink, r_0 is much smaller with respect to what one experienced at the satellite during uplink.

Equation 8.56 implies that, for large seeing (small r_0 and strong turbulence) values, the received signal spot size at the focal plane can be relatively large, with the consequence that all signal photons can be captured only using a larger surface area photodetector. Conversely, the use of a large surface photodetector can limit the receiver electrical bandwidth and, consequently, the detection of the digital signal at high bit rate. At the same time, if a single detector is used with a diameter smaller than d_{seeing}, the receiver will experience an additional loss of signal that for a number of applications (e.g., deep-space optical communications) cannot be sustainable.

However, analysis of the aberrations of the wave front phase with the help of adaptive optics can help limit the consequences of signal spot blurring at the focal plane. The blurring of the spot at the receiver focal plane (as indicated in Equation 8.56) refers to a long exposure when all the distortion components are integrated in time. Instead, when observing the time evolution of the signal spot at the focal plane, one would notice that the spot itself is moving (or dancing) randomly around a central position (centroid motion). This first-order effect is due to the random fluctuations of the angle of arrival of the signal wave front induced by the turbulence. The angle of arrival perturbation has a Gaussian distribution described by a zero mean and variance along the x and y axes as [22]:

$$\sigma_{\alpha x}^2 = \sigma_{\alpha y}^2 = 0.182 \left(\frac{D}{r_0} \right)^{5/3} \left(\frac{\lambda}{D} \right)^2. \tag{8.57}$$

Recalling that r_0 varies with the wavelength as $\lambda^{6/5}$, and noting Equation 8.57, it is simple to derive that the spot centroid motion at the focal plane is independent of the wavelength, while it represents a larger component of the spot blurring at longer wavelengths. With the help of Equations 8.42 and 8.51, Figure 8.7 plots the ratio between the angle arrival error (standard deviation) and the angular resolution of a telescope in function of D/r_0. Figure 8.7 illustrates that for $D/r_0 \sim 3$, the angle of arrival fluctuation is the more sizeable component of the enlargement of the spot size at focal plane and its removal can benefit the receiver operation. For larger D/r_0 values, instead, higher order wave front distortion (astigmatism, coma, etc.) contributes more heavily in the blurring of the signal spot.

Adaptive optics techniques can be used to reduce the size of the distorted spot size at the focal plane and to reduce it to a size closer to its diffraction limit [22].

Removal of the fluctuations of the angle of arrival was historically the first adaptive optics technique ever attempted [23], and it is conceptually illustrated in Figure 8.8. In essence, in Figure 8.8, a beam splitter removes a part of the incoming signal and redirects it to a wave front sensor that is able to (a) determine the instantaneous angle of arrival, and then (b) address a correction to a tip-tilt mirror that compensates the angle arrival offset and stabilizes the incoming beam.

The simplest form of this wave front sensor can be constituted by a quadrant detector (consisting of four element pixels) or a camera, both sensors able to determine the position of the centroids of the focused signal [22]. The acquired information of the centroid location on a focal plane can then be converted as tip-tilt angular information that drives the correction of the input tip-tilt mirror.

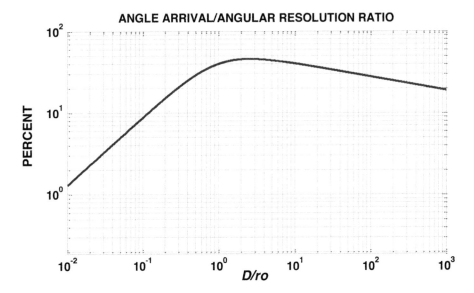

FIGURE 8.7 Ratio between the angle of arrival standard deviation and angular resolution of a telescope. The ratio peaks for $D/r_0 \sim 3$.

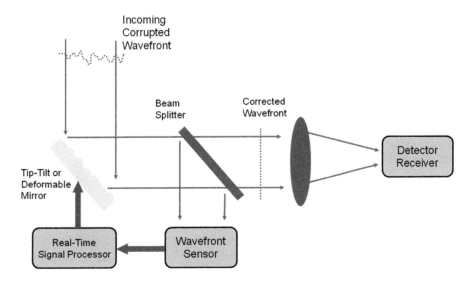

FIGURE 8.8 Block diagram for a generic adaptive optics system. The mirror on the path can represent either a tip-tilt system for angle of arrival compensation or a deformable mirror for higher order phase compensation.

Adaptive optics compensation of higher order distortions of the beam wave front is possible [22]. Under this scenario, a more complicated action is requested to correct the incoming wave front. Mainly, a spatially deformable mirror driven by the signal detected from a wave front sensor (e.g. a camera) is needed for this scope, of Figure 8.8. In this configuration, the spatially deformable mirror needs to be able to compensate for local aberration of the wave front [22, 24]. Temporal response of the correction of the actuator elements via the input deformable mirror is related to the turbulence time constant. The Greenwood frequency [25] is used to describe how fast the adaptive optics system needs to respond to the variation of the atmospheric turbulence. Related to turbulence strength and the transversal component of the wind speed [$v_T(h)$)] along the line of sight receiver-transmitter, the Greenwood frequency is given by:

$$f_G = \left[0.102k^2 \sec(\theta) \int_{h_0}^{L+h_0} C_n^2(h)v_T(h)^{\frac{5}{3}} dh \right]^{3/5} \text{Hz,} \qquad (8.58)$$

where f_G can vary from tens to hundreds of Hertz depending on the wind speed and turbulence strength. Generally, an adaptive optics system is required to have a close loop frequency at least four times f_G.

Conversely, in a number of applications, the signal level of the received beam can be too low for driving an adaptive optics system. In this case, sources other than the incoming signal can be used for the correction of the deformable mirrors. For instance, in astronomical telescopes, such external sources can be an angularly close star or a backscattering signal from a laser beacon [22, 24]. It is essential, however, that the spatial description of the atmospheric turbulence carried by an external source must represent that one experienced by the incoming signal. To do so, the angular separation between the signal beam and the external source must be less than the isoplanatic angle [26] given by:

$$\vartheta = \left[2.91k^2 \sec(\theta)^{8/3} \int_{h_0}^{L+h_0} C_n^2(h)h^{5/3} dh \right]^{-3/5} \text{rad.} \qquad (8.59)$$

Typical values of the isoplanatic angle seen by a ground telescope can vary from a few to tens of microradians.

8.1.9   BEAM EFFECTS

To complete our discussion of the effects of clear air turbulence on laser beam propagation in the atmosphere, the so-called beam effects need to be addressed. Generally, when a beam is propagating in the atmosphere, additional consequences of the action of clear air turbulence (besides scintillation and phase degradation of the wave front) are beam spreading and beam wander.

Beam spreading describes the broadening of the beam size at a target beyond the expected limit due to diffraction (or optical elements in the beam path) as the beam propagates in the turbulent atmosphere. Another effect of the turbulence on the beam propagation is relatively slow wandering of the beam around its central axis in a random fashion. A result of beam spreading and beam wander is that the long-term averaged beam irradiance at a given point is below an expected value, and manifests itself as a loss of signal. Beam spreading and beam wander particularly affect the propagation of uplink (ground-to-space) and horizontal atmospheric path link. To describe these effects for a Gaussian beam, we recall that given a collimated propagating Gaussian beam at a distance $\mathbf{z}$ from the source, the irradiance along a plane normal to the propagation direction is:

$$I(\mathbf{z},r) = \frac{2P_0}{\pi\omega^2(\mathbf{z})}\exp\left(-\frac{2r^2}{\omega(\mathbf{z})^2}\right)(\text{W/m}^2),\qquad(8.60)$$

where

P_0 is total (source) beam power in Watts.
r is the radial distance from the beam center.

$\omega(\mathbf{z})$ is the (radial) beam waist that after a propagation distance $\mathbf{z}$ (or in the case of an uplink beam is the slant path from ground to space) is given by:

$$\omega(\mathbf{z})^2 = \omega_0^2\left[1+\left(\frac{\mathbf{z}}{Z_0}\right)^2\right](\text{m}^2),\qquad(8.61)$$

in which ω_0 is the initial beam waist at $\mathbf{z}=0$ and Z_0 is the Rayleigh distance given by:

$$Z_0 = \frac{\pi\omega_0^2}{\lambda}(\text{m}).\qquad(8.62)$$

Beside the Rayleigh distance, the concept of the full divergence angle, θ_{DIV}, as the angle that comprise the 86.5% of the beam energy is defined as:

$$\theta_{DIV} = \frac{2\omega(z)}{z} = \frac{2\lambda}{\pi\omega_o}\qquad(8.63)$$

Again, Equation 8.60 refers to a generic Gaussian beam propagating in free space in the absence of optical turbulence. Conversely, when optical turbulence is present, the beam will experience a degradation in beam quality with a consequence that the long-term averaged beam waist in time will be $\omega_{eff}(\mathbf{z}) > \omega(\mathbf{z})$. Then, one can write the irradiance of the beam averaged in time as:

$$I(\mathbf{z},r) = \frac{2P_0}{\pi\omega_{eff}^2(\mathbf{z})}\exp\left(-\frac{2r^2}{\omega_{eff}(\mathbf{z})^2}\right).\qquad(8.64)$$

The effective waist, $\omega_{eff}(\mathbf{z})$, describes the variation of the beam irradiance averaged over long term at a distance $\mathbf{z}$ from the source. However, over short term, one notices that the beam will not only experience a spreading beyond what is expected in the absence of turbulence, but the beam spot itself will wander around a center position. This beam wandering, among other things, can be an additional cause of fluctuation (in addition to scintillation) of the signal irradiance at the targeted receiver. If we indicate as r_c the distance of the beam centroid from the center, one can express the effective beam waist as[27]:

$$\omega_{eff}(z)^2 = \omega_{st}(z)^2 + <r_c^2>$$ (8.65)

where $\omega_{st}(\mathbf{z})$ is the beam waist portion affected by (short term) beam spreading and $<r_c^2>$ is variance of the beam wander radial displacement that for an uplink beam is [27]:

$$\langle r_c^2 \rangle = 0.54 z^2 \left(\frac{\lambda}{2\omega_o} \right)^2 \left(\frac{2\omega_o}{r_o} \right)^{5/3}$$ (8.66)

in which r_o is atmospheric coherence length as Equation 8.50.

Evidently, due to the fact that $\omega_{eff}(\mathbf{z}) > \omega(\mathbf{z})$, the beam will experience an average reduction of the irradiance at the beam center (on axis) [27]:

$$\langle S \rangle = \frac{\omega(z)^2}{\omega_{eff}(z)^2} = \frac{1}{\left[1 + 5.65(\omega_o/r_o)^{5/3} \right]^{6/5}}$$ (8.67)

Equation (8.67) defines the term $<S>$ named as atmospheric Strehl ratio for the conditions from weak to strong turbulence. Conversely, the atmospheric Strehl ratio can be considered as another source of signal loss (with respect to the ideal propagation as in Equation 8.60), and therefore must to taken into consideration when budgeting the signal power (e.g., optical link budget). In a number of applications, bidirectional ground-to-space links (uplink and downlink) can be established, in this case a control loop can track the downlink beam, to correct the uplink tilt caused by the optical turbulence and remove the beam wander degradation for the uplink beam. In this case, it is shown [27] that the corrected atmospheric Strehl ratio, $<S_{tr}>$, can be expressed as:

$$< S_{tr} > \cong \frac{1}{\left[1 + \left(5.65 - \frac{4.86}{1 + 0.04(\omega_o / r_o)^{5/3}} (\omega_o / r_o)^{5/3} \right) \right]^{6/5}}$$ (8.68)

The atmospheric Strehl ratio is a good description of how the propagation of a Gaussian beam is affected by the optical turbulence averaged over long term. Generally, to understand the possible effects of the optical turbulence on Gaussian beam, the ratio between the beam waist at the source, ω_o, over the atmospheric coherence length as seen by the source on the ground, Equation 8.50, is a good metric. Essentially, one may consider three condition of operations: (a) $\omega_o/r_o \ll 1$, weak turbulence, where the beam is less affected by the optical turbulence, and the beam at target is not distorted by the turbulence; (b) $\omega_o/r_o \sim 1$ where beam is mainly affected by radial displacement that leads to beam wander in time around the optical axis; (c) $\omega_o/r_o \gg 1$, strong turbulence, where the beam wander is more detectable in addition to the fact the short-term realization of the beam at the target may break up showing a number of hot spots or "speckles." To better illustrate this concept, Figure 8.9 depicts realizations of beam propagation under three different condition of turbulence. The figure refers to the case of a propagation of a Gaussian beam at a wavelength of 1064 nm with a beam waist $\omega_o = 7$ cm. The beam is propagating in the atmosphere with at zenith angle of 60° with instantaneous images of the beam after a path of 60 km from the source. The images are produced using a wave optics code and a Hufnagel-Valley profile of the turbulence at different turbulence strength as in Figure 8.9b and c. Notice, that the figures represent just a single realization of the effect of the turbulence for a short exposure, while Strehl ratio of Equation 8.67 (and 8.68) refers to a long-term average of the beam shape affected by the turbulence. Particularly, Figure 8.9a depicts the beam at 60 km of distance from the source in (hypothetical) absence of turbulence, the beam is centered around the central axis and the beam size is as indicated in Equation 8.64. Instead, Figure 8.9b describes a realization for $\omega_o/r_o = 0.87$, with $r_o = 8.19$ cm at 1064 nm along the line of sight. The beam shape is still generally Gaussian experiencing some distortions, however, due to the turbulence the beam is moving around the central axis, in fact for this specific image the centroid coordinates are off center of 12 and 4 cm in the x and y directions. The beam will continue to wander around the central axis, and the final integrated long-term beam waist, $\omega_{eff}(z)$, can be calculated using Equation 8.67.

Finally, in Figure 8.9c is the depicted realization for $\omega_o/r_o = 2.75$, with $r_o = 2.55$ cm at 1064 nm along the line of sight. In this case, the beam experiences a strong degradation losing its spot-like shape and breaks up in 'speckles'. In Figure 8.9c, the coordinates of the beam centroid are −59 and −16 cm, respectively, along the x and y axes. Of course, in time the beam continues to wander around the central axis, breaking up in different shapes, while the long-term average shape of the beam is described by Equation 8.67. Finally, Figure 8.9d is comparing the long-term average distribution of the beam intensity for the three cases here described according to Equation 8.67. The atmospheric Strehl ratio greatly changes for the case for the case $\omega_o/r_o = 0.87$, and $\omega_o/r_o = 2.75$, with loss 8.7 and 18 dB, respectively, when compared to the ideal case of Figure 8.9a.

8.2 ATMOSPHERIC TRANSMISSION LOSS AND SKY BACKGROUND NOISE

8.2.1 ABSORPTION AND SCATTERING

Earth's atmosphere, via absorption and scattering, attenuates a propagating electromagnetic wave. Considering a collimated beam of initial beam intensity $I(0)$, after a

Uplink Beam: No Optical Turbulence

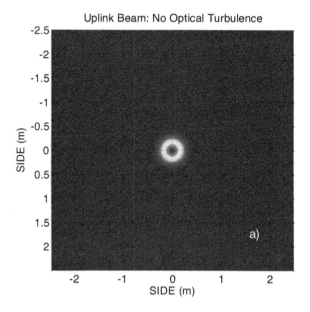

Uplink Beam: wo/ro=0.87

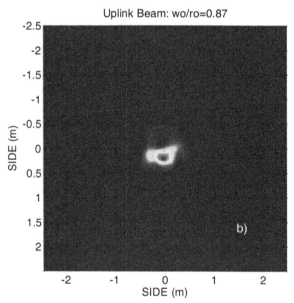

FIGURE 8.9 Three different realizations of a Gaussian beam spot for 60-km path in the atmosphere with a zenith angle of $\theta = 60°$. The initial beam waist is of 7 cm, while the wavelength is 1.064 μm: (a) absence of turbulence, (b) weak turbulence with $r_o = 8.19$ cm, (c) strong turbulence with $r_o = 2.55$ cm, (d) long-term averaged radial distribution of beam intensity after 60 km.

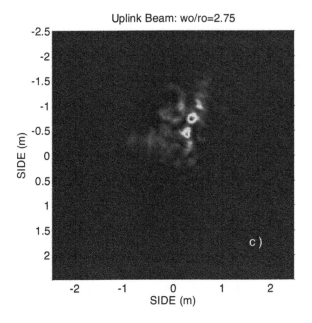

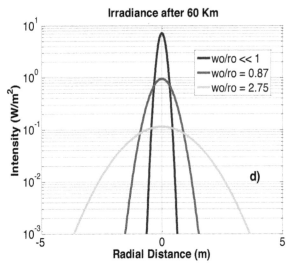

FIGURE 8.9 *(Continued)*

path of length L, the atmospheric transmittance T_{atm} related to atmospheric attenuation may be described by Beer's law as:

$$I(L) / I(0) = T_{atm} = \exp\left(-\int_0^L \alpha(z)dz\right) = \exp(-\tau), \qquad (8.69)$$

where

  τ is the optical depth of the path.
  α (dimensionally km^{-1} or m^{-1}) is the extinction or attenuation coefficient.

In case of constant attenuation coefficient (e.g., horizontal path in the atmosphere) one finds:

$$\tau = \alpha L. \tag{8.70}$$

The optical depth is related to the atmospheric loss, L_{ATM}, that can be expressed in dB as:

$$L_{ATM} = -10\log_{10} T_{atm} = 4.34\tau \; (dB). \tag{8.71}$$

The extinction coefficient is greatly dependent on the wavelength of operation, and can be considered the sum of the absorption coefficient α_A and scattering coefficient α_S. At the same time, both α_A and α_S have a contribution from the molecular and aerosol constituents of the atmosphere. In summary, one can write:

$$\alpha = (\alpha_{AM} + \alpha_{AA}) + (\alpha_{SM} + \alpha_{SA}). \tag{8.72}$$

Molecular absorption of the gases manifests itself with narrow absorption lines due to the resonance of photon energy, with molecules (and atoms) of the gas and water vapor present in the atmosphere. Of course, due to the variable density of the atmosphere constituents with altitude, the molecular absorption will be of different strengths at different elevations above sea level.

Molecular scattering, otherwise indicated as Rayleigh scattering, has a spectral dependence as [28]:

$$\alpha_{SM} = C_R / \lambda^4, \tag{8.73}$$

in which, C_R is a term depending on the different particle cross sections and concentrations, while the wavelength dependence indicates that Rayleigh scattering is stronger at shorter wavelengths (i.e., visible wavelengths), and it is less important at wavelengths in the near-infrared or longer.

Conveniently, for a ground-to-space link, one can approximate in a first order the optical depth (ground-to-space) due to Rayleigh scattering only as [29]:

$$\tau_{SM}(\lambda) = \frac{p}{p_o} 0.00877\lambda^{-4.05} \tag{8.74}$$

where p is the atmospheric pressure at location of interest, p_o is the atmospheric pressure at sea level, and the wavelength of interest λ is in microns.

Aerosols are particles suspended in the atmosphere with different concentrations. They have diverse nature, shape, and size. Aerosols can vary in distribution, constituents, and concentration. As a result, the interaction between aerosols and light can have a large dynamics, in terms of wavelength range of interest and magnitude of the atmospheric scattering itself. Because most of the aerosols are created at the earth's surface (e.g., desert dust particles, human-made industrial particulates, maritime droplets, etc.), the larger concentration of aerosols is in the boundary layer (a layer up to 2 km above the earth's surface). Above the boundary layer, aerosol concentration rapidly decreases. At higher elevations, due to atmospheric activities and the mixing action of winds, aerosol concentration becomes spatially uniform and more independent of the geographical location.

Scattering is the main interaction between aerosols and a propagating beam. Because the sizes of the aerosol particles are comparable to the wavelength of interest in optical communications, Mie scattering theory is normally used to describe aerosol scattering [30, 31]. Such a theory specifies that the scattering coefficient of aerosols is a function of the aerosols, their size distribution, cross section, density, and wavelength of operation.

For horizontal link, one can use a semiempirical method to describe and scale the scattering extinction coefficient. In a horizontal path, when the aerosol scattering is constant, it is defined as the visual range, V, for a wavelength of $\lambda = 550$ nm, as the distance (in km) where the total atmospheric loss is approximately 17 dB, or $\alpha V = 3.91$. Founding V, one can approximate the aerosol scattering extinction coefficient as [30]:

$$\alpha_{SA} = (3.91/V)(550/\lambda_c)^\delta \tag{8.75}$$

in which the wavelength of operation λ_c is expressed in nm, and the visual range is V indicated in km. The exponent δ is (approximately) related to V as:

$$\delta = 0.585 V^{1/3} \tag{8.76}$$

The value of V for a typical clear day can be $V = 23$ km (corresponding to a $\delta = 1.6$) or better, while for a hazy day the visual range may have a value of $V = 5$ km ($\delta = 1$). The values the exponential number assumes in Equations 8.73 and 8.74 (i.e., $\delta \ll 4$) indicates that aerosol scattering can substantially contribute to the overall extinction of the optical beam in the visible as well as in the near-infrared when compared to Rayleigh scattering.

The mechanism of absorption and scattering have been understood and characterized with a large degree of accuracy, and therefore a number of atmospheric radiative transfer software simulation packages have been developed to describe the effects of atmospheric absorption and scattering under different atmospheric conditions and over a large wavelength range. Historically, one of the first software programs describing the atmospheric effects on a laser beam was LOWTRAN (acronym for **LOW** resolution atmosphere **TRAN**smission) [32] developed by the US Air Force. LOWTRAN can be used to describe atmospheric transmittance and radiance in the wavelength range 0.2 to 28.5 μm (i.e., 350–40,000 cm^{-1}) with a resolution of 20 cm^{-1}. During a simulation, the software divides the atmosphere between sea level and 100-km altitude (or less) up to 33 layers. Each atmospheric layer is described by a variation of temperature, pressure, gas composition, gas mixing ratio, and aerosol distribution. The overall path atmospheric transmission (or optical depth) is calculated by compounding all the possible layer contribution components along the path through the different atmospheric layers.

Concerning aerosol concentration, LOWTRAN provides a number of models apt to properly describe the nature of the aerosols (e.g., rural, urban, maritime, desert, etc.) and their distribution within the atmospheric profile (i.e., boundary layer, troposphere, stratosphere, etc.). Other meteorological conditions affecting the beam propagation (such as fog, clouds, cirrus clouds, and rain) can also be simulated. As an improvement to LOWTRAN, the simulation program MODTRAN (acronym for **MOD**erate atmospheric **TRAN**smission) [33] was developed, allowing a narrower spectral resolution up to 0.2 cm^{-1} in its more recent version MODTRAN 6 [34]. However, the requirement to properly model the propagation of laser beam with a narrow linewidth in the sub-angstrom range has inspired the development of software programs such as FASCODE (acronym for **FAS**t atmospheric **S**ignature **CODE**) [35] which can compute the atmospheric transmittance with line-by-line resolution. Concerning the remainder of this section, we will present a number of simulation results of atmospheric transmittance (and sky radiance) obtained using MODTRAN software program.

As an illustration of the effects of atmospheric interaction with an optical beam, Figure 8.10 depicts the case of atmospheric transmittance over a horizontal path of 10 km at the sea level. The wavelength range is between 400 and 2500 nm, the visual range is 23 km (clear sky). In this simulation, the rural model is used for the

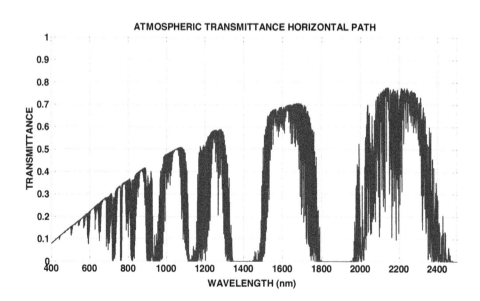

FIGURE 8.10   Atmospheric transmittance over 10 km at sea level after a MODTRAN simulation. A model of rural aerosol with visual range of 23 km, corresponding to relatively clear sky, is assumed.

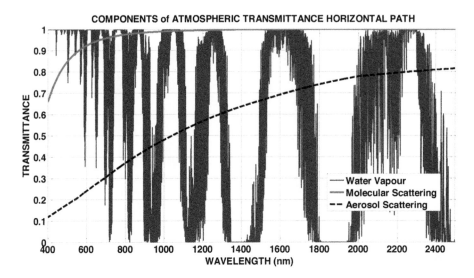

FIGURE 8.11   Contribution of Rayleigh scattering, Mie scattering, and water vapor to the atmospheric transmittance as in Figure 8.1. Notice that the band structure is mainly due to water vapor. Effects of molecular scattering are not noticeable over 1000 nm, while aerosol scattering is noticeable all over the spectrum.

aerosol distribution. The rural model is generally applied to describe locations (distant from urban centers and industrial areas) where aerosol particles are generated by vegetation, dust, etc. Specifically, the rural model assumes that the aerosol particles suspended in the atmosphere mainly consist of organic material mixed with water droplets.

Analyzing the transmittance spectrum, one can clearly notice that the spectrum itself is modulated between transmission bands with high transmittance and bands where the transmittance is close to zero (forbidden bands). One of the main contributors to this modulation of atmospheric spectrum is water vapor, as seen in Figure 8.11, where the effects of water vapor are shown along with Rayleigh scattering and Mie scattering (each plotted distinctively). In the example represented in Figure 8.11, other molecular components actively contribute to modulate the transmittance spectrum, such as CO_2, oxygen, etc. Usually, these gas elements' transition lines are very narrow with linewidth well below 1 Å. The effects of molecular and aerosol scattering to the transmittance spectrum appear more like a continuum, with molecular scattering disappearing at wavelength greater than 1 μm.

Interestingly, while the aerosol concentration is not very high in comparison to the other atmospheric constituents, aerosol scattering still has a sizeable effect over the transmittance spectrum, resulting in one of the main limiting factor in the high transmittance bands in the infrared.

To complete the discussion on the atmospheric transmittance, Figure 8.12 illustrates the transmittance spectrum for a slant path, Earth-to-space. In this example, two ground levels are considered, 0 km (sea level) and 2 km (e.g., a mountaintop).

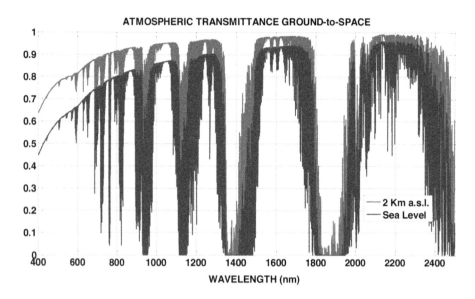

FIGURE 8.12 Atmospheric transmittance in an Earth-to-space path at zenith. A rural aerosol composition with a surface visual range of 23 km is considered. The data refers to the case of an observer located at two elevations: sea level (lower transmittance) and 2 km above sea level.

Again, the wavelength range considered is 400 to 2500 nm with clear sky conditions and a visual range, $V = 23$ km. Even if a ground-to-space path is in order of tens of kilometers, the related transmittance is higher when compared to the same distance horizontal link. The reason for this difference is the diminishing atmospheric density at higher elevation, which causes an overall smaller interaction with the propagating light beam. Moreover, on a mountain top, aerosol concentration is greatly reduced, and so, one should expect for aerosol extinction.

To confirm this statement, Figure 8.13 shows and compares the contribution to the transmittance of aerosol scattering at an observation altitude of 0 and 2 km. It must be noticed that the plot of Figure 8.12 refers to the shortest (vertical) ground-to-space path at the zenith. In the case of a slanted path, Earth-to-space (or vice versa) with an angle θ from the zenith, the path will be longer and the related atmospheric loss larger. However, if the optical depth at the zenith $\tau(\lambda)$ is known, one can compute the atmospheric loss in dB at the zenith angle θ as:

$$L_{ATM}(\theta) = -10log_{10}T_{ATM}(\theta) = 4.34\tau(\lambda)m(\theta) \qquad (8.77)$$

where $m(\theta) \geq 1$ is the relative air mass that takes into account the different contribution of a path length with respect to zenith, corresponding to $m(\theta = 0) = 1$. Under the parallel plane approximation of the atmosphere (ignoring Earth's curvature), one can express the air mass as:

$$m(\theta) = \sec(\theta) \qquad (8.78)$$

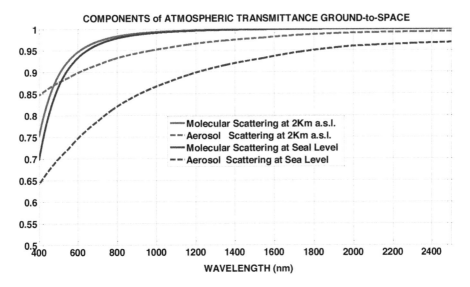

FIGURE 8.13 Atmospheric transmittance in an Earth-to-space path at zenith. The observer is at two different altitudes (sea level and 2 km above sea level): (a) molecular scattering contribution at 2 km, (b) molecular scattering contribution at sea level, (c) aerosol scattering contribution at 2 km above sea level, and (d) aerosol scattering contribution at sea level. Data obtained after MODTRAN simulation.

with this expression valid up to $\theta = 60°$. At larger zenith angle, $60° < \theta < 90°$, there are a number of approximations in the Literature to express the relative air mass, including the one in Ref. [36]:

$$m(\theta) = \frac{1}{\cos(\theta) + 0.505572(96.07995 - \theta)^{-1.664}} \quad (8.79)$$

in which the angle θ is expressed in degree.

8.2.2 BACKGROUND RADIATION AND SKY RADIANCE

In a number of applications, the receiver encompasses a source whose angular extension is larger than the receiver field of view. In this case, the extended source power captured by the optical receiver can be calculated when the source radiance B_λ is known, usually expressed dimensionally in W cm^{-2} sr^{-1} μm^{-1}. The resulting optical power, P_R, captured in presence of an extended source is therefore:

$$P_R = B_\lambda A_R \Omega_R \Delta\lambda L_T, \quad (8.80)$$

where

A_R is the area of the receiver (or telescope) in cm^2.
Ω_R is the receiver field of view in solid angle.

$\Delta\lambda$ is the system wavelength band of interest in µm.

L_T is the optical system loss.

In the case of space-to-ground optical communications, an extended source can be represented by planets, the Moon, or (event not recommendable) the Sun. Those astronomical bodies can be (depending on the applications) either a signal source or source of background noise when they are in line of sight with the transmitter. In the latter case, the background radiance is received in addition to the optical signal and it negatively affects the receiver performances.

Another common case where the background radiance constitutes a source of unwanted noise (for an optical communications receiver) occurs during daytime operations. Daytime sky radiance is originated by the scattering of the Sun irradiance. Spectrum of daytime sky radiance is correlated to Sun spectrum (Figure 8.14), which has its peak wavelength around 475 nm according to the properties of a blackbody emitter at a temperature $T = 5800$ K.

As illustrated for the case of atmospheric transmittance, the diffused sunlight of the sky can be originated by molecular and aerosol scattering, with the latter being the prominent source of scattering (and sky radiance) in the near-infrared region. Because of the different concentrations of aerosols in the atmosphere, sky radiance is larger near ground and smaller at higher elevations. Moreover, sky radiance will depend on the angular proximity of the observer line of sight to the Sun, fast decreasing in a nonlinear fashion, as the angle separation from the Sun increases. As an example of the discussion, Figure 8.15 presents two cases of daytime spectral sky radiance. The Sun is at 45° from the zenith, the observer is at sea level looking up in the sky (as in the case of a downlink) at 40° and 70° from the zenith (Sun and downlink path are in the principal plane). The data are generated by a MODTRAN simulation, which uses a rural model to represent the aerosol concentration with

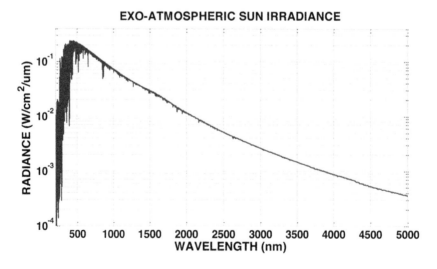

FIGURE 8.14   Exo-atmospheric spectral irradiance of the Sun at a distance of 1 AU.

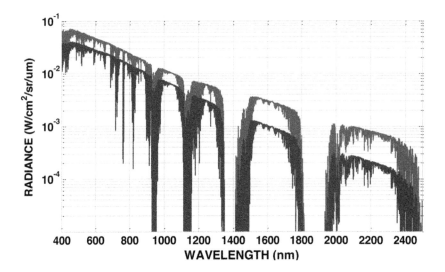

FIGURE 8.15 Daytime sky radiance at sea level. The Sun zenith angle is 45°. Two cases (radiance curves) are shown: (1) the observer zenith angle on the ground is at 40° (higher radiance curve) and (2) the observer zenith angle on the ground is at 70° (lower radiance curve). The rural aerosol model with a visibility of 23 km at sea level was used. Data obtained after MODTRAN simulation.

a visual range of 23 km at the ground level. The figure clearly shows that for the direction closer to the Sun (5°), a receiver will experience the larger amount of sky radiance which will peak around 470 nm (peak of the Sun spectral irradiance).

The action of the atmospheric transmission manifests itself on the spectral radiance by modulating the values of radiance and causing forbidden bands (similar to the spectral transmittance of the atmosphere), where the sky radiance (and sunlight in general) is virtually zero. For wavelengths of 1 μm and above, aerosol scattering is the main cause of sky radiance and therefore one should expect that at locations with low aerosol concentration, this portion of the radiance spectrum will be smaller in magnitude. As a demonstration of this concept, Figure 8.16 shows sky radiance data, *ceteris paribus*, when the observer (or receiver during a downlink) is located at 2 km above sea level. At 2 km, the concentration of aerosol is lower than sea level with consequent lower sky radiance values with respect to those indicated in Figure 8.14.

Clouds are another meteorological factor that contributes daytime sky radiance (besides the atmospheric opacity). Particularly noteworthy is the action of subvisual cirrus clouds that are composed of miniscule ice particles suspended high in the troposphere. These clouds affect the sky radiance in a mechanism similar to that of aerosols. MODTRAN and other similar software programs can also determine daytime sky radiance amount related to subvisual cirrus clouds. On the other hand, lower clouds have a more complex scattering mechanism including the fact that they can redirect sunlight irradiance in different directions in the sky, with the result that sky radiance is greatly dependent on the observation direction with respect to the Sun and clouds position/distribution in the sky.

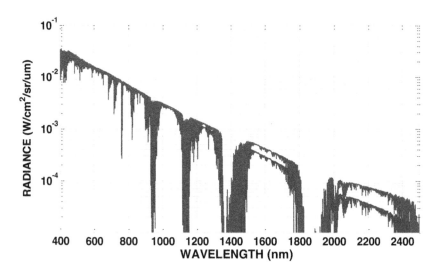

FIGURE 8.16 Daytime sky radiance at 2 km above sea level. The Sun zenith angle is 45°. Two cases (radiance curves) are shown: (1) the observer zenith angle on the ground is at 40° (higher radiance curve) and (2) the observer zenith angle on the ground is at 70° (lower radiance curve). The rural aerosol model with a visibility of 23 km at sea level was used. Data obtained after MODTRAN simulation.

8.3 CONCLUSIONS

The purpose of this chapter is to help a system engineer consider the effects and consequences of the atmospheric channel in an optical link. The resulting system parameters (e.g., losses) could then be quantified and used in a link budget that hypothetically approaches design.

Concerning the effects of clear air turbulence of the atmospheric channel, it is observed that these are consequences of the random nature of the refractive index of air. Some of the key effects to consider are the fade (and surge) of the signal due to scintillations, additional losses due to beam spreading and beam wandering, and the focal plane losses experienced when a (larger) spot size is imaged on the detector. Expressions were provided for different link configurations (uplink, downlink, etc.) to allow an easy calculation of the relevant parameters that could be used in a possible control design table. However, it is worthy to repeat that all these effects (due to clear air turbulence) are related in magnitude on the strength of the turbulence itself as described by the profile of the structure parameter of the refractive index $\left[C_n^2(h)\right]$. Therefore, great care must be taken in the choice of $C_n^2(h)$ that describes the turbulence. A number of parametric models of $C_n^2(h)$ have been provided in this chapter, more are available in the literature [2, 3]. Due to the random nature of the problem, it is quite difficult to properly represent all the possible variations of $C_n^2(h)$ in time ($C_n^2(h)$ changes can be very large during the 24 h of the day and during the year) and space (some locations can have more/less favorable $C_n^2(h)$ profile). One way to partially solve (or bound) the problem is to assume a possible profile of $C_n^2(h)$ (e.g., HV5/7), and then calculate the different parameters related to the clear air turbulence effects (e.g., beam spreading loss, probability of fade, etc.) while the profile is varied by a multiplicative factor (e.g., $0.5 \times HV5/7$, $5 \times HV5/7$, etc.). In this manner, one can predict how different turbulent strengths can affect the link. The multiplicative factor can be determined either by some heuristic approach, or scaled by measuring some turbulence parameters such as the C_n^2 close to the ground [37] or atmospheric coherence length [38].

Concerning atmospheric transmittance and sky radiance, these parameters can be obtained in principle by software simulation or by measurement in situ. However, after a number of measurement campaigns developed in recent years, one can also use the large number of data that already exists for the determination of atmospheric transmittance and sky radiance. For instance, the NASA AERONET program has deployed a number of sun photometers around the world for the characterization of aerosol concentrations with measurements of the atmospheric transmittance and radiance. The AERONET Program [39] makes readily available a multiyear data bank of atmospheric transmission and radiance for a number of wavelength channels collected by sensors distributed on locations all over the world.

REFERENCES

1. N. Kolmogorov, The local structure of turbulence in an incompressible viscous fluid for very large Reynolds numbers, *C. R. (Dokl.) Acad. Sci. USSR* 30, 301–305 1941.
2. L. C. Andrews and R. L. Phillips, *Laser Beam Propagation through Random Media*, SPIE Optical Engineering Press, Bellingham, WA, 1998.
3. R. R. Beland, Propagation through atmospheric optical turbulence, in *The Infrared and Electro-Optical Systems Handbook*, Vol. 2, F. G. Smith (Ed.), SPIE Optical Engineering Press, Bellingham, WA, 1993, Chapter 2.
4. G. Y. Jumper, H. M. Polchlopek, R. R. Beland, E. A. Murphy, P. Tracy, and K. Robinson, Balloon-borne measurements of atmospheric temperature fluctuations, 28th Plasmadynamics and Lasers Conference, AIAA-1997-2353, Atlanta, GA, June 23–25, 1997.
5. G. C. Valley, Isoplanatic degradation of tilt correction and short-term imaging systems, *Appl. Opt.* 19, 574–577, 1980.
6. R. J. Hill and S. F. Clifford, Modified spectrum of atmospheric temperature fluctuations and its application to optical propagation, *J. Opt. Soc. Am.* 68(7), 892–899, 1978.
7. V. I. Tatarskii, *The Effect of the Turbulent Atmosphere on Wave Propagation (Trans. for NOAA by)*, Israel Program for Scientific Translations, Jerusalem, 1971.
8. D. L. Fried, Aperture averaging of scintillation, *J. Opt. Soc. Am.* 57(2), 169–175, 1967.
9. J. H. Churnside, Aperture averaging of optical scintillations in the turbulent atmosphere, *Appl. Opt.* 30(15), 1982–1994, 1991.
10. H. T. Yura and W. G. McKinley, Aperture averaging of scintillation for space-to-ground optical communication applications, *Appl. Opt.* 22, 1608–1610, 1983.
11. G. Parry, Measurements of atmospheric turbulence-induced intensity fluctuations in a laser beam, *Opt. Acta* 28, 715–728, 1981.
12. E. Jakeman and P. N. Pusey, The significance of K-distribution in scattering experiments, *Physics. Rev. Lett.* 40, 546–550, 1978.
13. E. Jakeman, On the statistics of K-distributed noise, *J. Phys. A* 13, 31–48, 1980.
14. J. H. Churnside and R. J. Hill, Probability density of irradiance scintillations for strong path-integrated refractive turbulence, *J. Opt. Soc. Am. A* 4, 727–733, 1987.
15. J. H. Churnside and S. F. Clifford, Log-normal Rician probability-density function of optical scintillations in the turbulent atmosphere, *J. Opt. Soc. Am. A* 4, 1923–1930, 1987.

16. L. C. Andrews, R. Phillips, and C. Y. Hopen, *Laser Beam Scintillation with Applications*, SPIE Press, Bellingham, WA, 2001.

17. D. M. Strong, E. P. Magee, and G. B. Lamont, Implementation and test of wave optics code using parallel FFT algorithms, *Proc. SPIE* 4167, 34, 2001.

18. S. Coy, Choosing mesh spacings and mesh dimensions for wave optics simulation, *Proc. SPIE* 5894, 589405, 2005.

19. N. Roddier, Atmospheric wavefront simulation using Zernike polynomials, *Opt. Eng.* 29, 1174–1180, 1990.

20. D. L. Fried, Optical resolution through a randomly inhomogeneous medium, *J. Optic. Soc. Am.* 56, 1372–1379, 1966.

21. F. Roddier, The effects of atmospheric turbulence in optical astronomy, in *Progress in Optics XIX*, E. Wolf (Ed.), North-Holland, New York, 1981.

22. J. W. Hardy, *Adaptive Optics for Astronomical Telescopes*, Oxford University Press, New York, 1998.

23. D. P. Greenwood and D. L. Fried, Power spectra requirements for wavefront compensative systems, *J. Opt. Soc. Am.* 66, 193–206, 1970.

24. R. K. Tyson, *Principles of Adaptive Optics*, 2nd ed., Academic Press, Boston, MA, 1998.

25. D. P. Greenwood, Bandwidth specifications for adaptive optics systems, *J. Opt. Soc. Am.* 67, 174–176, 1976.

26. D. L. Fried, Anisoplanatism in adaptive optics, *J. Opt. Soc. Am.* 72, 52–61, 1982.

27. L. C. Andrews, R. L. Phillips, R. J. Sasiela, and R. R. Parenti, Strehl ratio and scintillation theory for uplink Gaussian-beam waves: Beam wander effects, *Optical Engineering* 45(7), 2006.

28. E. J. McCartney, *Optics of the Atmosphere*, Wiley, New York, 1976.

29. B. A. Bodhaine, N. B. Wood, E. G. Dutton, and J. R. Slusser, On Rayleigh optical depth calculations, *J. Atmos. Ocean. Tech.* 16, 1854–1861, 1999.

30. W. E. K. Middleton, *Vision through the Atmosphere*, University of Toronto Press, Toronto, 1963.

31. H. C. van de Hulst, *Light Scattering by Small Particles*, Wiley, New York, 1957.

32. J. E. A. Selby and R. A. McClatchey, Atmospheric transmittance from 0.25 to 28.5 μm: Computer code LOWTRAN 2, AFCRL-TR-72–0745, AD 763721, 1972.

33. A. Berk, L. S. Bernstein, and D. C. Robertson, MODTRAN: A moderate resolution model for LOWTRAN 7, GL-TR-89–0122, Phillips Laboratory, Geophysics Directorate, Hanscom Air Force Base, Bedford, MA, 1989.

34. A. Berk and F. Hawes Validation of MODTRAN 6 and its line-by-line algorithm, *J. Quant. Spectrosc. Radiat. Transf.* 203, 542–556, December 2017.

35. H. J. P. Smith, D. B. Dube, M. E. Gardner, S. A. Clough, F. X. Kneizys, and L. S. Rothman, *FASCODE—Fast Atmospheric Signature Code (Spectral Transmission and Radiance)*, AFGLTR-78–0081, ADA 057359, 1978.

36. Kasten, F. and Young, A. T., Revised optical air mass tables and approximation formula, *Appl. Opt.* 28, 4735–4738, 1989.

37. F. S. Vetelino, B. Clare, K. Corbett, C. Young, K. Grant, and L. Andrews, Characterizing the propagation path in moderate to strong optical turbulence, *Appl. Opt.* 45, 3534–3543, 2006.

38. L. Wang, M. Schöck, G. Chanan, W. Skidmore, R. Blum, E. Bustos, S. Els et al., High-accuracy differential image motion monitor measurements for the thirty meter telescope site testing program, *Appl. Opt.* 46, 6460–6468, 2007.

39. http://aeronet.gsfc.nasa.gov.

9 Optical Ground Station: Requirements and Design, Bidirectional Link Model and Performance

Marcos Reyes García-Talavera,
Zoran Sodnik, and Adolfo Comerón

CONTENTS

9.1   INTRODUCTION

When compared to traditional approaches using RF systems, there are well-known advantages of using optical wavelengths for communication between satellites including a large increase in bandwidth (and thus data rate), reduced terminal mass, reduced power consumption, very high antenna gain, no bandwidth and frequency restrictions, and no electromagnetic interference. The main disadvantage is that extreme pointing, acquisition, and tracking (PAT) accuracies are required.

As compared to intersatellite links, ground-based satellite links are peculiar because part of them cross the Earth's atmosphere. While intersatellite links have theoretically 100% access probability, ground-based satellite optical links cannot meet these requirements due to the possibility of cloud coverage. Additionally, constituent gases and particles suspended in the atmosphere will produce extinction of the electromagnetic power. Because of the short wavelengths employed in the optical links, the inhomogeneities of the index of refraction of the atmosphere arising from atmospheric turbulence may produce sizable distortion (usually negligible at microwave frequencies) of the wave fronts as they propagate through the link atmospheric section. These dynamic distortion effects induced by the atmospheric turbulence may have a considerable influence in the link performance.

In downlinks, for which the turbulence is close to the receiver, it will give rise to several effects. The first effect is the scintillation—random fluctuations of the power collected by the receiving aperture. The other effects can be called image effects. The image motion, also called tip/tilt or angle-of-arrival fluctuations, is an apparent fluctuation of the transmitter position around the nominal one, originated from the fluctuations of the average inclination of the wave front reaching the receiver aperture. The blurring is an apparent random spread of the transmitter source, coming from the distortion of the wave front across the aperture.

In uplinks (for which the turbulence is concentrated in the path section close to the transmitter), in addition to scintillation, the so-called beam effects may also occur. Beam effects can be considered as the duals of the image effects. In particular, large-size index of refraction inhomogeneities may result in deflecting the whole beam; smaller inhomogeneities crossed by the beam will give rise to beam distortion.

This chapter is focused on the requirements that an optical communications terminal on ground has to fulfill to establish a bidirectional link with a terminal in space, on providing a design reference for such a ground terminal (called optical ground station [OGS]), and finally on modeling and analyzing the performance of real ground-to-satellite optical links crossing the turbulent atmosphere.

9.2   GROUND TERMINAL REQUIREMENTS

The main purpose of an OGS is to establish a bidirectional communications link with an optical terminal onboard a satellite in the presence of the atmospheric turbulence. Considering the expected disturbances in the optical link due to the turbulence, built-in capabilities to monitor propagation effects are also looked for, to be capable of characterizing the link performance. These aims impose several top level requirements on OGS. These requirements can be classified into different groups.

9.2.1   OGS SITE

To mitigate (to the maximum possible extent) the atmospheric turbulence effects on the beams, the OGS has to be built on a high site, to reduce the optical air mass as much as possible, and thus obtain good-to-excellent atmospheric turbulence conditions. For this reason, the best astronomical sites in the world are the perfect candidates for an OGS, sites with low humidity, low occurrence of clouds and dust episodes, and low-strength atmospheric turbulence.

Additionally, common-purpose infrastructures are required to support the operation of an OGS. A very isolated site will impose additional costs on an OGS, which might make it unfeasible. In this case, facilities of an astronomical observatory provide the required capabilities for reliable maintenance and operation of the OGS.

Considerations related to the visibility of spacecrafts will also affect the site. High elevation angles will be preferred for the link to reduce the equivalent air mass and atmospheric turbulence effects. In the case of geostationary orbit, proximity of the site to the equator contributes to minimize the zenith angle and thus the air mass, which needs to be passed by a laser beam from/toward a geostationary satellite.

To optimize its sky coverage and to make possible on ground demonstrations with horizontal pointing, the OGS design shall allow it to cover the following range of telescope pointing directions: azimuth from 0° to 360° and elevation from 90° (zenith) to 0°. This requirement also has an impact on the OGS site selection, and other buildings on the site should minimize obstruction of these angles.

9.2.2   OGS TELESCOPE

9.2.2.1   Telescope vs. Link Budget

In the case of free-space optical communications and direct (incoherent) detection, the OGS telescope basically acts as a light collector (photon-bucket). The specification about the required diameter will be given by the link budget for a space-to-ground link. The main input parameters for the link budget will be the output power of the transmitter terminal onboard the satellite, its divergence, and the satellite location (distance to Earth). To achieve a certain power level at the OGS receiver detector, it is clear that the performance will be improved if a telescope with a larger aperture is used.

A large aperture also provides the advantages of having several small apertures. Several apertures have additional technical problems related to light combination like optical coupling to fibers, apertures matching, and coalignment.

In the case of coherent detection, the main objective is not only to collect as many photons as possible but also to obtain a good image resolution. The problem is like in astronomy, the telescope cannot be considered only as photon-bucket, special attention has to be paid to the optical quality. The phase of the wave brought to the receiver detector surface should be retrieved.

If coherent detection is foreseen, the case of having several small apertures has also intrinsic disadvantages. Special attention has to be paid to compensate for the optical delay in the different optical paths before combining the light at the detector.

The intrinsic disadvantage of having one large aperture is the additional complexity of the facility. But the factor that makes the difference obvious is the performance under atmospheric turbulence conditions.

9.2.2.2 Mitigation of Atmospheric Turbulence Effects with the Telescope

It is well known from the field of astronomy that light crossing atmospheric turbulence suffers from the inhomogeneities in the atmosphere produced by fluctuations of the index of refraction in the air. Several effects can be distinguished in the image quality degradation produced by the turbulence, and the telescope aperture plays an important role in the contribution to the different effects to the total image quality degradation.

Image motion is the result of global inclination of the wave front at the entrance of the receiver aperture (also called angle-of-arrival fluctuations). From the image point of view, it is observed as the movement around its nominal position of the sharp image points. Image motion is averaged with a larger aperture and, therefore, a large telescope diameter is preferable to reduce this effect.

Blurring is the spreading of the airy disk of a point source, while the center of the disk is maintained. It is observed as the loss of image sharpness. The blurring effect is related to the coherence diameter of the atmospheric turbulence r_0 (Fried's parameter). For aperture diameters much smaller than r_0, the blurring is determined by the aperture; while for diameters much bigger than r_0, the blurring is determined by the atmosphere. We are interested in having aperture diameters bigger than r_0 to reach the blurring limit imposed by the atmosphere.

Scintillation is the fluctuation of intensity of the wave front points, due to the fact that individual light rays are deviated differently by the atmospheric inhomogeneities, producing changes in the light collected by the telescope aperture. Again, the use of a large aperture reduces the effect of the scintillation in the receiver.

It can thus be concluded that a large aperture at the OGS telescope will provide a better performance from the receiver point of view. The respective sizes of the transmitting and receiving apertures need not be the same. From the transmitter point of view, the beam aperture should be adjusted to the specific link scenario (atmospheric turbulence conditions, link distance, etc.); the telescope diameter should not be a limit to adjust the transmitting aperture.

9.2.3 OGS Instrumentation

The OGS shall include all the instrumentation required to establish and test bidirectional optical communication links with satellites, including capabilities to determine the impact of turbulence and communication performance. Instrumentation of the OGS shall also be designed and operated to reduce the turbulence effects to the maximum possible extent.

There are several requirements imposed on the focal plane instrumentation for the laser communication system (CS). The instrumentation shall be located in a stable focal station, mainly for the laser system. The optical bench shall have flexibility of access to different systems. The diversity of instrumentation also requires a large available space. These arguments again favor the selection of a large facility with a large telescope provided with infrastructure around.

Two main systems may be distinguished in the focal plane instrumentation required for an OGS: the focal plane optics (FPO) and the focal plane control (FPC, including electronics and software).

The main performance requirements these systems have to fulfill are summarized below.

To provide flexibility to the OGS, a generic optical path shall be provided for receiving and transmitting signals. These generic optics shall be able to work in a large wavelength range (from visible to near-infrared). Therefore, the OGS design shall include a generic optoelectronic equipment and some satellite-specific components that could be easily exchanged.

The effect of atmospheric turbulence on the received signal shall be mitigated by means of a PAT system.

Demodulation capabilities (configurable, to be adapted to satellite transmitter format) shall be provided in the receiver system.

Apart from retrieving the received signal data, an additional purpose of an OGS is to characterize the satellite signal from the propagation point of view. Because of the importance of atmospheric propagation effects on an optical link performance, an OGS should also be equipped to determine these effects. Therefore, additional instrumentation shall be provided in the receiver path, capable of measuring effects of propagation through atmospheric turbulence.

With respect to the transmitter path, the effect of free-space propagation delay shall be compensated by means of a point ahead angle (PAA) to the transmitted optical signal.

The beam divergence of the transmitted optical signal shall be adjustable with good resolution, in order to be capable of coping with different link situations, mainly atmospheric turbulence conditions and satellite position information accuracy.

Various techniques have to be considered to avoid having a far-field pattern like a speckle field with strong intensity fluctuations in short sampling times. A far-field with high temporal stability is required to achieve a reasonable bit error rate (BER) for high data rates. Multiple apertures in the transmitter path shall be considered, together with introducing delays larger than the longitudinal coherence length of the transmitter laser in the transmitting paths of each aperture. This point will be specifically addressed in the design and performance section.

The radiant intensity transmitted by the OGS shall be adjustable, i.e., transmitter laser power shall be controllable in a large range (up to a factor of 100), and with good resolution.

The wavelength of the transmitted signal shall be adjustable (within the potential range of laser technology used), to be capable of working with satellites with different optical systems and receivers.

Modulation capabilities must be provided in the transmitter system. Special attention has to be paid to the transmitted signal polarization format, to be compatible with the optical payload onboard the satellite.

9.2.3.1 Focal Plane Optics

The first requirement of the FPO is that it should be mounted on an optical bench, rigidly connected to the OGS structure or foundation, as a means to avoid additional disturbances coming from differential vibrations between the telescope and the FPO.

The beam path from the telescope to the optical bench shall be as isolated as possible. A vacuum feedthrough should be considered.

Active elements (mirrors) are required to implement the acquisition and tracking. An active (steering) mirror shall be located at the plane of the reimaged telescope aperture, the internal pupil. To avoid lateral beam displacements, the center of rotation of the active mirror shall be located in the plane of the mirror surface.

The steering mirror should be capable of partially compensating atmospheric turbulence (low Zernike modes) to improve beam pointing and stability.

Capabilities shall be provided to have an image of the receiver/telescope field of view (FOV), as an auxiliary equipment for alignment and troubleshooting. The best way is to use the part of the received spectrum not used for any optical communication. Depending on the link budget, it has to be evaluated whether the entire satellite light will be directed to the rest of the system, or if a small percent of that satellite light will also be directed to this auxiliary camera.

The same telescope will be used to transmit and receive. The possibility of having a separate transmitting aperture has two main drawbacks: a second telescope with turbulence mitigation system will be required for launching the transmitter beam (implying additional costs); and guaranteeing the beam alignment between the receiver and transmitter apertures will be difficult. Therefore, an optical isolator (OI) shall be installed, to isolate the transmitter path from the receiver path. This is necessary to avoid part of the high transmitted power to enter into the sensitive detectors of the received beam.

The transmitter optics shall include a system to change the beam divergence. A beam expander (BE) with mechanisms to control the output beam diameters should be the appropriate option.

An optical subsystem to divide the beam into several apertures shall also be included. An additional goal will be to provide a means to select the number of subapertures to be used.

Another active mirror with high resolution shall be required to provide the PAA. PAAs for typical low earth orbit (LEO) or geostationary earth orbit (GEO) are in the order of a few arc seconds.

Finally, a beam rotation system shall be provided to avoid that the beams provided by the different transmitting subapertures, small in diameter and out of axis in the pupil plane, are vignetted by any structure in the path, mainly the secondary spider.

9.2.3.2 Focal Plane Control

To establish an optical link between the OGS and an optical payload onboard a satellite (and to characterize it), the OGS instrumentation has to fulfill three major functions: to acquire and track the satellite optical signal, to transmit a reliable optical beam to the satellite, and to measure the characteristics of both the received and transmitted signal. The FPC shall provide those functionalities.

To acquire and track the satellite signal (apart from controlling the telescope), the FPC has to include the control of all the mechanisms installed in the FPO (beam focus, beam divergence beam rotation, beam splitters, polarization plates); and a PAT system as one of its main subsystems. To generate the transmitted signal, the FPC has to include a control for the transmit laser system (TLAS) and its main parameters (wavelength and power). Finally, for the characterization of the signals and link, the FPC has to include (mainly) a CS with data transmitter and receiver (modulator and demodulator), a receiver front end (RFE) (with its proximity electronics), communications (BER) analyzer, and many other general-purpose instrumentation for beam characterization such as a spectrometer, polarimeter, or wave front sensor (WFS).

External communications capabilities shall be included to allow establishing a link with the different payload and satellite control stations, mainly to obtain orbit and performance information. Ancillary equipment shall also be available, to have information on conditions that will affect the link performance (like seeing monitor and meteorological station). These last elements are used for check-out, calibration, and monitoring, and will not interfere with the normal operation of FPC, although they will provide assistance and fundamental information to the OGS operators.

To optimize the operational cost and resources of the ground station, all the controls of the instrumentation must be centralized in a main control computer (MCC). It should be the FPC main computer and serve as the only interface between the OGS operator and the FPC. The basic architecture of the FPC is shown in Figure 9.1.

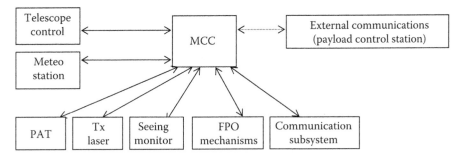

FIGURE 9.1 FPC block diagram.

9.2.3.3   FPC: General Requirements

The basic design of the FPC is driven by the following considerations.

- The FPC system will operate in an astronomical-type observatory. High altitude, low temperature, and humidity conditions have to be taken into account in the design of the control system.
- The PAT operation must be independent of the rest of the system. The acquisition sequence has very tight time constraints, and therefore, an independent real-time control system will perform this task. It will interface with the MCC, accepting commands and sending status flags, but it will maintain the closed loop tip/tilt control of the incoming laser beam, regardless of the tests being performed.
- Some of the measurements must be performed all the time during an experiment. Therefore, a set of instruments (such as the PAT, the BER analyzer, the spectrometer, or the polarimeter) must have the capability of taking data continuously.
- The sequence of acquisition and test must be precisely timed. For this reason, a high accuracy time signal shall be distributed independently to the different elements, and the intervention of a human operator shall be minimized at least during acquisition.
- The MCC will have to perform several precisely timed tasks simultaneously. It requires a system with good real-time and parallelism capabilities.

9.3   OGS DESIGN EXAMPLE

In the framework of the Data Relay and Technology Mission, the European Space Agency (ESA) has undertaken the development of Optical Data Relay payloads, aimed at establishing free-space communication links between satellites. The first of such systems put into orbit is the semiconductor-laser intersatellite link experiment (SILEX) project [1, 2].

SILEX has demonstrated the world's first free-space optical communication links between two spacecraft, namely the French low-orbit earth-observation satellite SPOT-4 and the ESA's geostationary satellite ARTEMIS (Advanced relay and technology mission). The link is being used to relay earth observation data. ESA's SILEX project has proven that state-of-the-art technology can meet the stringent pointing accuracy requirements of optical wavelengths.

In order to perform in orbit testing of the optical communications terminals involved in SILEX as well as for other future missions, ESA and the Instituto de Astrofísica de Canarias (IAC) reached an agreement to build the OGS in the IAC Teide Observatory. The OGS consists of a 1 m telescope and the suitable instrumentation for testing optical communications, capable of carrying out ground ↔ satellite bidirectional optical experiments.

The present section is based on the OGS design. The design [3, 4] is presented as an example fitting the general requirements mentioned in Section 9.2.3.3.

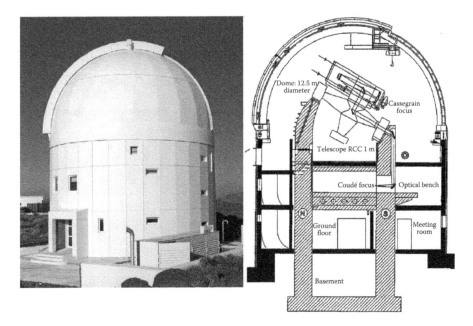

FIGURE 9.2 Left: Picture of the OGS building. Right: Section of the building, showing the upper floor with the telescope and the optical laboratory below with the Coudé focus.

9.3.1 BUILDING AND TELESCOPE

The OGS consists of an observatory building with a dome and the associated infrastructure, a telescope including control system and the optical payload test instrumentation. Figure 9.2 shows a picture of the ESA OGS and the section of the OGS building. In Figure 9.3, the telescope inside the dome can be seen.

The telescope is a 1 m Ritchey-Chrétien/Coudé telescope from Zeiss. The telescope is installed in the OGS upper floor, supported on an English mount. The telescope has two different optical configurations.

FIGURE 9.3 Picture of the OGS telescope inside the dome.

TABLE 9.1
OGS Building and Telescope Main Characteristics

OGS

Building location

Place	Teide Observatory of the Instituto de Astrofísica de Canarias (IAC), Tenerife Island, Spain
Longitude	16.5101° West
Latitude	28.2995° North
Altitude	2.393 km
Telescope	
Technology	Coudé system, SITAL mirrors, steel structure
Entrance pupil diameter	1016 mm
Coudé focal length	36.5 m
Telescope pointing mechanism	Linear motors with tacho-generator, coarse, and fine angular encoders
Open loop pointing accuracy	±50 μrad (anywhere on the sky)

The Ritchey-Chrétien system basic configuration has a focal length at the Cassegrain focus of 13.3 m (f/13), and a FOV of 45 arcmin. This configuration is used for standard astronomical use and Space Debris observations.

The Coudé configuration has a focal length of 39 m, f/39, and a FOV of 8 arcmin. This focus is the one used for optical payload testing. The main characteristics are summarized in Table 9.1.

The Coudé focus is located in a dedicated laboratory one floor below the telescope (see Figure 9.2). In order to minimize telescope and internal seeing effects on the optical signal, the optical path is guided from the upper telescope floor to the Coudé laboratory via a vacuum beam feedthrough. The Coudé laboratory is thermally stabilized by passive means, using only the thermal inertia of the concrete mass in connection with an appropriate external thermal insulation. Air over pressure is provided to prevent dust in the optical laboratory.

The telescope and Coudé optical bench constitute one separated concrete unit (inner building), that is, vibration isolated from the external building. This is a typical double foundation structure of astronomical optical telescopes facilities, to decouple the wind load on the building from the telescope, keeping it much more stable.

9.3.2 FPO Main Characteristics

All the instrumentation is installed on an optical bench in the Coudé laboratory. The bench comprises the optics, the actuators, and detectors for the PAT system, the instrumentation for optical signal analysis, and the laser transmitter. The technology used for the optical bench is commercial, standard honeycomb optical bench size 5×2 m², and off-the-shelf mounts, aligned during assembly.

A diagram of the FPO is shown in Figure 9.4. It comprises several light paths. First is described the receiver path, and then the transmitter path.

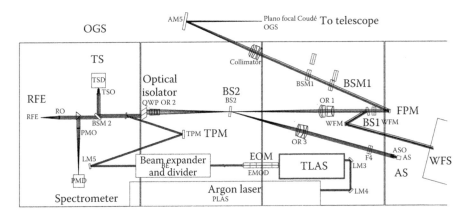

FIGURE 9.4   Optical diagram of the OGS FPO.

The light coming from the telescope and reaching the Coudé focus is collimated and routed via a beam splitter (BSM1) to reimage the entrance pupil onto the fine pointing mirror (FPM). BSM1 is used to send the visible spectrum, which is outside the wavelength band for communication to a large format camera, called Coudé Camera, capable of observing the full Coudé FOV. The purpose of this camera is direct observation, pointing verification (in some cases, looking at the sunlight reflected from the satellite), and background monitoring.

The FPM is the steering mirror necessary to implement the acquisition and tracking, and to compensate tip/tilt from atmospheric turbulence effects. It is a fundamental component, and combines a large aperture with a high bandwidth. From the FPM the light passes a beam splitter (BS1), which can direct part of the light to a WFS, and is focused by an optical relay system onto the acquisition sensor (AS) beam splitter (BS2). BS2 is a geometrical beam splitter, with an aperture in its center representing the FOV of the tracking sensor (TS). This way the AS will receive only the light reflected by the BS2, if the incoming signal is not centered on-axis. The light is reimaged onto the AS by another optical relay system. In Section 9.3.3.1, the purpose of the AS and the TS is explained in more detail.

The light that passes through the central hole of BS2 is collimated again and passes an OI, which isolates the transmitted and received beams by about 120 dB. The light is then routed via a filter and another beam splitter (BSM2) toward the TS and the beam analysis equipment. At this point, the received beam can be switched by a mirror between different analysis equipment: a RFE, a spectrometer, and a power meter/polarimeter.

The transmitter path starts at the TLAS. It is based on a solid-state titanium–sapphire laser, pumped by an argon laser. The argon laser provides a pumping power of 28 W. The Ti:Sap laser was chosen for several reasons: it can provide up to 7 W of output power at the required wavelength (847 nm), and its cavity can be tuned via a birefringent filter between 750 and 890 nm, offering a good potential to cover also other wavelength schemes. The laser output goes to a beam splitter (BS3) that provides a small fraction of the signal to be sent to analysis instrumentation.

This instrumentation is in charge of monitoring and controlling the output power and the wavelength, in closed loop with the two lasers.

The transmitted light passes through an electro-optical modulator (EOM), in charge of modulating the light in intensity, with data rates up to 100 Mbps (NRZ). For security reasons, a mechanical shutter is also provided at the output of the EOM, to switch off the beam during operation or in emergency cases. This is part of the standard laser safety equipment required for high-power lasers operation.

The light goes to a BE (zoom system), to provide the appropriate beam divergence out of the telescope. It allows to adjust the beam diameter between 40 and 300 mm at the output of the telescope. The beam also goes to a half-wave plate to convert the laser polarization. The light is then fed into a rotating beam divider, with a delay line network. This subsystem is one of the key elements for the uplink feasibility, and will be described in more detail later in the description of the FPC—transmitter.

The light reaches the transmit pointing mirror (TPM), which allows to control the beam direction to align the transmitted beam with the received one, and also to adjust the PAA required for satellite communications in any orbit. The transmitter beam is then combined with the received beam in the spectral OI; and from there on it follows the same optical path as the received beam. The only difference is that at BS1 a small part of the transmitted signal is directed toward the WFS to analyze the optical quality and the alignment of the transmitted beam.

9.3.3   FPC DESIGN

As it has been mentioned, the main function of FPC is to control all the different elements of the focal plane instrumentation. To this end, it performs the following main tasks:

1. *An initialization sequence*: Prior, and with the required anticipation, to the execution of a test sequence, FPC will initiate all the different elements and check their normal behavior.

2. *Acquisition*: FPC is responsible for reading the signal from the AS, and once a beacon is detected, to generate the correction signal to move the FPM so that the beacon impinges within the FOV of the TS, and the outgoing beam is correctly pointed in due time. The different phases of this process are described in Section 9.3.3.1.

3. *Tracking*: Once the signal is detected by the TS, the FPC refines the pointing in successive approximations (rallying, see Section 9.3.3.1), and once the beam is within the four central pixels of TS, it closes a fast tracking loop, reading data from TS and sending position correcting commands to FPM.

4. *Measuring and data recording*: As soon as it is possible to do so, and even during the acquisition phase, FPC reads and stores data from the spectrometer and the polarimeter. Once the communication channel is securely established, it also detects whether the signal clock is being recovered from RFE and begins taking BER measurements. It must be noticed that all these beam analysis measurements cannot be simultaneous, and switching off elements of the FPO is necessary to execute the sequence.

5. *TLAS control*: The FPC switches on and off the high-power laser beam as required by the operations, and sets it to the correct power and wavelength. The transmitted beam diameter is controlled as well. It also sets the TPM to the required position (PAA).

6. *Telescope control*: The FPC commands the telescope to point to the fore-seen satellite position and tracks it, with orbit information obtained from standard spacecraft trajectory data message (STDM) files. The telescope focus is also commanded. In case the fine pointing mechanism (FPM) approaches its range limit or the satellite position is offsetting with respect to the telescope axis, the FPC sends the required commands to the telescope control system to off-load the FPM. FPC is also able to query about the status and position of the telescope.

7. *Human–machine interface*: The FPC is finally the main control of the whole system, and therefore it provides a suitable man–machine interface, giving comprehensive information to the operator, and accepting commands from the person.

8. *External communications*: The FPC provides a communications channel with the Payload Test Laboratory (PTL) in Redu, Belgium, so the schedule of the tests, the ephemeris of the satellite and the test results can be communicated.

9. *Data retrieving and preprocessing*: During a single test sequence, a big amount of data will be taken. Once the test has been executed and the link has been released, the FPC is capable of retrieving the data from the different subsystems, for analysis and distribution.

10. *Meteorological and seeing data recording*: These data are shown to the operator and sent to PTL, in real time if it is so required.

11. *Other sensors and actuators*: Besides the aforementioned functions, the FPC will read, record, and actuate in other elements and sensors of the optical bench, such as movable mirrors, in order to change the configuration of the experiments, housekeeping, etc.

In Table 9.2 are summarized the main characteristics of the receiver and transmitter main subsystems of the FPC, including the PAT components.

In the following sections, the main FPC subsystems are described in more detail.

9.3.3.1 PAT Overview

The small divergence of the optical beams makes the PAT procedure a critical issue in the design of the OGS and of the link. Particularly, it makes necessary the use of two sensors with different requirements in what concerns FOV, accuracy, and response time. The first one, called AS, is in charge of performing the first detection of the incoming signal, and generating a correction that can be used to deflect the outgoing laser beam so that it reaches the satellite. This deflection also brings the incoming signal within the FOV of the second sensor, called TS, which in turn is used to keep this pointing as stable as possible during the experiments at high speed. The central pixels of the TS work as a quadrant detector (QD), and can be sampled with high resolution in order to be able to achieve a higher sensitivity in the determination of the spot position. A diagram showing the configuration of the different sensors within the FOV is

TABLE 9.2
Characteristics of the OGS FPC Transmitter and Receiver Main Subsystems

Transmitters

Communication laser	Argon laser pumped Ti:Sap laser
Laser power (average)	300 mW (out of aperture)
Laser beam diameter ($1/e^2$)	40–300 mm, four Gaussian beams
Communication wavelength	847 nm, range 843–853 nm
Communication polarization	Left-hand circular
Communication modulation	NRZ, 49.3724 Mbps (fixed data rate)
Beacon laser	—
Beacon beam intensity	—
Beacon beam divergence	—
Beacon wavelength	—
Beacon polarization/modulation	—

Receivers

Data receiver	Silicon avalanche photodiode (APD)
Data receiver FOV (diameter)	87.3 µrad
AS	Dalsa CA-D1–0128 frame transfer CCD chip, 128 × 128 pixel, pixel size 16 µm × 16 µm, 100% fill factor
AS readout rate	30–400 Hz
FOV per pixel	20.5 µrad × 20.5 µrad
AS useful FOV	2327 µrad (diameter)
TS	Thomson TH7855A frame transfer CCD chip, 14 × 14 pixel, pixel size 23 µm × 23 µm, 100% fill factor
TS frame rate	1000 or 4000 Hz (changeable)
FOV per pixel	21.8 µrad × 21.8 µrad
TS useful FOV	262 µrad (diameter)
Quadrant detection sensor	Four center pixels of TS
Mission telemetry recording	Tracking error and irradiance: 1 kHz
Point ahead mechanism	Single two-axis mirror, piezoelectric actuators, capacitive position sensors
Fine pointing assembly	Single two-axis mirror, electromagnetic actuators, inductive position sensors

included in Figure 9.5. For the deflection of both the incoming signal and the outgoing beam, a fast steering mirror will be used, the already mentioned FPM. Its bandwidth, while using the TS signal, allows to compensate also for the image motion component of the atmospheric turbulence effects on the received signal.

As an example to illustrate the open loop pointing, the characteristics of the SILEX experiment onboard the ESA satellite ARTEMIS will be used. The aperture of the optical terminals used in SILEX is 25 cm. At a 800 nm wavelength, the diffraction-limited beam divergence is 0.43 in. The mechanical accuracy required to point such a narrow beam makes it virtually impossible to control the terminals in an open loop. In order to overcome this difficulty, a cooperative acquisition procedure is necessary. The procedure chosen in SILEX appears in Figure 9.6.

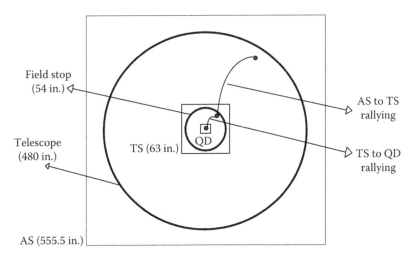

FIGURE 9.5   Different fields of view required by the PAT sequence and cooperative strategy.

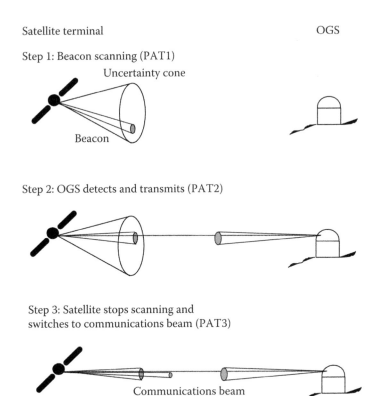

FIGURE 9.6   Acquisition sequence.

The ARTEMIS satellite emits a broad beacon, called acquisition signal, and scans it across its uncertainty cone (step 1, named phase PAT1), which is mainly given by its attitude control accuracy. It performs this scanning in such a way that the power is above the detection limit enough time (depending on the satellite beacon scanning speed) at each ground location.

Simultaneously, the OGS terminal surveys continuously a FOV wide enough to make sure that the satellite is within it (several arc minutes, 8 arcmin in the case of the OGS). When the scanning beacon reaches a position in which enough power is detected in the OGS's wide field sensor (AS), OGS determines its position and corrects the pointing of the outgoing signal in order to reach the satellite. At the same time, OGS centers the signal within the TS FOV (step 2, named phase PAT2).

When the GEO satellite detects a signal, it considers its beacon to be correctly pointed, and stops the beacon scan (step 3, named phase PAT3). Once this initial signal acquisition is successful, both terminals optimize their pointing by keeping the incoming signals as close as possible to the center of their respective TSs. Once both pointing control loops are stable enough, the system is ready to transmit data in both directions, and the satellite terminal turns on its communications beam (narrower, although brighter than the beacon). OGS is then ready to begin the measurements on the incoming beam, although previous measurements on the beacon can be recorded as well.

Once the signal is acquired, PAT keeps the incoming signal in the center of TS, by moving the steering mirror (the FPM). The main source of perturbation is the seeing effects due to atmospheric turbulence; these effects can be separated in three components, namely image motion, blurring, and scintillation. The design of OGS allows to correct the first of these components, image motion, by measuring the centroid position with the TS QD and generating a correction signal to control the FPM.

Notice that the pointing of the terminals is not aligned with the geometrical line of sight. Due to the relative movement of the satellite terminal and OGS, the pointing must correct the so-called PAA. This angle is compensated with a second steering mirror, which is the already mentioned TPM, that does not require a high bandwidth like the mirror to compensate for the image motion (i.e., like the FPM), but that requires a high positioning accuracy to establish the PAA continuously.

The PAT design shall be driven by the requirements of high-speed actuation and flexibility. The architecture should allow implementation of most PAT functions by using a software, without constraining the real-time performance.

9.3.3.2 PAT Performance

9.3.3.2.1 Signal Levels and Acquisition Probability

The level of the beacon and communication signals of ARTEMIS seen from OGS can vary, taking into account the atmospheric effects, the transmission losses of the optics, and the satellite position-elevation. The same holds for any satellite and ground station. PAT detectors must handle incoming signals in a wide range, in the case of the OGS from 100 to 7000 pW; and a spot size (blurring) between 0.5 arcsec (best site conditions) and 12 arcsec (satellite at horizon) full width at half maximum (FWHM). The frame rate, output gain, and detection threshold are controlled in both AS and TS to cope with this range and to achieve an acquisition

TABLE 9.3
AS Characteristics

AS Specifications and Performance

CCD array size	128×128 pixels
Pixel size	$16 \times 16\ \mu m$
Effective FOV (total FOV)	480 in. $\times$ 480 in. (540 in. $\times$ 540 in.)
FOV per pixel	4.22 in. $\times$ 4.22 in.
Acquisition probability	0.999
Acquisition algorithm	Pixel of maximum intensity
Acquisition position accuracy	± 0.66 pixels in both axis
Frame rate	1–400 Hz
Input signal power	100–3000 pW
Maximum acquisition delay	66.9 ms (specified 100 ms)

probability of 0.999. The main specifications and performance of the two sensors are summarized in Tables 9.3 and 9.4.

9.3.3.2.2 Timing Performance

The reaction time is the maximum duration of the acquisition procedure. PAT must ensure that the outgoing beam is correctly pointed toward the satellite within the time that the onboard terminal beacon is still pointing to it. In the case of the OGS, the time specified is 350 ms (maximum 384 ms) after the arrival at OGS of the beacon signal from the satellite. Otherwise, the satellite will keep on with the beacon scanning and the link will not be closed. This time is divided into two:

1. Acquisition duration (time to detect the signal in the AS and compensate the offset to move it within the TS FOV): <200 ms, 8 arcmin FOV
2. Rallying duration (time to center the signal in the TS QD): <150 ms, 54 arcsec FOV

TABLE 9.4
TS Characteristics

TS Specifications and Performance

CCD array size (pixel size) Pixel size	14×14 pixels $23 \times 23\ \mu m$
Effective FOV (total FOV)	54 in. $\times$ 54 in. (63 in. $\times$ 63 in.)
FOV per pixel	4.5 in. $\times$ 4.5 in.
Rallying algorithm	Pixel of maximum intensity
QD	Central 2×2 pixels
Tracking algorithm	Centroid of QD (subpixel)
Signal gains	1–2–4–8
Frame rates	1000–4000 Hz
Input signal power	94–7400 pW

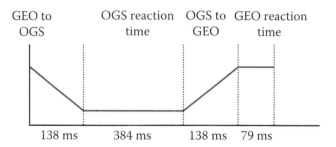

FIGURE 9.7 Acquisition sequence timing budget.

To this time we have to add the signal propagation time from satellite to ground and back (138 ms in the case of the geostationary orbit). The timing budget is represented in Figure 9.7. The timing performance achieved with the OGS design is summarized in Table 9.5.

9.3.3.2.3 Close Loop Control Performance

The performance of the OGS PAT closed loop control, in charge of satellite signal fine tracking and atmospheric turbulence compensation, is shown in Figure 9.8. The image motion spectrum is represented by solid line (perturbation). The residual image motion, when the PAT loop between the TS QD and the FPM is closed, is represented by dashed line (residual error). The perturbation is concentrated in the low frequencies, where the attenuation achieved by the PAT control loop is maximum, higher than 40 dB. The figure of merit is the perturbation rejection factor (rms attenuation in the whole spectrum), defined as the ratio between the root mean squared image motion without correction and the residual perturbation with closed loop (see the formula below). An attenuation factor higher than 10 is achieved; this perturbation attenuation is crucial for the link performance.

$$\text{Attenuation rms factor} = \frac{\sqrt{\sum_{f=0}^{f=ff} \text{Perturbation power}}}{\sqrt{\sum_{f=0}^{f=ff} \text{Residual errors power}}}$$

9.3.3.3 Optical Verification System

One fundamental consideration for an OGS design is to count with the capability of verifying the PAT performance before the satellite is available. Otherwise, the mission will assume strong risk. For this purpose, a verification system shall be included in the optical design, capable of simulating the satellite signal, simulating the atmospheric turbulence conditions, and feeding the signal through the same optical path as the real beam.

In the OGS, the optical verification system (OVS) is in charge of these tasks. The image motion is simulated using a two-tilt axes piezoelectric mirror, with controllable frequency and amplitude movement. Blurring can also be generated.

TABLE 9.5
PAT Timing Performance

	Specified	Performance
Detection delay (ms)	n.a.	Max: 33
Acquisition (ms)	200	Mean: 66
		Max: 166
Rallying (ms)	150	Mean: 34
OGS response (ms)	350	Mean: 116
		Max: 233

The position of the spot within the FOV shall also be adjustable. The flux of the power meters, the polarimeter, and the spectrometer are well calibrated with the OVS.

The OVS optical signals shall be seen through the rest of the FPO as if they were coming from the telescope. To comply with that requirement, a telescope–collimator set was designed simulating the real OGS telescope and collimator. Finally, the FPO bench shall have space available for the installation of this OVS, with an easy optical path to couple the signal at the entrance of the FPO.

9.3.3.4 Transmit Laser System Overview

TLAS is a coherent light source, whose power and wavelength need control. As in any other control problem, there is a set of outputs to be controlled that can be changed by means of a set of inputs. The outputs to be controlled are the wavelength

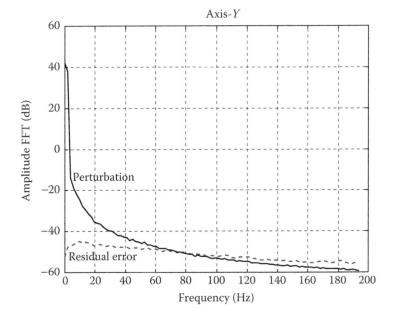

FIGURE 9.8 Atmospheric image motion and residual error with control loop closed.

and the power, and the inputs to be controlled are the pump power and the wavelength of the TLAS cavity gain (i.e., the minimum cavity loss).

The wavelength is read by means of a Fourier spectrometer and controlled by rotating a birefringent filter that is placed inside the Ti:Sap laser cavity. This rotation is achieved by means of a linear displacement and a lever; in turn, the linear displacement is achieved by means of a DC motor and a micrometric screw.

The power is read by means of a photodiode (or alternatively, a thermopile) and controlled by means of the pump power supplied by the pump (argon) laser. Notice that the output power depends not only on the pump power but also on the wavelength, since the gain spectrum of the Ti:Sap is wavelength-dependent. Therefore, power and wavelength are partially coupled.

In order to implement the control system, different instruments are controlled from a PC-type computer by means of a general-purpose interface bus (GPIB). Therefore, from the transmit laser control computer, the wavemeter and the power are read, and the pump power and the position of the tuning screw, set by means of GPIB, command the different instruments. This also provides monitoring of the status of the pump laser in terms of alarms and internal variables, interesting to know from the point of view of safe operation.

The specifications of this subsystem established the key parameters that define the performance of the wavelength and power controls. They are summarized in Table 9.6. The overall system block diagram can be depicted as shown in Figure 9.9.

9.3.3.5 Transmitter Configuration

The distribution of the intensity of the transmitted laser at the satellite orbit does not have an ideal Gaussian profile, but is much closer to a moving speckle field, generated by the effect of the atmospheric turbulence eddies that the beam crosses in its path to the free space. Statistics show that the most probable intensity at any position within a speckle field is zero. Therefore, such a far-field pattern, with the very short exposure times associated with optical communications (rather high data rates) will produce strong scintillation in the satellite receiver that would result in a very high BER. The specific propagation effects in the uplink have been already summarized, and they are explained in detail in Section 9.4.

TABLE 9.6
TLAS Specifications

Wavelength Control	Long-Term Wavelength Stability	Better than ± 0.1 nm
	Wavelength tuneability range	750–890 nm
	Maximum wavelength tuning speed	2.7 nm/s
	Wavelength resolution	0.1 nm
Power control	Long-term power stability	Better than ± 0.1 W
	Power range	0–6 W
	Maximum power adjustment speed	1 W/s
	Power resolution	0.1 W

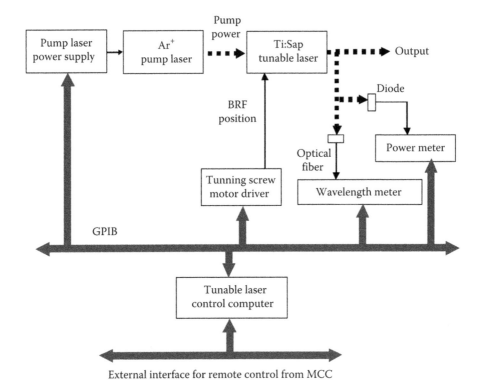

FIGURE 9.9 TLAS general layout diagram.

To keep the BER at acceptable levels, it is necessary to produce an uplink far-field pattern with high temporal stability at the satellite receiver terminal. To achieve this, a special transmitter optical system was conceived for the OGS. Once the transmit laser beam goes through the BE, it is split into four individual beams. The beams are arranged in the telescope aperture such that they do not hit the supporting spider of the secondary mirror, i.e., each beam goes through one of the quadrants of the primary mirror. To compensate for the image rotation inherent to the Coudé configuration, the beam splitting device also has the capability of rotating and tracking the telescope motion.

The BE allows to cope with different atmospheric turbulence conditions. The smaller apertures (40 mm) generate a higher divergence angle, which is necessary in the case of good seeing conditions to provide a large far-field pattern.

However, splitting the beam into four is not enough to solve the speckle problem in the far field. The four apertures synthesize a large aperture, and the speckle field would be identical to the one generated by a single large aperture (assuming an equivalent size). It can be argued that the beams coming from the different subapertures are spatially decorrelated, crossing different turbulent cells on their way through the atmosphere and thus generating four different speckle fields. But these four speckle fields will interfere again and synthesize a speckle field identical to the one generated by one large equivalent aperture.

To overcome this problem, an incoherent superposition of the four subapertures was realized. For this purpose, the four transmitted beams are fed through individual delay lines, which exceed the coherence length of the laser, in order to make them incoherent with respect to each other. Then each beam passes along separate paths through most of the turbulence in the lowest atmosphere, and generates an individual speckle field. However, these four speckle fields are mutually incoherent and can therefore not interfere with each other. This approach is possible because the coherence length of the Ti:Sap laser is below 20 mm at an output power level of 5 W. It decreases when increasing the output power, starting at about 50 mm at the laser threshold. The three delay lines required were designed

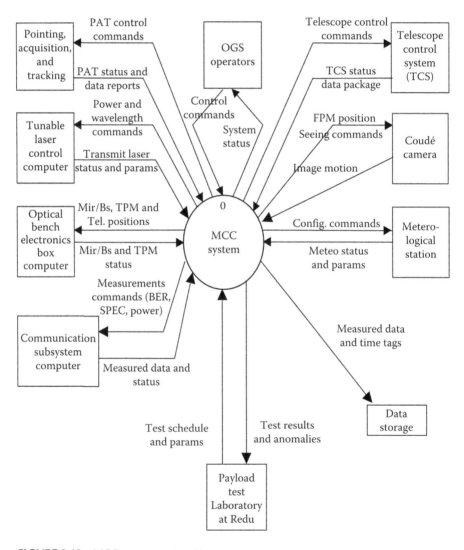

FIGURE 9.10 MCC system context diagram.

as bulk optical delay lines, with a total length below 300 mm. This way, the effect in the communication signal rise/fall time is negligible.

9.3.3.6   MCC Overview

The MCC system will work under the environment defined in Figure 9.10. All the external interfaces and the inputs and outputs the system must handle are identified there.

The MCC system is composed of a set of functional or logical modules (the word "module" is used here as a logical component of the system rather than as a software module) that can be combined in different ways to perform all the different functions of the entire MCC System. These logical modules are generic and can be applied to any OGS design. Figure 9.11 shows a diagram of the MCC system logical model.

The main components of the logical model perform the functions of

- Test planning and test results
- Test performing

9.3.3.6.1   Test Planning and Test Results

This logical component covers all the modules involved in the preparation of an OGS operations test, and in the interaction with the OGS operator either to show him/her test results or system status and as a means for him/her to enter commands into the system.

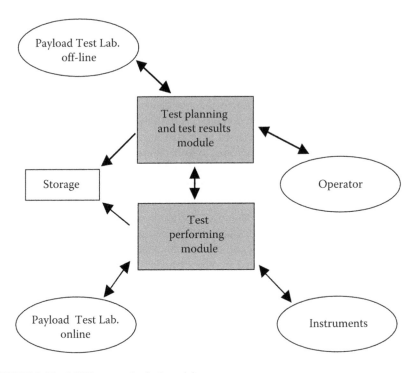

FIGURE 9.11   MCC system logical model.

9.3.3.6.2   Test Performing

This logical component covers all the modules directly involved in the operations test. This includes the control and setup of all instruments needed to carry out the optical communication tests and, the measurements needed to be taken for the analysis of such communications and logs of the entire system.

9.4   PERFORMANCE PREDICTION MODEL FOR LASER BEAM PROPAGATION THROUGH ATMOSPHERIC TURBULENCE

Free-space propagation theoretical models are obviously not valid to predict the performance of laser communications through the atmosphere. Specific models have to be developed for this case, taking into consideration the peculiarities related with the direction of propagation. The models presented hereafter have been reviewed and fed back with the real link performance. Nevertheless, the validation of the models is to some extent not complete, due to the limited amount of experiments and their difficulty.

For the downlinks, a plane wave model is assumed to carry out the propagation calculations and performance predictions. To cope with intensity-scintillation saturation effects, which can appear for low-elevation angles, the so-called heuristic theory has been applied [5]. For high elevation-angle paths, the results found relating to intensity scintillation coincide quite well with those predicted by the Rytov's weak-disturbance theory [6]. The spatial covariance function of the intensity fluctuations is computed and used to take into account the aperture-averaging effect of the 1 m diameter OGS telescope. Also, the spatial covariance function for the total collected-power fluctuations is calculated to assess, through the Taylor hypothesis, the temporal spectra of the collected-power log-amplitude fluctuations for several transverse-wind velocities. This is used, together with the assumed lognormal probability law for the collected-power fluctuations, to determine probabilities of exceeding a given fade level and the mean continued durations of fade above a given level. Angle-of-arrival rms values and spectra have also been calculated to provide estimates to be used for the OGS fine-pointing system adjustments. The results given are essentially an extension of those presented in Ref. [7].

As for the uplinks, significant effort has been devoted to model the dynamic behavior of beam-wander related effects, which, on one hand, does not seem to have very often been addressed in the technical literature, and on the other hand, may have a significant impact in the performance of the communications links. An admittedly simplified model that considers the power fluctuations in the satellite receiver to stem from the turbulence-induced wander of a Gaussian beam (with divergence increased with respect to vacuum-propagation diffraction effects by the short-term turbulence-induced beam spread) to which spherical-wave wave front intensity scintillation effects are independently superimposed, has been used. Given the extremely long path section between the last atmosphere disturbing layers and the satellite, the wave reaching the satellite receiver is expected to present an extremely high degree of spatial coherence, so that no aperture-averaging effects are considered. A probability density function for the total log-amplitude of the power received in the satellite is derived, which allows the computation of the probability of fade and surge occurrences. Also, a spatial covariance function for the log-amplitude fluctuations of the total received power in the satellite is constructed. This allows the calculation of temporal spectra of these log-amplitude fluctuations through Taylor's hypothesis. A heuristic approach is then used to estimate the average continued times during which a given fade will be exceeded.

9.4.1 DOWNLINK DYNAMIC MODEL

The propagation effects of the atmosphere on the downlinks are essentially the same as on starlight, since the wave reaching the upper layers of the atmosphere is to a very good degree of approximation a plane wave. Starting from a model of the profile of the refractive index structure parameter $C_n^2(Z)$ along the propagation path, we establish a method to obtain predictions about the link behavior. Using the so-called heuristic theory [5], we compute the spatial covariance function of the log-amplitude fluctuations $C_\chi(\vec{\rho})$. Assuming the log-amplitude is a normal random variable, the relative spatial covariance function of the intensity fluctuations $C_I(\vec{\rho})$ can be computed from that of log-amplitude fluctuations as

$$C_I(\vec{\rho}) = e^{4c_x(\vec{\rho})} - 1 \tag{9.1}$$

Since the receiver will be using a finite-size aperture to collect the optical wave, we will need to estimate the collected-power covariance function $C_P(\rho)$, which is related to the intensity covariance function $C_I(\vec{\rho})$ through its convolution with the aperture optical transfer function $K_W(\vec{\rho})$

$$C_P(\vec{\rho}) = \int_S K_W(\vec{\rho}')C_I(\vec{\rho}' + \vec{\rho})d\vec{\rho}' \tag{9.2}$$

Once the spatial covariance function of the collected-power fluctuations is obtained, the Taylor's hypothesis will allow us to compute the temporal covariance function of the collected-power fluctuations for different transversal wind velocities whose Fourier transform yields the collected-power fluctuation spectrum $S_P(f)$. The spatial covariance function of the collected-power log-amplitude fluctuations $C_{\chi P}(\vec{\rho})$ is easily computed from the spatial covariance function of the collected-power fluctuations $C_P(\vec{\rho})$ assuming the first one is a normal random variable. Its value at zero displacement gives the collected-power log-amplitude variance $\sigma_{\chi P}^2$, and the Fourier transform of the covariance function in turn will yield the collected-power log-amplitude fluctuations spectrum $S_{\chi P}(f)$.

Considering the three C_n^2 height-dependent models adjusted after measurement campaigns carried out between the two IAC observatories (150 km interisland optical link experiments) [7], the log-amplitude variance of collected-power fluctuations computed for the 1 m OGS aperture is shown in Figure 9.12.

The deformation of the focal spot and the jitter of its position around the focus (angle-of-arrival fluctuation) can be taken into account by means of the Fried's coherence diameter, that for a plane wave takes the form $r_0 = \left[0.42k^2 \int_0^L C_n^2(z)dz\right]^{-3/5}$.

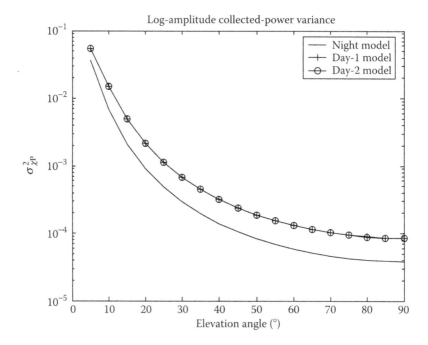

FIGURE 9.12   Variances of the log-amplitude of the collected-power calculated for the 1-m diameter aperture of the OGS, for the three models (night, day-1, and day-2) [7] and for different elevation paths.

The estimation of the image motion dynamical behavior can be attempted from the knowledge of the angle-of-arrival spatial covariance function plus Taylor's hypothesis. Working out the expression for the image centroid position in a receiving system found in Ref. [7], one obtains the spatial covariance function of the angle-of-arrival fluctuations on the same direction u over two identical apertures of diameter D separated at a distance ρ as [8]

$$c_{1u2u}(\rho) = \frac{32\pi^2}{D^2} \times 0.033 \int_0^L C_n^2(z) \int_0^\infty K^{-8/3} J_1^2\left(\frac{D}{2} K\right) J_0(\rho K) dK\, dz \qquad (9.3)$$

where

K is the spatial frequency
$C_n^2(z)$ is the function defining the structure constant of the index-of-refraction fluctuations along the path between the OGS ($z = 0$) and the satellite ($z = L$)

Substituting $v\tau$ for ρ in Eq. (9.3), with v the transverse component of the wind velocity along the path, a temporal covariance function $c_u(\tau)$ for the angle-of-arrival fluctuation on a direction u is found, from which the temporal spectrum of the angle-of-arrival fluctuation a along that direction can be computed as

$$S_{a_u}(f) = \int_{-\infty}^{\infty} c_u(\tau) e^{-j2\pi f t} d\tau \qquad (9.4)$$

9.4.1.1 Single-Beam Uplink Model

To assess the turbulence-induced collected-power fluctuation in the ground-to-satellite uplinks, a simplified model for a single-beam uplink has been considered in the first instance. The model considers two sources of fluctuation:

- *On-axis intensity scintillation*: Scintillation on the wave front of a beam not subject to beam wander.
- *Beam wander*: Intensity fluctuations due to the wander of the beam.

In turn, the following simplifying assumptions have been made (Figure 9.13):

- Fluctuations due to on-axis scintillation can be calculated as those of a spherical wave.
- Beam wander statistics can be dealt with as the dual of angle-of-arrival fluctuations in a downlink, substituting the diameter of the collecting aperture in the downlink by the beam diameter at the transmitter location [9, 10].

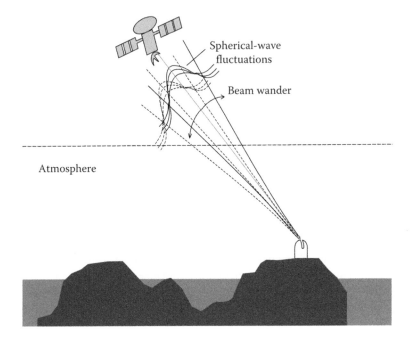

FIGURE 9.13 Outline of the uplink modeling. Spherical-wave intensity fluctuations are superimposed on a beam subject to wander effects.

The first assumption is a good one for very divergent beams. In the ranges of divergences the OGS-transmitted beams can be adjusted, for the less divergent settings it fails taking into account the decrease of the on-axis scintillation with respect to the spherical wave situation and the dependence of the scintillation strength on the offset with respect to the beam axis. With these assumptions, the instantaneous intensity on the satellite can be expressed as

$$I = I_0 e^{2(\chi_s + \chi_w)} \tag{9.5}$$

where

I_0 is the average power in the direction of the beam axis and
χ_s and χ_w are, respectively, the log-amplitude fluctuation due to wave front on-axis scintillation—without beam wander—and the log-amplitude fluctuation due to beam wander.

In turn, assuming a turbulence-affected Gaussian beam with root mean squared short-exposure angular radius at $1/e$ of the maximum intensity $\Delta\theta$, χ_w is related to the instantaneous angular beam deviation α through

$$\chi_w = -\frac{1}{2}\frac{\alpha^2}{\Delta^2\theta} \tag{9.6}$$

Assuming χ_s and χ_w are independent random variables, the probability density function $f_\chi(\chi)$ of the total log-amplitude fluctuation $\chi = \chi_s + \chi_w$ can be written as the convolution of the probability density functions of χ_s and χ_w [11]:

$$f_\chi(\chi) = \int_{-\infty}^{\infty} f_{\chi w}(\chi')f_{\chi s}(\chi-\chi')d\chi' \tag{9.7}$$

where $f_{\chi s}(\chi_s)$ and $f_{\chi w}(\chi_w)$ are the probability density functions of χ_s and χ_w, respectively.

A normal probability law is assumed for χ_s, so that

$$f_{\chi s}(\chi_s) = \frac{1}{\sqrt{2\pi}\sigma_{\chi s}} e^{\left(\left((\chi_s+\sigma_{\chi s}^2)^2\right)/(2\sigma_{\chi s}^2)\right)}.$$

If the turbulence-induced angular deviations, α_x and α_y, of the beam from the nominal pointing along two mutually perpendicular axes are taken as independent zero-mean normal random variables with variance α_α^2 [12], it is easily seen using Equation 9.6 that χ_w is an exponential-law random variable [11] with probability density function

$$f_{xw}(\chi_w) = U(-\chi_w)\left(\frac{\Delta\theta}{\sigma_\alpha}\right)^2 e^{(\Delta\theta/\sigma\alpha)^2\chi_w},$$

where $U(x)$ is the step function. By using these expressions for $f_{\chi S}(\chi_S)$ and $f_{\chi w}(\chi_W)$ in Equation 9.7, one finds

$$f_\chi(\chi) = \frac{1}{2}\left(\frac{\Delta\theta}{\sigma_a}\right)^2 e^{(\Delta\theta/\sigma_\alpha)^2\{\sigma_{\chi S}^2[1/2(\Delta\theta/\sigma_\alpha)^2+1]+\chi\}} \text{erfc}\left\{\frac{1}{\sqrt{2}}\sigma_{\chi S}\left[\frac{\chi+\sigma_{\chi S}^2}{\sigma_{\chi S}^2}+\left(\frac{\Delta\theta}{\sigma_\alpha}\right)^2\right]\right\} \quad (9.8)$$

The variance $\sigma_{\chi S}^2$ is computed as the value for $\rho = 0$ of the spherical-wave spatial covariance function of the log-amplitude fluctuations at the receiver $C_{\chi S}(\rho)$. The Kolmogorov's spectrum for the index of refraction fluctuations of the atmosphere and the heuristic theory [5] for the intensity fluctuations of a spherical wave propagating through a turbulent medium have been used to compute $C_{\chi S}(\rho)$.

Drawing on the mentioned duality between angle of arrival in the downlink and beam deviation in the uplink, the spatial covariance function of χ_W, $C_{\chi W}(\rho)$, is found to be

$$C_{\chi W}(\rho) = \frac{c_{lu2u}^2(\rho)}{\Delta^4\theta} \quad (9.9)$$

where $c_{lu2u}(\rho)$ is given by Equation 9.3 replacing D by the $\sqrt{2}W_0$, with W_0 the beam-waist radius of the assumed Gaussian beam at $1/e$ of its on-axis intensity. Obviously, when doing this substitution, $c_{lu2u}(0)=\sigma_\alpha^2$.

Because χ_S and χ_W are independent, the variance of χ is the sum of the variances of χ_S and χ_W, and the variance of χ_W being $\sigma_{\chi W}^2 + (\sigma_\alpha/\Delta\theta)^4$, we have $\sigma_\chi^2 = \sigma_{\chi S}^2 + (\sigma_\alpha/\Delta\theta)^4$.

The range of validity of the simplifying hypothesis stated at the beginning of the section has been checked against numerical simulation of the beam propagation using a split-step algorithm and fractal phase screen to simulate the turbulence effects. Figure 9.14 shows the check results for the nighttime turbulence profile model and two elevation angles. The continuous lines give the total variance as a function of the transmitted beam-waist radius (hence of the beam divergence). For small radii, the beam is very divergent and it is well approximated by a spherical wave; the effect of beam wander is virtually negligible for such divergent beams. As the beam-waist radius increases, beam wander effects begin to be noticeable. As the beam waist further increases, beam wander decreases because of averaging effects of the atmosphere on wide beams; its effect on the total variance diminishes; and eventually the total variance goes down again to the spherical-wave one. That decrease is not shown by the numerical simulation (Figure 9.14). Actually, when beam wander is significant, the satellite will be receiving an incoming beam with important random mispointing that will be seen as intensity fluctuations. It is likely that the numerical simulation is reproducing the off-axis increase in scintillation predicted by theoretical results of beam propagation in a turbulent medium [13]. Note that saturation effects are also noticeable in the simulation for the larger beam-waist radius values.

Calling v the transverse component of the wind velocity along the path, Taylor's hypothesis allows to calculate the temporal correlation functions from the spatial correlation functions, by the substitution [14] $\rho \rightarrow (L/z)v\tau$ in the computation of the

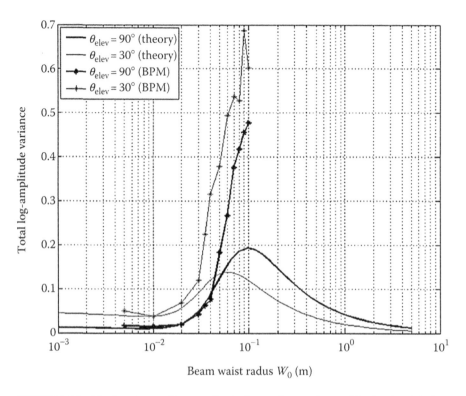

FIGURE 9.14 Total variance of log-amplitude fluctuations at the satellite for the nighttime turbulence profile model. Continuous lines: simplified model for uplink fluctuations. Points: simulation using beam propagation method (BPM) with fractal phase screens.

contribution $dC_{\chi S}(\rho)$ of the atmosphere elemental slice dz at z from the transmitter to the spatial covariance function of the spherical-wave log-amplitude fluctuations at the satellite range L, and the substitution $\rho \rightarrow v\tau$ in $C_{\chi W}(\rho)$. From the so computed temporal covariance functions of log-amplitude fluctuations due to wave front intensity fluctuation without beam wander and log-amplitude fluctuations due to beam wander, the temporal covariance function for the total log-amplitude fluctuation is found:

$$C_\chi(\tau) = C_{\chi s}(\tau) + C_{\chi s}(\tau) \tag{9.10}$$

From the above equation, the temporal spectrum of log-amplitude fluctuations can be computed as

$$S_\chi(f) = \int_{-\infty}^{\infty} C_\chi(\tau) e^{-j2\pi f\tau} d\tau \tag{9.11}$$

The beam wander temporal characteristics, considered so far, consider only the effect of the atmosphere. Nevertheless in the normal OGS operation mode, its PAT system is continuously compensating the turbulence-induced angle-of-arrival fluctuations. Because the cause of these fluctuations (large turbulent eddies) is the same that produces beam wander effects, a partial compensation of the beam wander effects is to be expected. This must appear as a reduction in the effective value of σ_α^2 with respect to the open loop value that can be calculated from Equation 9.3 making $\rho = 0$ and substituting $\sqrt{2} W_0$ for D, and in a modification of the temporal spectrum of beam-wander induced log-amplitude fluctuations

$$S_{\chi w}(f) = \int_{-\infty}^{\infty} C_{\chi w}(\tau)e^{-j2\pi f \tau}d\tau \qquad (9.12)$$

contributing to the total log-amplitude fluctuations. Further details can be obtained from Ref. [15].

9.4.1.2 Multiple Beam Uplink Model

The concept for the multiple beams (M-beam) uplink model is to consider them as an extended source formed by M point sources [16]. As in the one-beam model, χ_S and χ_w components for each beam are considered statistically independent. Assuming that the beams are identical, the expression for the spatial covariance function of on-axis total log-amplitude fluctuation is given by

$$C_{\chi T}(\vec{r}) = \frac{1}{M}\sum_{i=1}^{M} C_\chi^S(\vec{r}) + \frac{1}{M^2}\sum_{i=1}^{M}\sum_{j=1,j\neq i}^{M} C_\chi^{PR}(\vec{\rho}_i - \vec{\rho}_j) \qquad (9.13)$$

where

$C_\chi^S(\vec{r})$ denotes the spatial covariance function of the log-amplitude fluctuations for a spherical wave propagating from ground to the satellite,

$C_\chi^{PR}(\vec{r})$ denotes the spatial covariance function of the log-amplitude fluctuations for a plane wave that would propagate from the satellite to ground, and

$\vec{\rho}_k$ is the position vector of the kth source.

For the beam wander, it is assumed that the two perpendicular components (α_{ix}, α_{iy}) of the random mispointing (dual of the downlink angle-of-arrival fluctuations) for all i beams are jointly Gaussian random variables. This leads to the following expression for the joint probability density function.

$$f_\alpha(\alpha_{1x},\alpha_{1y},\alpha_{2x},\alpha_{2y},...,\alpha_{Mx},\alpha_{My}) = \frac{1}{(2\pi)^N |C|^{1/2}}e^{-\frac{\alpha^T C^{-1}\alpha}{2}} \qquad (9.14)$$

where C is the covariance matrix of the angle fluctuations. The total combined probability of fade is given by adding the probability of fade due to the on-axis

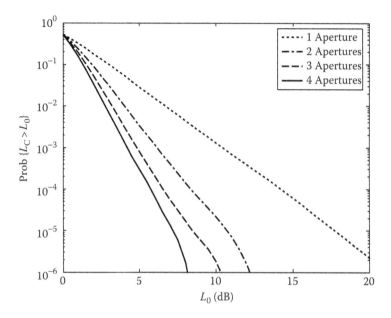

FIGURE 9.15 Probability of exceeding the loss in the abscissa for 40 mm beam-waist diameter and nighttime turbulence model.

scintillation component and the probability of fade coming from the beam wander, $L_C = L_T + L_W$, with a probability distribution function given by

$$F_{LC}(L_C) = \int_{-\infty}^{\infty} f_{LT}(L)F_{Lw}(L_C - L)dL = f_{LT} * F_{Lw} \tag{9.15}$$

In Figure 9.15 are presented the predictions of fade probability for 1, 2, 3, and 4 beams, for the 40 mm aperture case. For the distances between the beams in the OGS (distributed as a square within the 1 m aperture), it is found that the averaging of on-axis scintillation effects is negligible. Nevertheless, there is a noticeable averaging of scintillation related to beam wander, that is translated into a reduction in the probability of exceeding a given fade when increasing the number of beams. This effect is more significant for big fades. This method is expected to be limited to weak/moderate integrated refractive turbulence.

9.5   EXPERIMENTAL METHODOLOGY

Test procedures have to be established in order to be capable of obtaining the performance information from the optical communication experiments, and a rigorous methodology shall be followed to feedback the theoretical models [17].

The optical communication terminals have to be designed to measure and register at high data rates, the received power fluctuations (scintillation), and the image motion (angle-of-arrival fluctuations, only downlink). Spectral and statistical analysis of these data provide the required temporal spectra and fade probabilities and duration. However, special attention has to be paid not only to the outputs but also to the inputs of the models, the experimental conditions, which determine the degree of validity of the theory vs. experiment comparison.

To validate the link theoretical models, three types of inputs are necessary:

1. *Terminal characteristics*: Total transmitted power (P_T), transmitted wavelength (λ), transmitted polarization, transmitted divergence (or beam diameter, W_0), transmission up to the terminal aperture (or optical attenuation)
2. *Link characteristics*: Downlink/uplink, link distance (z), elevation (for atmosphere section), free-space losses (diffraction), and
3. *Atmospheric conditions*: C_n^2 profile, extinction, wind speed.

The terminal characteristics and link characteristics are defined and established in advance to the experiment to be carried out. The main uncertainty to be able to compare experimental results with the model outputs will be in the atmospheric conditions. Looking into the model described in the previous section, two parameters are identified as atmospheric turbulence inputs to the theoretical model: $C_n^2 (z)$, used to determine the spatial covariance function of the log-amplitude fluctuations and wind speed (mean value of transverse wind speed along the turbulence path), to obtain the temporal covariance functions of collected-power fluctuations and collected-power log-amplitude fluctuations, applying Taylor's hypothesis, $\rho \rightarrow \upsilon\tau$.

In the case of the OGS, several facilities were coordinated to provide the possibility of retrieving the necessary atmospheric information during the optical communication experiments. Three telescopes of the IAC Teide Observatory were put into operation simultaneously to establish the optical link and measure the turbulence conditions in parallel. The facilities are shown in Figure 9.16.

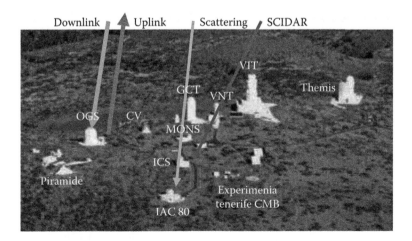

FIGURE 9.16   Teide Observatory facilities coordinated for simultaneous laser link experiments and atmospheric effects evaluation.

9.5.1 TURBULENCE PROFILES MEASUREMENTS

The alternatives to provide the $C_n^2(z)$ are

1. To have allocated a second telescope in the observatory (in the case of the OGS the 1.5 m Telescopio Carlos Sanchez [TCS]) during the optical communication experiments, to perform turbulence profiles measurements ($C_n^2[z]$) with a generalized Scintillation Detection and Ranging (SCIDAR) instrument. The measured profiles can be directly input to the theoretical models, providing the optimum way to evaluate the performance prediction model.
2. For daytime experiments (SCIDAR not available), a C_n^2 parameterized theoretical model has to be used, with the parameters adjusted to the observatory. Even an observatory tuned C_n^2 theoretical model would be a better case. The most significant parameters are calculated from laser link measurements, and the rest (more stable) are established according to the average values for the site.

Figure 9.17 shows a typical measured $C_n^2(h)$ profile, corresponding in this case to the OGS site. The peak of the ground layer (including dome seeing) is clearly distinguished (the height of the observatory is 2400 m). The profile of the C_n^2 mean is represented by the solid line and the profile of the standard deviation of the mean by the dashed line.

Fundamental image quality parameters like isoplanatic angle and seeing are also calculated from the $C_n^2(h)$ profiles.

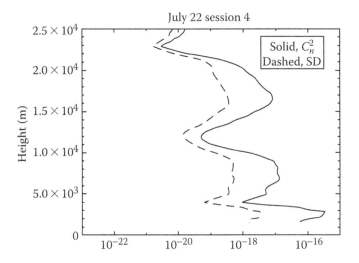

FIGURE 9.17 Averaged $C_n^2(h)$ profile measured during an optical link experiment.

9.5.2   Wind Speed Determination

1. Wind profiles can also be determined from the generalized SCIDAR instrument measurements. The mean transversal wind speed necessary for the models can be calculated from the wind profiles after SCIDAR data reduction by cross-correlation of the pupil scintillation distributions.
2. As an alternative for daytime experiments, the data of the meteorological institutes close to the site can be used, specifically the balloon wind profile measurements.

Instrumentation balloons launched by meteorological institutes can provide important information for the evaluation of the link performance. Balloons normally travel from sea level to an altitude of about 30 km, thus crossing the whole atmospheric turbulence. Among other meteorological parameters, the wind speed and direction are computed from information about the position of the balloon.

The main contributions of wind in the atmospheric turbulence effects are the intensity and the temporal frequency of wave front fluctuations. From the theoretical model point of view, the required parameter is the wind velocity v_0 to be used in the Taylor hypothesis, the temporal coherence of the atmospheric turbulence, or characteristic time of the turbulence changes. From the vertical wind profiles, v_0 can be determined.

The time at which the meteorological institutes launch the balloons has to be taken into account, as some assumptions about the stability of the wind in time have to be done to evaluate the performance of the link at times different from the balloon data.

The distance between the ground station site and the balloons launching place also plays a role in the validity of the data. The contribution of the high layers can be assumed the same as they are not influenced by the ground. It is not the case with the boundary layer where the winds depend strongly on the surface. Normally, the contribution of the high layers is larger in the wind velocity v_0.

9.5.3   Extinction

The attenuation and transmission coefficients have to be determined for all the experimental sessions from measurements done simultaneously at the same site.

Additionally, in the case of the OGS, a third telescope (the 80 cm IAC80) has been used in nighttime experiments to analyze the OGS outgoing beams (uplink) propagation through the turbulence, to validate static link propagation models. This telescope is also used for extinction measurement campaigns.

9.6   EXPERIMENTAL RESULTS

Experimental campaign plans and procedures have to be established in order to get the maximum return from the optical communication tests between the terminals onboard and the ground station. Simultaneous atmospheric turbulence measurements have to be as well scheduled, with the operational and instrumental specific constraints. The results presented in this section correspond to the OGS and ARTEMIS experiments [17–21].

9.6.1 DOWNLINK STATISTICAL RESULTS

A general result from the downlink analysis is that the intensity distribution in most of the experiments fits a lognormal distribution, and the residual errors in general follow a normal distribution.

The different experimental results obtained (mainly scintillation index, fade and surge probabilities and mean durations, log-amplitude power spectral density [PSD], and probability density functions) have been used for a comparative analysis, with the purpose of finding relationships between different parameters and with turbulence measurements. In Figure 9.18, the scintillation index vs. the probability of fade is represented. The same process has been followed in Figure 9.19, but comparing the scintillation with the probability of surge. A lognormal fit (solid line) has been included in the figures. The fade and surge levels used as reference for this analysis is 0.5 dB. In the case of the fade probability, the results obtained correspond to those expected from a lognormal distribution, as can be concluded from the lognormal fit. In the case of the surges, only a small deviation is observed in the cases of high scintillation index.

In Figure 9.20, fade probability vs. surge probability is represented. They should have a linear (1 to 1) relationship if the data from which they were calculated follow a lognormal distribution. This can be easily demonstrated taking into account that the log-amplitude χ follows a normal distribution.

While a clear relationship has been found between the scintillation and the fade and surge probability, it seems that this relation cannot be extended to the duration of fades and surges. A similar behavior is observed in the uplink.

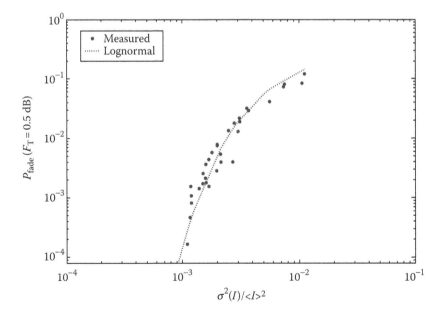

FIGURE 9.18 Downlink scintillation index vs. 0.5 dB fade probability.

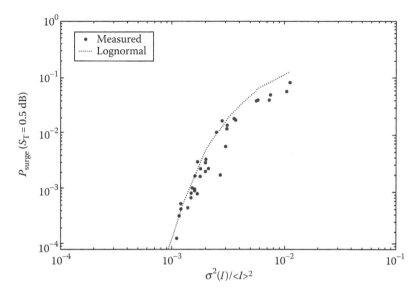

FIGURE 9.19   Downlink scintillation index vs. 0.5 dB surge probability.

The link performance results have been also analyzed in comparison with the atmospheric turbulence measurements performed simultaneously by the IAC SCIDAR. Key turbulence parameters have been calculated from the C_n^2 profile measurements: the isoplanatic angle, θ_0, and the seeing. In addition to this, the meteorological data from the National Institute of Meteorology of Spain has provided the

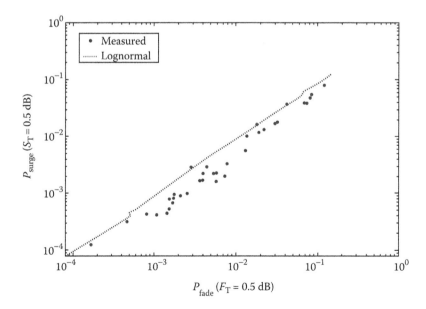

FIGURE 9.20   A 0.5 dB fade probability vs. 0.5 dB surge probability.

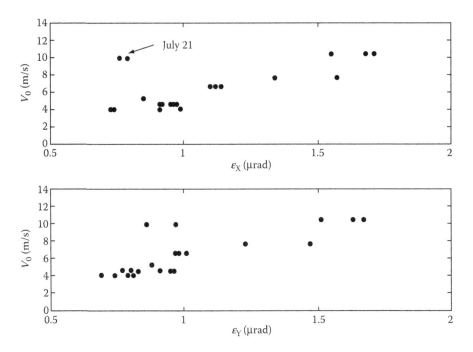

FIGURE 9.21 V_0 vs. residual tracking errors in X and Y axes.

wind speed and wind profiles, together with the rest of meteorological parameters. Seeing, θ_0, and the transversal mean wind speed v_0 (m/s) used for the Taylor hypothesis, have been used for the analysis. Only the latter presents a clear relationship with the tracking residual errors (remaining image motion after tracking loop), as shown in Figure 9.21 (only one experiment, indicated in the figure, deviates from that behavior). The correlation of wind speed with mean intensity and residual error in X and Y has been confirmed with the analysis of variance, delivering a strong significance.

With respect to the comparison of the predictions from the theoretical model and the experimental results, a reasonable approximation has been obtained with respect to the measured variances in the night downlinks. The fade depth obtained is small in both cases and also a similar tendency has been observed in the mean duration of fades. Figure 9.22 shows a good agreement between theoretical model predictions and experimental results, with respect to the fade probability. The four graphics are in the same scale.

9.6.2   DOWNLINK COMMUNICATION EXPERIMENTS RESULTS

In all the successfully established bidirectional link experiments, it has been possible to do BER measurements in the downlink for any atmospheric conditions (those that failed, the reason has been technical). ARTEMIS CS is configured to transmit pseudorandom sequence (PRS15) in 2PPM format at a data rate of 2.048 Mbps.

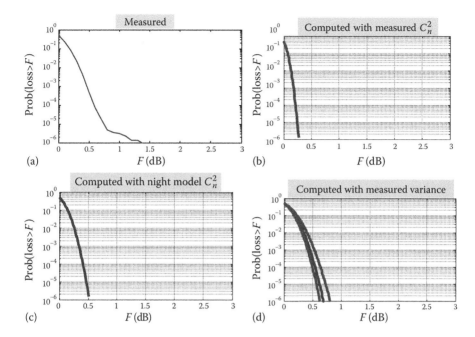

FIGURE 9.22 Comparison of fade probabilities; (a) experimentally measured. Theoretically predicted using (b) the measured simultaneous C_n^2 profile, (c) the theoretical C_n^2 model for nighttime at IAC observatories, and (d) the measured log-amplitude variance.

The BER and bit error counts (BEC) were recorded during all the sessions at the OGS with a gate period of 20 sec and a gap of approximately 1.8 sec between measurements. Table 9.7 summarizes all the BER measurements of all the successful sessions, recorded by the communication subsystem (CSS) of the OGS.

With respect to the uplink communication experiments, the performance results are very reduced, basically due to the poor quality of the modulation of the OGS-transmitted signal (Table 9.7).

It has to be pointed out that, during some of the experiments, there were periods of time with perfect links, without any bit error during the sessions. In these periods BER are <1.000e-8, which is the theoretical ARTEMIS feeder link quality limit. Figures 9.23 and 9.24 show two representative cases of typical evolution with time of BER during the links. Figure 9.23 corresponds to a stable behavior; Figure 9.24 corresponds to a case with oscillations during the links, varying from a perfect link (<1.000e-8) to 3.000e-5.

9.6.3 Uplink Statistical Results

Before entering into the operation of the SILEX project, limited results had been achieved in laser communication from ground to space. The uplink experiments of this project have provided for the first time statistical results in this field. As already pointed out, a big effort has been invested in the theoretical modeling of the beam

TABLE 9.7
Summary of Downlink Communication Experiments Results

Date (Session)	Mean BER	Max BER	Min BER	Comment
22/05/03 (1)	1.1877e-7	2.1973e-7	2.4414e-8	BER trend: stable, slightly decreasing
22/05/03 (2)	5.1467e-8	1.4648e-7	<2.4414e-8	BER trend: stable, slightly decreasing
22/05/03 (3)	1.4909e-7	5.1270e-7	<2.4414e-8	BER trend: stable, slightly decreasing
23/05/03 (1)	1.7332e-5	2.0410e-5	1.5698e-5	BER trend: stable, slightly decreasing
23/05/03 (2)	3.3470e-5	4.6509e-5	2.2656e-5	BER trend: variable ↑ ↓
24/05/03 (1)	7.4014e-6	2.6318e-5	4.1748e-6	BER trend: variable, monotonically decreasing
05/06/03 (1)	5.7422e-7	8.7891e-7	2.9297e-7	BER trend: stable
05/06/03 (2)	6.8917e-7	1.0010e-6	3.9063e-7	BER trend: stable, slightly decreasing
05/06/03 (3)	9.7233e-7	1.5381e-6	6.5918e-7	BER trend: stable, slightly decreasing
05/06/03 (4)	4.1829e-7	7.8125e-7	2.1973e-7	BER trend: stable
06/06/03 (1)	3.5795e-7	6.1035e-7	1.2207e-7	BER trend: stable
06/06/03 (2)	4.4491e-7	8.3008e-7	1.4648e-7	BER trend: stable, slightly decreasing
06/06/03 (3)	5.7431e-7	9.7656e-7	3.1738e-7	BER trend: stable, slightly decreasing
06/06/03 (4)	1.5668e-7	2.0264e-6	1.1475e-6	BER trend: stable
10/06/03 (1)	2.9447e-6	9.6924e-6	<2.4414e-8	BER trend: variable, conspicuous bump
10/06/03 (2)	2.4077e-7	4.3945e-7	1.2207e-7	BER trend: stable
10/06/03 (3)	1.3508e-6	9.8877e-6	1.4648e-7	BER trend: stable with a final increase
10/06/03 (4)	1.9198e-6	1.2549e-5	2.9297e-7	BER trend: stable with a final increase
21/07/03 (1)	1.1656e-7	2.6042e-7	2.4414e-8	BER trend: stable.
21/07/03 (2)	9.8969e-7	5.1758e-6	4.8828e-8	BER trend: stable with a final exponential increasing
21/07/03 (3)	7.5684e-7	7.6497e-7	7.4870e-7	Only two measurements
22/07/03 (1)	8.8781e-7	7.3893e-6	4.8828e-8	BER trend: exponential decreasing
23/07/03 (2)	<1.0739e-8	2.4410e-8	<1.000e-8	Perfect link: BER = 0
23/07/03 (4)	<1.0716e-8	4.8830e-8	<1.000e-8	Perfect link: BER = 0
24/07/03 (1)	3.2813e-6	1.4795e-5	3.0924e-7	BER trend: exponential decreasing, plateau, exponential increasing
24/07/03 (2)	6.2733e-6	1.3118e-6	<1.000e-8	BER trend: exponential increasing, plateau, exponential decreasing
24/07/03 (3)	4.1257e-6	1.4469e-5	7.3242e-7	BER trend: exponential decreasing, plateau, exponential increasing
24/07/03 (4)	1.4178e-5	2.5016e-5	2.1484e-6	BER trend: exponential increasing, plateau, exponential decreasing
06/08/03 (1)	7.5372e-6	2.0850e-5	6.6732e-7	BER trend: variable ↑ ↓
06/08/03 (2)	1.6650e-5	3.7988e-5	8.4635e-7	BER trend: variable ↑ ↓
06/08/03 (3)	9.6327e-6	1.4453e-5	6.7383e-6	BER trend: exponential decreasing
06/08/03 (4)	1.4233e-5	4.1406e-5	1.2044e-6	BER trend: variable ↑↓↑
07/08/03 (1)	5.9010e-6	1.4827e-5	1.3021e-7	BER trend: variable ↑ ↓
07/08/03 (2)	1.8224e-5	3.6230e-5	5.3711e-7	BER trend: variable ↑ ↓
07/08/03 (3)	4.6516e-6	1.9857e-5	1.1882e-6	BER trend: variable ↓ ↑

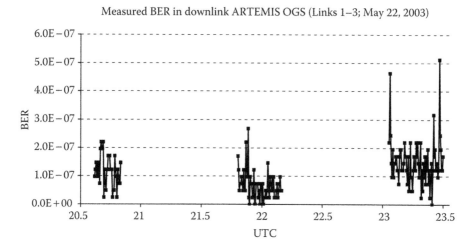

FIGURE 9.23 Example of BER measurements recorded, highly stable during the links.

wander effects, as well as in the multibeam performance prediction. As part of the OGS design, the transmitter system includes the feature of having up to four mutually incoherent beams (see previous sections), conceived initially with the purpose of reducing the irradiance fluctuations and thus reducing the scintillation, fade probability, and surge probability in ARTEMIS receiver.

The results obtained show that there is a strong dependence between the transmitting aperture diameter (beam divergence) and the effect of using several apertures on the scintillation, fade and surge probability, and fade and surge mean duration. In Figure 9.25a, the statistics of 1 dB fade probability using different numbers of beams (4, 3, 2, and 1 beams) are presented, for a transmitting subaperture diameter of 40 mm. The same statistics but for the 1 dB fade mean duration (in ms) are shown in Figure 9.25b.

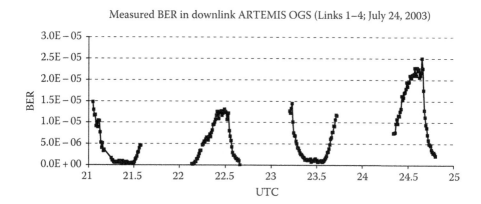

FIGURE 9.24 Example of BER measurements recorded with strong oscillations.

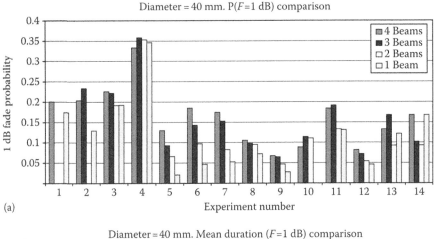

(a)

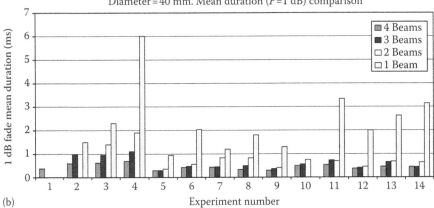

(b)

FIGURE 9.25   Uplink performance with 40 mm diameter subaperture, using different number of transmitted beams (a) 1 dB fade probability and (b) 1 dB fade mean duration.

It can be observed that, instead of reducing the fade probability, in most of the cases increasing the number of beams increases the fade probability. The same happens with the scintillation index and the surge probability. Other possible reasons, like the lack of dynamic range in ARTEMIS receiver for low level signals or a small misalignment between the transmitting beams, have been preliminarily analyzed to discard any instrumental effect, and no clear conclusions have been obtained till now, so in principle this can be due to atmospheric turbulence. However, the behavior of the fade mean duration (and surge mean duration) is predicted by the theory, it is reduced dramatically when using more than one transmitted beam.

Figures 9.26a and b represent the same results as Figure 9.25, but for a larger beam diameter, 100 mm (smaller beam divergence). In this case, the behavior of the scintillation index (fade probability and surge probability) is, in most cases, the same as the fade (and surge) mean duration. Therefore, increasing the number of beams improves all the link performance parameters. The averaging effect of multiple subapertures

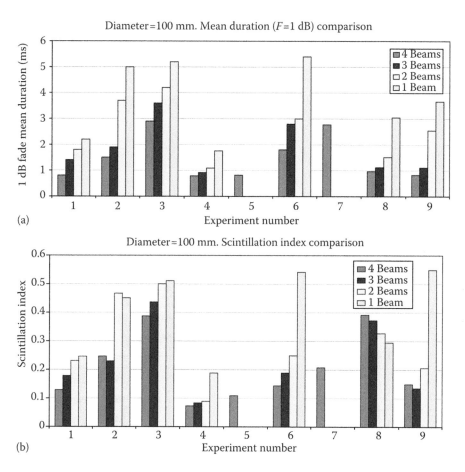

FIGURE 9.26 Uplink performance with 100 mm diameter subaperture, using different number of transmitted beams (a) 1 dB fade probability and (b) 1 dB fade mean duration.

for larger beam diameters (200 and 300 mm) is also confirmed, although with limited number of data, that does not allow to extract definitive conclusions.

As an example, in Figures 9.27 and 9.28 are presented the comparison of the probability density functions (normalized to the mean intensity) with different number of beams, for two experiments: one with 40 mm aperture and one with 100 mm aperture. In the 40 mm case, the distribution width increases slightly with the number of beams (Figure 9.27a), which indicates that the scintillation index and the fade and surge probabilities also increase (Figure 9.27b). In the 100 mm case, the distribution width decreases when increasing the number of beams (Figure 9.28a), implying that the scintillation index and the fade and surge probabilities also decrease (Figure 9.28b).

From the point of view of the log-amplitude fluctuations spectra, a good agreement is also found between the experimental results and the predictions of the uplink model for one beam, using the simultaneously measured C_n^2 profile (Figure 9.29). This good agreement is observed in all the one-beam uplinks analyzed.

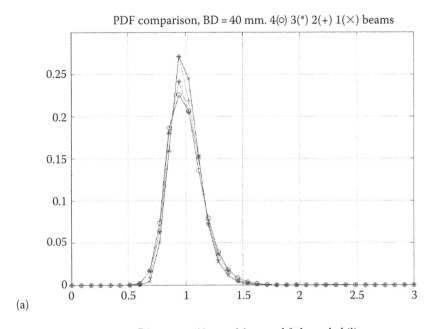

(a)

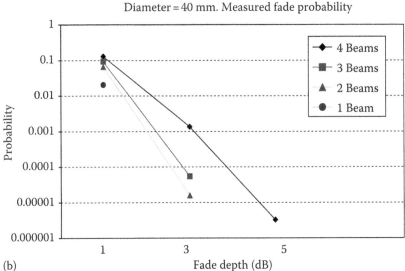

(b)

FIGURE 9.27 Comparison of a 40 mm experimental case with respect to the number of beams. (a) Probability density function and (b) fade probability for different fade levels.

9.6.3.1 Comparative Analysis of Uplink Results with Theoretical Predictions (Wave Optics)

To complete the analysis of the experimental results, numerical methods for beam propagation (wave optics) have also been used in the framework of this project to validate the analytical models. In Figure 9.30 are summarized the results obtained

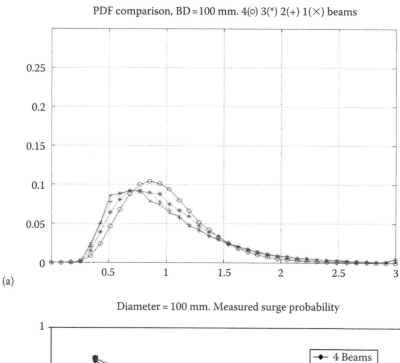

(a)

(b)

FIGURE 9.28 Comparison of a 100 mm experimental case with respect to the number of beams. (a) Probability density function and (b) fade probability for different fade levels.

with respect to the scintillation indexes measured during all the experiments, for one-beam uplink. The asterisks represent the measured scintillation index values, for different beam diameters (transmitting apertures):

1. Light solid lines represent the predicted scintillation using two measured C_n^2 profiles measured by IAC.

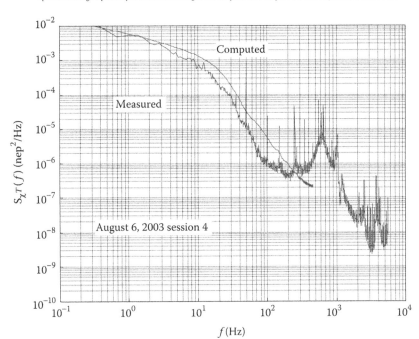

Uplink Overall log-amplitude spectrum. Model: Ch2ago6s4. Wind profile:vent 06ago vt=4.6 ms.Tx Ap. diameter 0.070711 m.

FIGURE 9.29 Comparison of measured log-amplitude fluctuations spectrum with the theoretical predictions for one beam using simultaneously measured C_n^2 profile.

2. Light dashed lines correspond to the same predictions as case 1 but without the beam wander effect (only on-axis scintillation).
3. Upper dark solid line represents the numerical scintillation predictions using the C_n^2 night model profile (adjusted at IAC observatories).
4. Dark dashed line corresponds to the same predictions as case 3 but without beam wander.
5. Lower dark solid line represents the predictions using the Andrews and Phillips [13] Gaussian beam model (only on-axis scintillation), using the C_n^2 night model profile as the input. Similar result is obtained using the measured C_n^2 profiles.

Several conclusions can be extracted from these results. The measured scintillation index values for large apertures (over 50 mm) are well within the range of the numerical simulation results. For small apertures (40 mm), some of the experimental values seem to be too small, not predicted by any model nor C_n^2 profile. While a good approach is found in long-term beam width theoretical models and experimental results, the propagation models that do not consider the beam wander phenomenon are underestimating the short-term atmospheric turbulence effect in the uplink beam propagation.

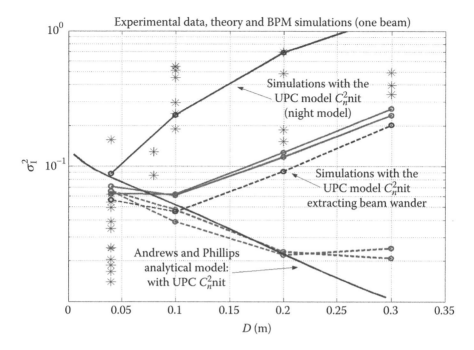

FIGURE 9.30 Comparison of scintillation indexes, with the ones predicted by theoretical models and by numerical simulations measured.

9.7 CONCLUSIONS

This chapter aimed at specifying and designing a reference OGS, as well as characterizing the results of free-space laser communication links between satellites and the ground, developing and validating tracking algorithms, and theoretical models that allow to predict the performance depending on the meteorological and atmospheric turbulence conditions. The detailed analysis of the laser communication terminals involved in the ARTEMIS—OGS project allowed to determine their right configuration to obtain the maximum return from all the experiments. The experimental campaigns have proven that ARTEMIS—OGS laser communication links in night and daytime and under difficult atmospheric seeing conditions can be established; these links are being routinely demonstrated.

Theoretical models have been developed to predict the laser link performance, based on the atmospheric turbulence conditions. While a plane wave model (widely accepted) has been used for the downlink, for the uplink the model considers two sources of scintillation: on-axis scintillation, of a spherical wave; and beam wander. Theoretical models have also been developed for multiple transmitting apertures. According to them, the scintillation averaging effect using multiple beams is noticeable only for the beam wander component.

In the downlink, the preliminary comparative analysis of experimental results with atmospheric turbulence did not show any correlation with the seeing and isoplanatic angle parameters. However, there is a confirmed correlation between tracking residual errors (residual angle-of-arrival fluctuation or image motion) and scintillation with mean transversal wind speed in the turbulence layers.

The scintillation reduction when increasing the number of beams, predicted by low turbulence regime theory, is not clearly confirmed experimentally for the uplinks, at least for large beam divergences. With large divergences, the scintillation does not decrease when rising the number of subapertures. The same holds for the fade and surge probability, they even increase. With small divergences, a reduction in scintillation (and fade and surge probabilities and durations) is observed when increasing the number of subapertures. On the other hand, for all divergences, the fade and surge mean durations decrease (improve) clearly when the number of subapertures is increased. This behavior is predicted by the theoretical model.

The contribution of the beam wander to the total scintillation and fade statistics depends on the beam divergence and the turbulence conditions. For large beam divergence, lower fade probabilities are obtained with weak turbulence. In turn, for small beam divergence, lower fade probabilities are obtained with strong turbulence.

The comparison of multiple beam uplinks with theoretical predictions still needs additional work. Though the measured C_n^2 profiles have been used for the comparison, no clear conclusions have been obtained till now. The nonstationary behavior of the atmosphere can contribute to the difficulty in getting to generic conclusions.

ACRONYMS

APD	Avalanche photodiode
ARTEMIS	Advanced relay and technology mission
AS	Acquisition sensor
BEC	Bit error counts
BER	Bit error rate
BPM	Beam propagation method
BPSK	Binary phase shift keying
CSS	Communications subsystem (of the OGS)
ESA	European Space Agency
FPC	Focal plane control
FPM	Fine pointing mirror
FPO	Focal plane optics
FWHM	Full width at half maximum
GEO	Geostationary earth orbit
IAC	Instituto de Astrofísica de Canarias
LEO	Low earth orbit
MCC	Main control computer (of the OGS)
OGS	Optical ground station
OI	Optical isolator
PAA	Point ahead angle
PAT	Pointing, acquisition, and tracking (system)
PSD	Power spectral density
RFE	Receiver front end (of the OGS)

SCIDAR	Scintillation Detection and Ranging
SILEX	Semiconductor-laser intersatellite link experiment
STDM	Spacecraft trajectory data message
TLAS	Transmit Laser System
TPM	Transmit pointing mirror
TS	Tracking sensor
UPC	Universidad Politécnica de Cataluña

REFERENCES

1. G. Oppenhäuser et al., The European SILEX project—concept, performances, status and planning, *Free Space Laser Communication Technologies II*, SPIE 1218, 27–37, 1990.
2. T. Tolker-Nielsen et al., In orbit test result of an operational optical intersatellite link between ARTEMIS and SPOT4, SILEX, *Free Space Laser Communication Technologies XIV*, SPIE 4635–4730, 2002.
3. R. H. Czichy et al., Design of an optical ground station for in-orbit check-out of free space laser communication payloads, *Free Space Laser Communication Technologies VII*, SPIE 2381, 26–37, 1995.
4. M. Reyes et al., Design and performance of the ESA optical ground station, *Free Space Laser Communication Technologies XIV*, SPIE 4635–4704, 2002.
5. S. F. Clifford et al., Saturation of optical scintillation by strong turbulence, *Journal of the Optical Society of America*, 64(2), 148–154, Feb. 1974.
6. V. I. Tatarskii, The effects of the turbulent atmosphere on wave propagation, *National Technical Information Service*, Springfield, VA, 1971.
7. A. Comerón et al., ASTC Inter-island measurement campaign final report, Data analysis, Atmospheric modeling, Fade and bit-error-rate statistics, July 1996.
8. A. Rodríguez-Gómez et al., Temporal statistics of the beam-wander contribution to scintillation in ground-to-satellite optical links: An analytical approach, *Applied Optics*, 44, 4574–4581, 2005.
9. R. J. Sasiela, *Electromagnetic Wave Propagation in Turbulence: Evaluation and Application of Mellin Transforms*, Springer-Verlag, New York, 1994.
10. Robert, K. Tyson, *Principles of Adaptive Optics*, Academic Press, Boston, MA, 1998.
11. A. Papoulis, *Probability, Random Variables and Stochastic Processes*, 2nd ed., McGraw-Hill, New York, 1984.
12. R. E. Hufnagel, Variations of atmospheric turbulence, *Digest of Technical Papers*, Topical Meeting on Optical Propagation through Turbulence: Topical Meeting, Boulder, CO, July 1974, p. Wa1-1.
13. L. C. Andrews and R. L. Phillips, *Laser Beam Propagation through Random Media*, SPIE Press, Bellingham, WA, 1998.
14. S. F. Clifford, Temporal-frequency spectra for a spherical wave propagating through atmospheric turbulence, *Journal of the Optical Society of America*, 61(10), 1285–1292, Oct. 1971.
15. F. Dios, J. A. Rubio, A. Rodríguez, and A. Comeron, Scintillation and beam-wander analysis in an optical ground station—satellite uplink, *Applied Optics*, 43(19), July 1, 2004.
16. A. Comeron, F. Dios, A. Rodríguez, J. A. Rubio, M. Reyes, and A. Alonso, Modeling of power fluctuations induced by refractive turbulence in a multiple-beam ground-to-satellite optical uplink, *Free Space Laser Communications V*, SPIE 5892, 2005.
17. M. Reyes et al., Ground to space optical communication characterization, *Free Space Laser Communications V*, SPIE 5892, 2005.

18. M. Reyes et al., Analysis of the preliminary optical links between ARTEMIS and the optical ground station, *Free Space Laser Communication and Laser Imaging II*, SPIE 4821, 33–43, 2002.

19. J. Romba, Z. Sodnik, M. Reyes, A. Alonso, and A. Bird, ESA's bidirectional spaced to ground laser communication experiments, *Free Space Laser Communications IV*, SPIE 5550, 2004.

20. M. Reyes et al., Propagation statistics of ground-satellite optical links with different turbulence conditions, *Optics in Atmospheric Propagation and Adaptive Systems VII*, SPIE 5572, 2004.

21. A. Alonso, M. Reyes, and Z. Sodnik, Performance of satellite to ground communications link between ARTEMIS and the optical ground station, *Optics in Atmospheric Propagation and Adaptive Systems VII*, SPIE 5572, 2004.

10 Reliability and Flight Qualification

Hamid Hemmati

CONTENTS

10.1   INTRODUCTION

The vacuum environment of space along with constant bombardment by moderate to elevated levels of cosmic background radiation imposes unique challenges on the reliable operation of a spaceborne optical and optoelectronic assembly. A spaceborne component or system must endure ten to hundreds of g's of mechanical shock, depressurization during launch, pyroshocks, as well as rapid temperature change (~5°C/min) and depending on the satellite's location moderate to high levels of radiation. Storage temperature specifications of ±60°C and controlled operational temperatures of ±30°C are common in such environments. Examples of extreme temperature environments in space are the lunar surface temperature ranges from −233°C to 123°C and the Venus surface temperature averaging at +464°C. Electrons trapped in the Earth's Van Allen belt maintain velocities approaching the speed of light (Chen et al., 2007). Furthermore, radiation environments at Jupiter approach 10 Grad, posing major shielding and instrument lifetime challenges for any exploration mission to these environments.

    Reliability of a device or a subsystem for spaceflight depends on the ability to consistently meet the requirements before and after launch and throughout the life of the mission. Failure modes, mechanisms, causes, estimates, risks, and remedies are conceived, established, and investigated by the reliability engineers working together with subsystem engineers (Andrews and Moss, 2002; Hemmati, 2006). To quantify a subsystem's performance under the qualification environment, fairly accurate knowledge of the environmental conditions anticipated during and after launch is required for such parameters as shock, vibration, thermal, and radiation. Certain reliability and qualification issues may be mitigated via modeling and component design, while others would require verification through testing. The development of a systematic qualification approach is critical to the reliability of a given part or subsystem.

    Environmental tests performed at the assembly and actual flight unit levels are likely to yield significant information about the flight worthiness of equipment. These tests could consist of the following:

Acoustics	Nondestructive inspection
Creep/stress-relaxation	Pressure cycling
Corrosion	Pyroshock
Electrical arc tracking resistance	Random vibration

(continued)

Electromagnetic compatibility	Radiation (ionizing, solar, atomic oxygen)
Electrostatic discharge (ESD) susceptibility	Radiated and conducted emissions
Fatigue	Sinusoidal vibration
Drop and mechanical shock	Static charge and ESD sensitivity
Galvanic corrosion	Thermal soak/bake/aging/dwell
Lifetime	Thermal cycling
Metal migration	Thermal vacuum
Magnetic acceptance	Thermal gradients
Meantime between failure	Welded parts

Key reliability factors for optoelectronic devices include

Alloy electrodes	Metal diffusion (e.g., aluminum)
Bonded parts	Particle impact
Charge build-up	Photodarkening
Contamination	Point defects
Dislocations	Radiation damages
Displacement damage	Soft solders (e.g., indium)
Facet oxidation	Surface degradation
Fiber connection	Upsets, latch ups, burnout
Hermiticity	Whisker formation
Impurity of the material	Workmanship

A reliability engineer's partial list of primary activities includes:

1. Fault-free analysis, where electrical and mechanical elements are affected
2. Worst case analysis (WCA) of power supply transients
   a. Voltage/temperature or frequency margin testing may be used as an alternative to WCA
3. Stress analyses
   a. Electrical and electromechanical parts
   b. Corona discharges (pressure cycling)
   c. Structural (e.g., fatigue and pressure cycling)
   d. Thermal
4. Failure mode and effect analysis of each system's external interfaces including ground test equipment interfaces to flight hardware

Description of the general qualification tests including test standards, qualification of COTS components such as individual optoelectronic and photonic devices, parts screening and radiation effects on these devices follows.

10.2 QUALIFICATIONS TESTS

The qualification of a particular design may include three primary stages: (1) prototype qualification, that is, qualification of nonflight units (2) qualification of proto-flight units often tested at levels well beyond the specification and (3) qualification of the actual flight unit frequently tested at the less stringent qualification requirement levels. The alternative is to design and fabricate the components based on environmental specifications. But, that is a very costly process. A component that passes the qualification requirements and standards is qualified for the very specific environment defined by the standard only.

10.2.1 QUALIFICATION STANDARDS

Established qualification standards are intended for a particular set of applications of specific devices. These standards are typically based on the state-of-the-art practical knowledge to make a device or a system ruggedized for a given environmental condition. Different aerospace and government organizations have developed different standards depending upon mission criticality. For instance, The NASA/Jet Propulsion Laboratory (JPL) qualification procedures for deep-space missions and NASA/Goddard Space Flight Center (GSFC) procedures for near-Earth missions and constellations of LEO satellites consisting of hundreds of rapidly developed spacecraft for broadband communications delivery. A few of the many common standards are described below.

10.2.1.1 MIL Standards

Military (MIL) standards provide a comprehensive list of methods and procedures to qualify electronic devices. The MIL standard 883, in particular, is applicable to optoelectronic devices for military applications. Methods within the MIL-STD-883 for testing of lasers, detectors, and other optical or optoelectronic devices used in a flight laser communication system are as follows:

Tests within MIL-STD-883

Accelerated life test (ALT) (e.g., method 1005)
Burn-in (e.g., method 1015)
Constant acceleration (e.g., method 2001)
Destructive physical analysis (e.g., method 5009)
External visual (e.g., method 2009)
Leak (e.g., method 1014)
Internal visual inspection (e.g., method 2017)
Mechanical shock (e.g., method 2002)
Nondestructive bond pull (e.g., method 2023)
Particle impact noise detection (e.g., method 2020)
Solderability (nonoperational, e.g., method 2003)
Temperature cycling (e.g., method 1010)
Thermal shock (nonoperational, e.g., method 1011)
Total dose radiation (e.g., method 1019)
Radiographic (e.g., method 2012)
Vibration (nonoperational, e.g., method 2007)

10.2.1.2 Telcordia Qualification

Telcordia qualification guidelines on optics and optoelectronic components are established primarily to address the fiberoptic industry needs (Telcordia.com). Telcordia requirements and standards specify performance, testing, and evaluation criteria for vibration, temperature cycling, mechanical shock, electromagnetic interference (EMI)/ESD, accelerated aging, and other qualification-related effects. In this case, only component samples are screened per batch under the Telcordia specifications. Telcordia standard requirement for lifetime specifies a minimum of 10-year reliability performance. Although Telcordia-qualified parts may exceed spacecraft lifetime requirements, the actual environmental specifications need to be compared to show relevancy to space and the launch environment. Requirements may exceed the launch and space environment or be significantly more benign, such as submarine-based telecommunication systems. Even though Telcordia qualification requires COTS components to pass shock, vibration, elevated and low-temperature extremes it does not ensure reliability for the specific components integrated into the system. Telcordia generic requirements (GR), pertaining to laser communication systems are as follows:

GR-20	Optical fiber and cables
GR-326	Optical connectors
GR-468	Optoelectronic devices
GR-910	Optical attenuators
GR-1209	Passive optical components
GR-1221	Branching component reliability
GR-1312	Optical fiber amplifiers
GR-2853	AM/Digital video laser transmitters and receivers
GR-2883	Optical filters

10.2.1.3 NASA Standards

The NASA Standard NASA-STD-8739.5 specifies requirements on fiberoptic terminations, cable assemblies, and installation (https://standards.nasa.gov/standard/nasa/nasa-std-87395). This standard prescribes a process and requirements for reliable fiberoptic assemblies, cables, terminations, and installations. It pertains to passive components and specifically, the implementation of fiber cables. Besides, the NASA Electronic Parts and Packaging program provides guidelines for qualifications and reliability assessment (https:nepp.nasa.gov).

10.2.1.4 IEEE Standards

Certain IEEE standards specify environmental conditions for survivability of photonics and microelectronic systems in the space environment. Based on physics, failure of semiconductor devices, test levels for radiation, vacuum, temperature, and humidity have been identified. Pertinent standards include

IEEE 1156.4	Environmental specifications for spaceborne computer modules
IEEE 1393	Spaceborne fiberoptic data buses

10.2.2 TEST AND CHARACTERIZATION METHODOLOGY

This step includes prescreening tests, initial qualification testing, identification of possible failures and failure mechanisms, failure mitigation and requalification, acceptance testing, and Waiver procedures.

Some of the tests mentioned in the introduction section are briefly described here. These are the main sources of failure and are the most common tests, which can readily be done and applied to most environments.

10.2.2.1 Mechanical Vibrations and Shock

Flight hardware typically experiences vibrations during the ground test, transportation, launch, and flight operations. The vibration frequency may be a sine wave or random. A significant source of vibrations is acoustic noise often particularly intense at the first minute of a satellite launch or airplane takeoff. Different launch vehicles and host platforms generate different vibration patterns. Steady-state and random accelerations are imparted from the rocket motors to the rocket and its payloads. Once in orbit, deployment of antennas, solar panels, or other structures induce shock waves to spaceborne payloads. For an example of test specifications regarding vibration and acoustics testing, refer to the NASA document "Vibroacoustic qualification testing of payloads, subsystems, and components" (NASA Goddard Space Flight Center, https://llis.nasa.gov/lesson/817). Table 10.1 provides the vibration profile of a typical spacecraft mission.

10.2.2.2 Electromagnetic Compatibility and Electromagnetic Interference

The emissions and conduction of electromagnetic emission from one instrument to the other must be avoided in all flight missions. The electromagnetic compatibility (EMC)/EMI stems from one or more of the following sources: conduction, radiation, and power. The MIL-STD-462 test standard provides guidelines for EMC implementation.

10.2.2.3 Thermal Vacuum Stability and Outgassing

Devices can outgas or degrade in the vacuum environment. Use of materials with very low outgassing properties and condensation to avoid contamination of critical optical surfaces is recommended.

During the satellite launch, orbit transfer events and cycles in and out of the Earth's shadow, the spacecraft experiences large temperature variations. before deployment of solar panels, the payload's electrical power is supplied merely through batteries, limiting the capacity of electrical heaters. Phase-change materials are being considered for use on spacecraft to mitigate thermal variations. For an Earth-orbiting payload, optoelectronic systems are typically designed for $-10°C$ to $+55°C$ operation and $-20°C$ to $+70°C$ survival temperature range.

TABLE 10.1
Example of Frequency Ranges of Vibration Experienced by Flight Payloads

Vibration Types	Frequency (Hz)	Response
Random vibration	10–2000	0.01–1 g^2/Hz
Acoustic vibration	20–10,000	100–140 dB
Pyrotechnic shock	30–10,000	5–10,000 g

TABLE 10.2
Environmental Thermal Loading Conditions

Parameters	Values	Notes
Solar energy flux	1.32–1.43 kW/m^2	Above the atmosphere
Reflected solar energy	0.45 kW/m^2	Albedo (maximum)
Outgoing infrared	0.1–0.5 kW/m^2	
Cold sky temperature	2.725 K (dart sky)	
Temperature extremes	−65°C to +125°C	Uncontrolled
	−10°C to +55°C	Controlled

Thermal vacuum stability and outgassing behavior of materials used in flight should be compatible with the mission environment but should not adversely affect mission performance. The tests should include operational, survival, and hot/cold start modes of operation. The thermal balance test typically involves closed-loop control during testing. It is advisable to consider utilizing those organic materials with an aggregate mass loss of less than 1.0%, and a collected volatile condensable mass that does not exceed 0.1%. Extensive thermal-vacuum bake-outs prior to integration may be necessary for some materials utilized to eliminate contamination of optics. Table 10.2 summarizes thermal loading conditions for low Earth orbit (LEO).

10.2.2.4 Static Charge and ESD Sensitivity

ESD characteristics should be evaluated to ensure they meet the requirements. ESD can cause damage that is not even visible under very high optical magnification but affects performance and lifetime. To define the extent and complexity of the static control program, testing and analysis should be performed, thereby establishing static charge sensitivity of the spaceborne electronic devices (NASA 4002 for IESD and NASA TP-2361 for surface charging).

10.2.2.5 Electrical Arc-Tracking Resistance EBER

The possibility of arc tracking due to electrical discharges should consistently be avoided. One approach is to prevent the electrical wire insulation, wire accessories, and materials to enter contact with electronic circuits. Anodized aluminum surfaces may exhibit arcing when biased to sufficiently high voltages (Hillard and Ferguson, 1995). Selective regions of Kapton films, when overlapped, have exhibited pyrolization (Stueber and Mundson, 1993).

10.2.2.6 Galvanic Corrosion

The MIL-STD-889 provides guidelines for controlling dissimilar metals that are in intimate contact and can result in galvanic corrosion. One approach utilizes compatible materials.

10.2.2.7 Metal Migration and Whisker Growth

Diode lasers are among some of the parts in a lasercom system that utilizes silver and gold for contact bonding. Gold, silver, tin, and copper used in integrated circuits and circuit boards have demonstrated metal migration. Pure tin, zinc and cadmium finish or solder are known to be susceptible (without warning) to the formation of needle-like protrusions (strands), or whiskers, capable of electronic short-circuiting. Therefore, their use should be prevented.

10.2.2.8 Welding

The MIL-STD-1595 provides guidelines for automatic, semiautomatic, or manual welding. Any weld rods or wires used as a filler metal on structural parts should be certified and documented for composition, type, and supplier to provide traceability to the end-use item. Nondestructive inspection of all critical welds is always prudent.

10.2.2.9 Nondestructive Inspection

The MIL-I-6870 standard provides guidelines for nondestructive inspection and evaluation. These tests help reduce risks associated with highly stressed, mission/safety-critical items.

10.2.2.10 Radiation

Energetic particles can degrade or cause permanent damage to detectors and lasers. Accelerated tests based on the expected radiation environment from mission orbit parameters and extrapolated to mission lifetime are an integral part of the radiation hardness qualification procedure. In electronics, absorption of energetic particles, cosmic radiation absorption, and micrometeorite strike can cause degradation, permanent damage, or single-event effects (SEEs).

Proton-induced displacement damage is the primary radiation damage mechanism in injection laser diodes (Barnes et al., 2002). Total ionizing dose and displacement damage reflected as an increased dark current for detectors is also a primary degradation mechanism for photodiodes. Increased attenuation due to photodarkening is a common effect of radiation on optical fibers.

10.2.3 Qualification of Custom Off-the-Shelf (COTS) Components

It is unlikely that a flight project can rely entirely on COTS commercial components that are qualified to the exact requirements of the mission. Qualification of COTS components for space use, where samples of existing devices or assemblies are qualified to the required specifications, reduces time and cost while increasing risk. In comparison, qualification from the source approach is lower risk but at a significantly higher cost. One should keep in mind that most COTS optical and optoelectronic components are not designed for the space environment, have minimal flight history, and their failure-in-time values and failure mechanisms are not well established. This is particularly true for plastic parts. Certain COTS and most plastic parts pose some of the biggest unknowns in terms of test and evaluation. Typically, detailed information on the device construction and origin of its constituent components is not available; therefore, resulting in increased risk levels. Quantities in tens or even hundreds of a given part may be necessary (often from multiple vendors) to quantify a given part with low qualification heritage (Hemmati and Lesh, 1989).

Devices manufactured in different lots at the same foundry can also vary in radiation susceptibility (Dodd et al., 2010). Device heritage may be applied only to the part that is exactly the same as that flown in space.

COTS devices that are Telcordia (or similarly) qualified are much preferred, as these devices comply with a great deal of the requirements for flight qualification. However, some important specifications such as materials for use in the vacuum environment (e.g., outgassing properties), radiation tolerance, vacuum (zero humidity), and zero-gravity environment are likely absent from devices that are qualified for terrestrial applications. As a minimum, in qualifying COTS components the program begins by developing the following methodologies: selection, screening, risk reduction, qualification, test and characterization (Berthier et al., 2002). These procedures are explained below.

10.2.3.1 Selection Methodology

Assuming that the overall specifications and performance characteristics of the device meet the preset requirements, multiple factors influence the selection criteria, including:

- Commercial availability/affordability in quantities and in a timely manner,
- Prior qualification (e.g., for fiberoptic industry—Telcordia),
- Availability of lifetime/meantime between failure (MTBF) and other reliability data,
- Willingness and adequate experience level of the manufacturer to perform minor upgrades, and
- Degree of obsolescence prior to the actual selection of flight components.

10.2.3.2 Screening Methodology

It is prudent to develop a set of environmental limitations, both survival and operational. An example is MIL-STD-883, with standards and specifications downloadable at http://dscc.dla.mil, which provides various screening methods and their pertinent documents include those listed earlier.

10.2.3.3 Risk-Reduction Methodology

Several implementation approaches help in minimizing some of the risk factors. These include:

- Procurement of small quantities of the same device from several different manufacturers,
- Preliminary testing for down-selection,
- Procurement of large quantities (10–100 sec) of the best-performing devices, ideally manufactured under identical conditions—traceable to flight devices,
- Implementation of redundancy,
- Derating of the device during operation (e.g., applying lower operating current to laser diodes typically used by the manufacturer), and
- Allocation of margin to end-of-mission life performance specification.

10.2.3.4   Qualification Methodology

The qualification methodologies based on Standards, as described in the earlier segments of this chapter apply equally well to the COTS components.

10.3   QUALIFICATION OF INDIVIDUAL DEVICES

A lasercom transceiver consists of multiple subsystems and components. Brief descriptions of some principal constituents of a transceiver are provided below. In developing the flight article, it is critical to follow the "test as you fly and fly as you test" criteria (Kim et al., 2002; Ott et al., 2006a). The use of many of the components in terrestrial fiberoptic telecom systems enhances their chances of being subject to stringent requirements set by Telcordia standards for photonic devices. Therefore, if available, an excellent starting point for meeting flight qualification is to employ devices that adhere to this standard. Outgassing and radiation tolerance are among the specifications not included with this standard and can be addressed separately. A dedicated qualification testing procedure is essential for these requirements.

For radiation qualification of COTS devices, space radiation environment modeling software is used to first calculate the expected mission radiation environment. Simulation results can be used to determine mission radiation requirements. Radiation modeling results will provide expected dose levels based on aluminum shielding thickness as well as linear energy transfer (LET) spectrum for protons and heavy ions. Components can be selected based on the capability to survive the expected environment (capability to survive to expected dose levels), or radiation testing can be completed to evaluate the performance of the components. A subsystem cannot be more reliable than its constituents, and the semiconductor laser diode is a fundamental critical part.

10.3.1   Qualification of Semiconductor and Fiber Lasers/Amplifiers

As described in Chapter 4, diode-laser-pumped crystalline hosts doped with rare-earth ions (or doped fiberoptics) are the primary candidates for spaceflight lasers. These lasers typically consist of several components including semiconductor pump laser(s) in continuous wave (CW), quasi-CW, or pulsed mode of operation, fiber or diode-based master-oscillator laser for oscillator/ amplifiers modules, modulators (e.g., $LiNbO_3$), the host medium (crystal or fiber), fiberoptics, fiber connectors, optics (lenses, isolators, filters, beam multiplexers, harmonic generators, mirrors, and coatings), and laser driver electronics. Laser driver electronics and microcontrollers within TECs for lasers are particularly susceptible to radiation effects and are a concern when evaluating COTS lasers. Assuming available redundancy for the laser transmitter, future missions will be requiring lifetimes on the order of 10 to 15 years. To qualify a laser, each of the constituent elements will have to be qualified separately. In constructing a laser, proper attention should be given to contamination control. Prior to the environmental qualification of flight laser transmitters, the development of a detailed test plan is highly prudent. Some of these plans have been summarized in comprehensive publications and will not be elaborated further (Peri et al., 2003; Wright et al., 2005; Hemdow et al., 2006; Hovis, 2006; Ott et al., 2006b).

TABLE 10.3

Comparison of Radiation Sensitivity for Two Er-Doped Fibers

Er-Doped Fiber	Er Concentration $\times 10^{24}/m^3$	Ge (mol%)	Al (mol%)	Attenuation at 980 nm (dB/m-krad)	Attenuation at 1550 nm (dB/m-krad)
1	4.5	20	12	0.013	0.0025
2	16	23	10	0.012	No data

Source: Henschel et al. (1998).

Some of the semiconductor laser's failure mechanisms include infant mortality failure, catastrophic optical mirror damage (COMD), solder creep/debonding, wear out failure, and bulk failure (Fukuda, 1991). Infant mortality failure occurs early on and is primarily related to manufacturing defects. COMD occurs due to absorption at the semiconductor laser facet causing a major thermal gradient. Wear out failure is gradual degradation and can be attributed to the growth of material defects. Bulk failure results primarily from crystal defects in the region where high intracavity fields circulate.

An example of lasers qualified under Telcordia specifications is the 915-nm fiber-coupled single-emitter pump diode lasers (6390 series), currently offered commercially (jdsu.com). These lasers have an output power and efficiency of >5 W and >45%, respectively. An MTBF of 200,000 h (>22 years) has been quoted for certain versions of this laser.

Generally, the signal and pump diode laser's lifetime (used with diode-pumped lasers and laser amplifiers) are of primary concern to the overall reliability of the laser. Various programs and facilities have been developed to better understand the reliability of these devices (Ciminelli et al., 2003; Peri et al., 2003; Hendow et al., 2006).

Fiber lasers and amplifiers are typically doped with appreciable concentrations of rare-earth elements such as erbium (Er) or ytterbium (Yb). The dopant in the fiber alters the response to ionizing radiation relative to undoped fibers.

Ott surveyed the effect of the dopant type in fiber lasers and optical amplifier on radiation response (Ott, NEPP, 2004). This survey indicated that aluminum concentration in the fiber causes more radiation-induced damage to the fiber than Er or Yb concentrations. Tables 10.3 and 10.4 show the influence of Al concentration on radiation-induced attenuation in four different fibers (Barnes et al., 2005).

TABLE 10.4

Comparison of Radiation Sensitivity for Two Yb-Doped Fibers

Yb-Doped Fiber	Yb (mol%)	Ge (mol%)	Al_2O_3 (mol%)	P_2O_5 (mol%)	Radiation-Induced Attenuation (dB/m-krad)
1	0.13	5	1	1.2	0.07
2	0.18	0	4.2	0.9	0.86

Source: Rose et al. (2001).

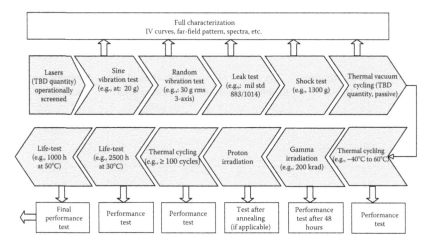

FIGURE 10.1 Example of a qualification process for semiconductor diode lasers, also applicable to other optoelectronic components.

Figure 10.1 shows an example of a qualification procedure for diode lasers and other optoelectronic components (Hemmati, 2002).

Besides vacuum issues and the zero-gravity operation, the primary unique space environment effect is radiation and will be discussed next.

10.3.2 QUALIFICATION OF PASSIVE FIBERS AND FIBER CONNECTORS

Besides photodarkening (due to repeated exposure to optical radiation and ionizing radiation accumulation) proper selection of fiber cladding to avoid outgassing and coefficient of thermal expansion (CTE) mismatch with the fiber's core requires special attention. Reliability of fiberoptic connections to active elements (e.g., diode pump lasers) under mechanical, thermal, and optical stresses has to be verified utilizing standard test procedures (e.g., MIL standards, discussed below). Failure modes include fiber and assembly damage from fiber flex/tension/twist and thermal cycling, and fiber end-face damage can result from vibration and shock.

Extensive qualification data on standard fibers exists in the literature. However, some of the new fibers, namely the large mode area fibers, and the photonic crystal fibers, have been studied in less detail (Limpert et al., 2003; Galvanauskas and Samson, 2004).

10.3.3 QUALIFICATION OF PASSIVE OPTICS

A wealth of information is available in the literature on the qualification of discrete passive optical components and optical coatings (Bruni et al., 1996). This would help in selecting certain substrate materials. However, as optical coatings (metallic, dielectric, etc.) could vary widely in the method of deposition, each coated material typically requires dedicated qualification tests. Failure modes include radiation-induced darkening and color-center generation. For fiberoptic components, the quality of couplers, connectors, filters, isolators, and other passive components need to be qualified separately. Again, a Telcordia-qualified version of these devices will be further, along the qualification path, relative to other versions. A detailed description of the qualification of lithium niobate crystal for optical modulators is given in Moyer et al. (1997).

10.4 PACKAGING FOR THE SPACE ENVIRONMENT

Facet oxidation, moisture, contamination, inappropriate packaging approaches such as CTE mismatch and use of indium solder, pressure gradients during launch and maneuvers, laser beam-induced deposition of hydrocarbons, electron, proton, and ion bombardment are among environments and manufacturing approaches harmful to optoelectronic components.

Hermetically sealed semiconductor laser devices may suffer from contamination, where laser-induced photodissociation of organics and outgassed hydrocarbons deposit on the laser's active facet (Sherry et al., 1996). The resulting laser beam absorption can lead to catastrophic damage or destruction of the semiconductor lasers. One solution is to introduce oxygen into the sealed package to react with carbon to act as a better material. Without proper precautions, similar facet contamination may occur in a vacuum, especially at the elevated temperatures encountered by the device.

MIL standard 883 may not sufficiently address the procedures for a hermetically sealed optoelectronic parts intended for use in space. Depending on the environment, proper test procedures should be established. A packaging qualification process may follow the steps of visual inspection of the housing, bake-in, assembly, vacuum bake-in, (laser) welding of the lid, performance verification, coupling tests, temperature storage, performance test, leak test, and final performance characterization and validation.

Some of the recommended approaches for diode laser packaging are as follows (Hemmati, 2006):

1. Avoid materials that can outgas (e.g., epoxies, organics, and insulators),
2. Avoid indium solder and utilize flux-free hard solders instead,
3. Avoid device fracture by preventing CTE mismatch with the submount,
4. Avoid materials that are prone to failure due to creep and fatigue,
5. Avoid thermal gradients and allow for proper heat sinking of the package. (Note that thermal convection does not occur in a vacuum, making the device more susceptible to damage),
6. Avoid glued fibers,
7. Utilize only laser welding, soldering, and glass feed-throughs, and
8. Utilize hermetic packages with a mixture of nitrogen and oxygen.

10.5 RADIATION EFFECTS

Over the duration of a mission, an Earth-orbiting lasercom transceiver will be subjected to the space radiation environment. Some radiation exposures are additive, such as total dose and displacement damage. The total radiation effect experienced by the flight terminal depends on the duration of exposure (mission lifetime), altitude and inclination of the orbit, and time-varying solar events. The combined ionization environment of protons and electrons typically defines the total accumulated dose, over the life of the mission. The overall effect of the radiation environments ranges from degradation in performance to complete failure.

Materials used for flight hardware should be able to withstand the radiation environment specified with typically less than 20% degradation in their applicable properties. In applications where the estimated radiation dosage exceeds the 20% degradation level or is greater than the available test data, shielding may be used. In assessing materials for space environmental resistance, the effects of vacuum-ultraviolet, ultraviolet, gamma rays, electron, and proton radiations are typically considered. In cases where there are no available data, testing may be required.

10.5.1 RADIATION ENVIRONMENTS

The space radiation environment consists primarily of trapped particles in the Van Allen Radiation Belts, solar energetic particles (SEPs), and galactic cosmic rays (GCRs) (Armstrong and Colborn, 2000; Huston, 2002). Energetic electrons and protons trapped by the Earth's magnetic field and its geomagnetic tail surrounds the Earth. This region is commonly referred to as the Van Allen radiation belts (Webb and Greenwell, 1998). The belts have two primary zones. The inner belt extends in altitude from about hundreds of kilometers to about 6000 km, while the outer belt extends to about 60,000 km. The South Atlantic anomaly comprises a region of lower magnetic field strength that allows energetic particles to penetrate down to about 100 km. While protons and electrons populate the inner belt, the outer belt consists primarily of energetic electrons. Coronal mass ejections and solar flares produce intense bursts of SEPs, including protons and heavy ions (electrons are also included but typically an order of magnitude lower in flux) over the polar caps. GCRs originate from outside the solar system and are formed from diffusive shock acceleration of supernova remnants. Primary GCRs consist of protons, alpha particles and heavy-ion nuclei. GCR particles move perpendicular to the Earth's magnetic field lines and can be deflected at the equator and funneled toward the poles. In this manner, GCRs are most relevant at high altitude and high inclination, polar orbits (Suparta, 2014).

Trapped protons and electrons from the radiation belts, protons from SEPs, and protons from GCRs cause total ionizing dose (TID) and total nonionizing dose (TNID) or displacement damage effects. TID effects occur when charged particles deposit ionizing energy into the target material, and the ionization alters the material by generating electron-hole pairs and inelastic Coulombic scattering (Alig and Bloom, 1975; Miroshnichenko, 2003). Displacement damage effects are caused when a nucleus is hit by an incident particle and displaced in the crystal lattice (Miroshnichenko, 2003).

Protons and heavy ions from GCRs and solar flares, as well as trapped protons in the radiation belts, cause SEEs in electronic devices. SEEs occur when a charged particle is deposited or passed through active components with electrical circuits, such as memory, power, and logic devices (Stark, 2011). SEEs cause a disruption

TABLE 10.5
Expected Total Ionizing Dose

Expected Total Ionizing Dose for 100 mils (2.54 mm) Aluminum Shielding Thickness			
Orbit	1-Year Mission	5-Year Mission	10-Year Mission
ISS	0.50 krad	2.48 krad	4.96 krad
1000 km, 0° Inc.	1.89 krad	9.46 krad	18.9 krad
800 km, Polar	2.94 krad	24.2 krad	44.3 krad

Expected Total Ionizing Dose for 15-Year GEO Mission			
Aluminum shielding	100 mils	200 mils	300 mils
Expected TID	710 krad	120 krad	44 krad

in electronic device operation and can experience destructive or nondestructive consequences. Common SEEs include single-event upsets (SEUs), which affect the logic state of a circuit and induce soft errors, and single-event functional interrupts (SEFIs), which are identified by device functional "hangs" (Baker, 2002). Single-event latch-ups (SELs) induce an elevated current state in the device and can result in loss of device functionality or permanent damage (Samaras, 2014). The effect of radiation on electronics components is well-documented and will not be further discussed here (Pease et al., 1988).

Accumulated radiation effects in different Earth orbits are vastly different. For example, for low, medium, and geo-synchronized Earth orbit (LEO, MEO, and GEO) missions, the accumulated dose over 7 years is 5 to 10, 20 to 200, and 50 krad beyond 100 mils of Aluminum, respectively. Mission duration, mission start date (solar cycle period), altitude, and inclination all contribute to the TID level. Table 10.5 lists the expected TID values at 100 mils aluminum shielding thickness for various LEO orbits and mission durations for mission start in the year 2020, modeled using OMERE (Outil de Modelisationde l'Ennvironment Radiatif Externe) radiation environment modeling software (Aniceto, 2017).

Table 10.6 summarizes the LEO radiation environment.

TABLE 10.6
LEO Radiation Environment Typical Characteristics

Characteristics	Values
Orbital altitude	10–1000 km
Dose rate/year	0.1 krad
Particle impact/m²/year	11–26
Ultraviolet radiation	0.12 kW/m^2
Charged plasma	0.1–0.2 eV
Atomic oxygen	4.5 eV
Solar protons	10^2–10^5 MeV

The estimated TID of radiation (rad), accumulated over a period of 1 year, beyond 100 mils of aluminum for the LEO and GEO is:

	LEO (1000 km)	GEO (38,000 km)
Inclination	60°	0°
Dose/year (krad)	17.3	6.6

Radiation types at different orbits are depicted in Figure 10.2.

A list of some of the particles and rays and examples of their radiation effect on the components that comprise an optical communication system, is given in Table 10.6.

For a given mission, the magnitude of the radiation level for the space environment can currently be predicted to high accuracy with statistical models using radiation environment modeling software such as OMERE, SPENVIS and NOVICE. Prudent design practices, strategic component placement, and shielding will have to be exercised to minimize the radiation exposure of a given component. Electronics and optoelectronic parts are the most radiation-susceptible components (Howard and Hardage, 1999).

For diode lasers and photodiodes, displacement damage and total ionizing dose damage can be calculated prior to assembly, to determine the required radiation hardness levels. For optics, radiation hard glasses such as cerium-doped glasses, fused silica, or UV-grade sapphire present good options for radiation hardness (Gusarov, 1999). For the mirror substrate, in particular, SiC is a radiation-resilient material. Meticulous attention has to be paid to thin-film coatings on glass or mirror substrates. For electronics, a limited number of radiation-hard components are available to choose from for circuit design.

Radiation effects on optics and optoelectronics components that constitute a laser communication terminal are less known and will be summarized here. Taylor (1999) provides an excellent review of the effects of space radiation on key photonic components. To generalize, semiconductor lasers and photodetectors are sensitive to displacement damage (due to protons and neutrons). Passive optical elements and fibers are sensitive to ionizing radiation, such as those caused by electrons, gamma rays and protons (TABLE 10.7).

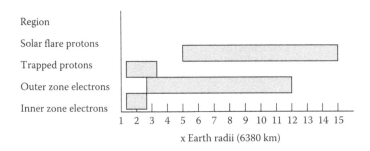

FIGURE 10.2 Altitude dependence of the particle types for total ionizing dose (TID).

TABLE 10.7
Radiation Effects of Particles and Rays on Lasercom Components

Effect	Trapped Electrons	Trapped Protons	Solar Protons	Cosmic Rays
Glass (and fiber) darkening	●	●		
Diode laser power degradation	●	●	●	
Detector noise increase	●	●	●	
Total ionizing dose damage	●	●	●	
Scintillation	●	●	●	
Single-event upsets		●	●	●
Displacement damage		●	●	
Bremsstrahlung	●			

Note: Blank areas indicate little to no occurrence.

10.5.2 Radiation-Induced Effects in Optics and Coatings

Refractive optical components experience a significant degradation depending on the radiation exposure levels and the particle type (Baccaro et al., 2005). Some of the optical materials with no appreciable degradation up to 100 krad include Si, Suprasil, Synasil, SiO_2, ZnS, and Al_2O_3. Table 10.7 summarizes radiation tolerances of some refractive materials.

Particular attention should be devoted to doublet and triplet lenses that may have used epoxies to bond lenses together. The lens material may be radiation tolerant, but not the epoxies. For both refractive and reflective optics, optical coatings (dielectric or metallic) deposited on optical elements can also experience radiation-induced damage (Fernandez-Rodriguez et al., 2004). Most dielectric coatings applied to mirrors and beam splitters are highly durable. Some of the optical coating materials showing a few percent degradation at 100 krad include TiO_2, MgF_2, MgO, and HfO (Table 10.8).

10.5.3 Radiation-Induced Effects in Semiconductor Lasers

Radiation effects on edge-emitting diodes and vertical-cavity surface-emitting lasers (VCSELs) are well-documented (Barnes, 1970; Shoone et al., 1997;

TABLE 10.8
Radiation Tolerance of Certain Refractive Materials

BK7 glass	Browning starts at 1 krad
	Significant darkening at 0.1 Mrad
K7G, GG-375G, K-5G, BK-1G glasses	Radiation tolerant to1 Mrad
Fused silica	No light loss up to 10 Mrad
Sapphire	No light loss up to 100 Mrad

Le Metayer et al., 2003; Troupaki et al., 2006). For a brief duration of Earth-orbiting missions, radiation effects on semiconductor diode lasers are minor and can be effectively shielded. Radiation-induced failure modes include displacement damage in the active region, excess minority carriers, decrease in minority carriers, and nonradiative recombination. However, change in the threshold current, a significant reduction of output power, and shift in central wavelength may occur for longer missions (on the order of 10 years). With a fluence of protons, the percentage change of threshold current is found to be about the same for 1550-nm InP distributed feedback lasers, the 1330-nm InP lasers, the 905-nm InGaAs strained quantum well lasers, and several times that of 850-nm AlGaAs VCSELs (Johnson, 2004). Annealing of laser damage occurs more rapidly when lasers operate above the threshold (Johnson, 2004).

10.5.4 RADIATION-INDUCED EFFECTS IN PHOTODETECTORS AND DETECTOR ARRAYS

Most detectors, particularly avalanche photodiodes (APDs), and especially Geiger-mode APDs are highly susceptible to radiation damage (Baccaro, 1999). A detector's dark current increases with proton- or neutron-induced displacement damage. Transient signal pulses are produced by photodetectors as a result of the proton or cosmic ray-induced damage. A detector's leakage current may increase as a result of TID-induced damage.

Displacement damage from proton irradiation leads to increased dark current and degraded responsivity in InGaAs photodiodes due to the generation of defect traps and recombination centers in the active region. A study on proton irradiation of InGaAs PIN photodiodes by Barde et al. (2000) observed an overall 10 to 1000-fold increase in dark current values from proton irradiation with energy levels between 9 MeV and 300 MeV and fluences between 2×10^9 p/cm^2 and 10^{10} p/cm^2. The study recognized a linear relationship between dark current and proton fluence. Troska et al. (1997) observed a nonlinear increase of 6 to 7 orders of magnitude in the dark current of InGaAs PIN photodiodes from 24-GeV proton irradiation with fluence levels of 4×10^{14} and 10^{15} p/cm^2.

Hamamatsu InGaAs PIN diodes were irradiated with 105-MeV proton radiation in reverse bias voltage configuration (Aniceto et al., 2017). Photodiode dark current at $1V_R$ linearly increased with increasing proton fluence levels. Based on OMERE space radiation environment simulations and dark current results from one of the irradiated devices, the average dark current value is expected to be between 10.71 nA and 26.73 nA at $1V_R$ for the 500-km orbit altitude, and the average dark current value is expected to be between 26.73 nA and 73.85 nA at 1VR for the 1000-km orbit altitude after 1-year mission durations.

Shielding could be used to mitigate proton radiation effects, but effective shielding material and thickness will be based on the environment, i.e., proton energy spectrum. Circuit level processing could potentially be configured to be tolerant of gradual dark current and responsivity degradation (Aniceto et al., 2017).

Charge-coupled devices (CCDs) are exceedingly sensitive to a variety of radiation-induced effects. High levels of shielding are typically necessary. CMOS-type active pixel sensors are significantly less sensitive to radiation (Hopkinson, 2000).

Detectors with II–V structure are considerably more radiation-tolerant than silicon detectors. Silicon phototransistors include an inferior tolerance for radiation-induced damage.

CMOS image sensors are known to be susceptible to SEEs, including single-event transients (SETs), SELs, and single-event disturbance effects (Hopkinson, 2000; Lalucaa et al., 2013). The dominant parametric degradation effects are an increase in supply current due to ionizing dose and creation of "hot" pixels (i.e., pixels with high dark current values) due to nonionizing interactions in the pixel sensitive volume (Bogaerts et al., 2002). Milanowski et al. (2017) tested a commercial 12-megapixel CMOS sensor, CMOSIS CMV12000, with 105-MeV protons. No SELs or SEFIs were observed. Hot pixel accumulation with fluence, consistent with well-understood displacement damage effects, was characterized for fluences up to 2×10^{11} protons/cm². An annealing period of 106 days yielded a decrease in the average hot pixel count by over two orders of magnitude (Milanowski et al., 2017).

10.5.5 RADIATION-INDUCED EFFECTS IN FIBERS (INCLUDING DOPED FIBERS)

Formation of color centers in the core region of the fiber has been reported in the literature (Williams et al., 1996). In doped fibers, the parasitic absorption by the color center (ions) can lead to significant throughput losses, a phenomenon known as photodarkening. Composition of the fiber, and in particular, presence of the impurities and particular dopants (e.g., Al), have been associated with enhanced photodarkening, while the introduction of certain other dopants (such as Ce ions) have been shown to reduce the susceptibility to ionizing radiation (Zellmer et al., 1999).

Depending on the type, optical fibers show varying degrees of sensitivity to radiation (Marshall et al., 1996). Fibers have a high nonlinear response to TID, while displacement damage is not an issue (Ott, 1988). Infrared fibers (1320 and 1550 nm) are considerably more radiation-resistant than shorter wavelength (700–850 nm) fibers. Doped fibers (e.g., Er-doped or polarization preserving) show enhanced susceptibility to radiation. Among Er^{3+}, Yb^{3+}, and Er^{3+}/Yb^{3+}-doped fibers, Er^{3+}-doped fibers were observed to be the most radiation-sensitive and Er^{3+}/Yb^{3+} co-doped fibers to be the most radiation-resistant (Fox et al., 2010). Pure silica fibers can tolerate hundreds of kilorad while some doped glasses include an exceptionally low tolerance to radiation. Annealing and optical bleaching, particularly at elevated temperatures, are often helpful to partially recover from darkening.

It has been determined that Si fibers and doped fibers exhibit a maximum transmittance loss of 10% for total gamma-ray exposure of 3–6 krad, typical for a seven-year LEO environment (Fox et al., 2007). A kinetic model developed by Ahrens et al. (2001) yields end-of-life attenuation of 0.12 dB/m for silica-based Er^{3+}/Yb^{3+} co-doped fibers after 100 krad of total ionizing dose.

10.5.6 OPTICAL MODULATORS

A NASA study by Ott et al. (2002) evaluated the reliability of commercial $LiNbO_3$-based optical fiber modulators for space flight environments. The study concluded that titanium-diffused and proton-exchanged waveguide $LiNbO_3$ devices are not

susceptible to radiation-induced effects up to 1 Mrad during high-dose-rate exposures greater than 0.1 rads/min. The NASA study claimed there is no definitive
indication that LiNbO$_3$ modulators and devices are susceptible to displacement damage, and gamma radiation exposure should be appropriate for simulating radiation-
induced effects. Based on the size and materials of most LiNbO$_3$ modulators, the
radiation testing has demonstrated these devices to be insensitive to typical space
flight environment total dose levels and dose rates [Ott et al., 2002].

Thomes et al. (2007) exposed a hermetically sealed, commercial LiNbO$_3$ modulator, Photline Technologies MXPE-LN, to gamma radiation to a total dose level of
52 krad at a dose rate of 13 rads/min. The study concluded that gamma radiation did
not affect the modulator performance. Le Kernec et al. (2017) conducted gamma and
proton radiation testing on commercial LiNbO$_3$ modulators and III-V semiconductor modulators. Gamma radiation testing revealed differences in X-cut and Z-cut
LiNbO$_3$ modulators. X-cut LiNbO$_3$ modulators were found to be significantly more
radiation tolerant than Z-cut LiNbO$_3$ modulators.

10.5.7 COHERENT DIGITAL SIGNAL PROCESSING ASICs AND FPGAs

Technological advancements from the terrestrial communications industry, such
as optical coherent digital signal processing (DSP) application-specific integrated
circuits (ASICs) and field-programable gate arrays (FPGAs) have enabled high-
capacity optical coherent communication systems of 100 Gbps and greater. There
has been recent interest in utilizing optical coherent COTS transceivers for high
throughput space applications (Robinson et al., 2018). Radiation assessments of commercial optical coherent DSP ASICs show potential for adopting this technology
for a space-based optical communications system. Radiation assessments of commercial 100 Gbps optical coherent DSP ASICs, Inphi CL20010A1, and Acacia DSP
ASIC within the AC100M transceiver were conducted by Aniceto et al. (2017, 2018).
The ASICs were constructed via 28-nm bulk CMOS technology, which can be sensitive to destructive SEEs.

Inphi CL20010A1 ASICs were tested with proton radiation at 64 MeV and 480 MeV
energy levels. No destructive SELs were observed. The ASIC survived and experienced
no performance degradation from a proton total fluence of 1.27×10^{12} p/cm^2 with an
equivalent TID exposure up to 170 krad(Si) while tested in the noise-loaded optical
loopback configuration with 64-MeV protons (Aniceto et al., 2017).

Aniceto et al. (2018) completed gamma and proton radiation testing of the
AC100M transceiver in an active noise-loaded optical loopback configuration.
Communication with the AC100M was lost at a TID level of ~13.7 krad from gamma
radiation testing. Power cycling of the AC100M module and the EVB did not clear
the module communication fault or restore nominal operation. Based on data and
fault monitoring logs of the AC100M and EVB, the loss of communication was
caused by a failure of a DC power converter on the AC100M PCB due to scattered
gamma radiation of less than 1 krad. During proton radiation testing, a total of
23 nondestructive SEEs were induced from a cumulative fluence of 5.25×10^{10} protons/
cm^2. The AC100M survived proton irradiation to a TID level of up to 66.7 krad.
Based on a single-unit test of a COTS AC100M, the study concluded that it could be

suitable to use the AC100M for some LEO missions (i.e., 10-year ISS orbit, 100-mil Al shielding, 2× margin). Replacing auxiliary components around the AC100M ASIC could lead to a more radiation-tolerant transceiver based on the COTS core.

10.6   ACCELERATED LIFE TESTING

For a multiyear mission lifetime, it is not always possible to test the life of a device for the actual period. Accelerated life testing (ALT) is an early identification method for some of the failure modes and the likely mechanisms to occur in a standard life test (Suhir, 2002). However, as ALT is performed under elevated stressing conditions, certain failures may occur that typically would not have happened. Examples include alteration of material properties at extreme temperatures, elevated stress-causing dislocations, or burn-in tests triggering failure mechanisms. A proper accelerated testing program generates degradation modes and failure mechanisms similar to those observed in actual operating conditions.

ALT applies excessive stress levels (e.g., significantly more elevated temperatures than specified), or higher frequency of stress cycle to induce the failure in a shorter amount of time. For any given device under test, one must design specific accelerated test conditions that relate to the initial requirements for environmental conditions.

A list of common accelerated test conditions includes:

- Thermal: cycling, gradients, shock, storage, and aging at high and low temperatures,
- Electrical: voltage extremes and power cycling,
- Mechanical: shock, fatigue, creep, stress, random, and sinusoidal vibration,
- Radiation: TID, displacement damage, and single-event effects, and
- Combination of the above, all within the allowable nondamaging ranges.

Accelerated and highly accelerated life testing can provide information on failure modes, mechanisms and statistics. Predictive and post-test failure and statistical analysis provide an insight adequate enough to mitigate the causes of failure. Broadly, ALT constitutes a basis for improving the quality of the qualification tests rather than a substitute. On the negative side, however, accelerated testing in some cases can be misleading compared to actual exposure rates in the natural environment. The so-called extreme low-dose radiation effects (ELDRS) have been found to result in significantly higher damage rates (factors of 10-100) when a dose is applied over extended periods. ELDRS testing is typically used merely on devices with bipolar technology oxide regions. Using dose rates above standards, such as IEEE-1156.4, can also induce more radiation damage than what could be experienced during the mission. This has called into question ALT for some device families.

10.7   QUALIFICATION EXAMPLE

As an example of qualification specifications for a flight worthy device, Table 10.9 provides the space qualification levels for a particular component of a fiber system for flight (FireFibe) (Chalfant et al., 2001).

TABLE 10.9

Specifications and Space Qualification Requirement Values for a Commercial Fiber System

Parameter	Nonoperating Conditions	Operating Conditions	Standards, References, and Test Methods
Thermal/ vacuum	−46°C to	−46°C to +71°C	IEEE-1156.4
	+81°C	1 cycle, 144 h hot, 24 h cold, $\Delta T/\Delta t = 10°C/min$	EIA RS-455-52 IEEE-1393
Thermal cycles	−46°C to	−46°C to +71°C	IEEE-1156.4 IEEE-1393
	+81°C	200 up to 1000 cycles $\Delta T/\Delta t = 10°C/min$ Attenuation rate <0.5 dB/km at 1300nm	EIA RS-455-52
Temperature shock		−46°C to +71°C	IEEE-1393
		$\Delta T/\Delta t = 30°C/min$	DOD-Std-1678, method4020, test condition A
Outgassing		Maximum volatile condensable material content of 0.1%	IEEE-1156.4 IEEE-1393
		Maximum total mass loss of 1.0%	SP-R-0022 when tested in accordance with ASTM-E-595
Pressure	Sea level to 5×10^{-6} Torr	Sea level to 5×10^{-6} Torr $\Delta P/\Delta t = 100$ Torr/s	IEEE-1156.4 IEEE-1393
Relative humidity	0–95% Noncondensing	0–95% Noncondensing	
Pyrotechnic shock		30 G at 100 Hz, 3000G from 1 kHz to 10 kHz	IEEE-1156.4
		3 shocks per axis	IEEE-1393
Random vibration		20 Hz–0.125 G^2/Hz	IEEE-1156.4
		50 Hz–800 Hz–0.8 G^2/Hz	IEEE-1393
		2000 Hz–0.125 G^2/Hz	EIA RS-455-11
		3 min in each axis. Attenuation rate does not increase by more than 0.5 dB/km at 1300 nm.	
		Peak acceleration must be at least 20 G	
		After a total ionizing radiation dose of 10 krad (Si) (dose rate of 1300 rads/min), the fiber attenuation rate does not increase by more than 20 dB/km at 1300 nm over the attenuation rate due to other effects. The system operates when exposed to a proton flux of 10^5 protons/sq. cm	IEEE-1393 EIA RS455-49

TABLE 10.9 (*Continued*)
Specifications and Space Qualification Requirement Values for a Commercial Fiber System

Parameter	Nonoperating Conditions	Operating Conditions	Standards, References, and Test Methods
Total radiation dose per year	Trapped e⁻ and p, heavy ion	30–200 krad(Si) per year Special testing required for Military	IEEE-1156.4 IEEE-1393
SEE rate	Nondestructive	$<3 \times 10^{-3}$ events per day	IEEE-1156.4
SEE rate	Destructive	$<3 \times 10^{-5}$ events per day 30–200 krad(Si) per year Special testing required for Military	IEEE-1156.4 IEEE-1156.4 IEEE-1393

Source: Courtesy of Space Photonics, Inc.

10.8 SUMMARY

- The technology of space qualification of optoelectronics and photonics components and subsystems appropriate for space-based optical communication is still developing and is defined by implementers of flight projects.
- Appropriate flight-qualified components are generally not available commercially.
- Qualification of COTS components results in lower cost but is a higher risk approach.
- Qualification from the source is at a lower risk but typically higher cost approach.
- To quantify a subsystem's performance under the flight qualification environment, a fairly accurate knowledge of the environmental conditions during and after launch is required for thermal, vibration, vacuum, outgassing, radiation, etc.
- Certain reliability and qualification issues may be mitigated via modeling, while others would require verification through testing.
- Development of a systematic qualification approach is critical to the reliability of a given part or subsystem.
- Quantities in tens to hundreds of a given part may be necessary for qualification, often from multiple vendors, to quantify a given part with low qualification heritage.
- Semiconductor lasers and photodetectors are sensitive to displacement damage due to protons and neutrons. Passive optical elements and fibers are sensitive to ionizing radiation, such as those caused by electrons, gamma rays, and protons.
- Derated operation of active elements (e.g., diode laser's operational current) improves reliability of the device.

- Heritage may be applied only to the part that is exactly the same as that flown in space.
- Use of gold-indium contact should be avoided in active components.
- Understand single-event upset behavior and causes in photodetectors and laser diodes.
- Analyze possible pathways for spacecraft charging.
- Selection of Telcordia-qualified components is a good starting point in the qualification process.
- COTS devices may require special qualification testing - in particular, plastic parts.

REFERENCES

Ahrens, Robert G., Jaques, J., LuValle, M., DiGiovanni, D., and Windeler, R., "Radiation effects on optical fibers and amplifiers." Testing, Reliability, and Applications of Optoelectronic Devices (Vol. 4285). International Society for Optics and Photonics, 2001.

Alig, R. C. and Bloom, S., Electron-hole-pair creation energies in semiconductors, *Phys. Rev. Lett.*, 35(22), 1522, 1975.

Andrews, J. D. and Moss, T. R., *Reliability and Risk Assessment*, 2nd ed., PE Publishing, London, 2002.

Aniceto, R., Evaluation of the performance of coherent optical communications commercial DSP ASICs in low earth orbit radiation environments. Diss. Massachusetts Institute of Technology, 2017.

Aniceto, Raichelle J., Milanowski, R., Moro, S., Cahoy, K., and Schlenvogt, G., "Proton Radiation Effects on Hamamatsu InGaAs PIN Photodiodes." Radiation Effects Conferences on (RADECS), Data Workshop, 2017.

Aniceto, Raichelle J., Moro, S., Grier, A., Milanowski, R., and Cahoy, K., "Assessment of Gamma and Proton Radiation Effects on 100 Gbps Commercial Coherent Optical Transceiver." International Conference on Space Optics (ICSO), Data Workshop, 2018.

Aniceto, Raichelle J., Moro, S., Milanowski, R., Isabelle, C., Hall, N., Vermeire, B., and Kerri, C., "Single Event Effect and Total Ionizing Dose Assessment of Commercial Optical Coherent DSP ASIC." Nuclear and Space Radiations Effects Conference (NSREC), Data Workshop, 2017.

Armstrong, T.W. and Colborn, B.L., NASA publication # NASA/CR-2000-210072, Evaluation of trapped radiation model uncertainties for spacecraft design, 2000.

Baccaro, S., Radiation damage effect on avalanche photodiodes, *Nucl. Instrum. Methods A*, 426, 206–211, 1999.

Baccaro, S., Piegari, A., Di Sarcina, I., and Cecilia, A., Effect of gamma radiation on optical components, *IEEE Trans. Nucl. Sci.*, 52, 1779–1784, 2005.

Baker, D. N., How to cope with space weather, *Science*, 297, 1486–1487, doi: 10.1126/science.1074956, 2002.

Barde, S., Ecoffet, R., Costeraste, J., Meygret, A., and Hugon, X., Displacement damage effects in InGaAs detectors: experimental results and semi-empirical model prediction, *IEEE Trans. Nucl. Sci.*, 47(6), 2466–2472, 2000.

Barnes, C. E., Effect of Co60 irradiation on epitaxial GaAs laser diodes, *Phys. Rev. B.*, 1, 4735–4747, 1970.

Barnes, C., Ott, M., Becker, H., Wright, M., Johnson, A., Marshall, C., Shaw, H., Marshall, P., LaBel, K., and Franzen, D., NASA electronics parts and packaging (NEPP) program assurance research on optoelectronics, *Proc. SPIE Conf. Proc.*, 589707, 2005.

Barnes, C., Ott, M., Johnston, A., LaBel, K., Reed, R., Marshall, C., and Miyahira, T., Recent photonics activities under the NASA electronic parts and packaging (NEPP) program, *Proc. SPIE Conf. Proc.*, 4823, 189–195, 2002.

Berthier, P., Laffitte, D., Perinet, J., Goudarad, J. L., Boddaert, X., and Chazan, P., New qualification approaches for optoelectronic devices, *Proceedings of the IEEE Electronic Components and Technology Conference*, San Diego, CA, pp. 551–557, 2002.

Bogaerts, J., Dierickx, B., and Mertens, R., Enhanced dark current generation in proton-irradiated CMOS active pixel sensors, *IEEE Trans. Nucl. Sci.*, 49(3), 2002.

Bruni, R. J., Clark, A. M., Moran, J. M., Nguyen, D. T., Romaine, S. E., Schwartz, D. A., and Van Speybroeck, L. P., Verification of the coating performance for the AXAF flight optics based on reflectivity measurements of coated witness samples, *Proc. SPIE Conf. Proc.*, 2805, 301–310, 1996.

Chalfant, C., Orlando, F., and Parkerson, P., Photonic packaging for space applications, Space Photonics, Inc., IMAPS OE Workshop, Bethlehem, PA, 2001.

Chen, Y., Reeves, G. D., and Friedel, R. H., The energization of relativistic electrons in the outer Van Allen radiation belt, *Nat. Phys.*, 3, 614–617, 2007.

Ciminelli, C., Armeniese, M. N., and Passaro, V. M. N., Reliability test procedures for tunable lasers, *SPIE Conf. Proc.*, 4944, 83–96, 2003.

Dodd, P. E., Shaneyfelt, M. R., Schwank, J. R., and Felix, J. A., Current and future challenges in radiation effects on CMOS electronics, *IEEE Trans. Nucl. Sci.*, 57(4), 1747–1763, 2010.

Fernandez-Rodriguez, M., Ramoos, G., del Monte, F., Levy, D., Alvarado, C. G., Núñez, A., and Álvarez-Herrero, A., Ellipsometric analysis of gamma radiation effects on standard optical coatings used in aerospace applications, *Thin Film Solids*, 455–456, 545–550, 2004.

Fox, Brian P., Simmons-Potter, K., Thomes, W. J., and Kliner, D. A. V., Gamma-radiation-induced photodarkening in unpumped optical fibers doped with rare-earth constituents, *IEEE Trans. Nucl. Sci.*, 57(3), 1618–1625, 2010.

Fox, B. P., Schneider, Z. V., Simmons-Potter, K., Thomes Jr., W. J., Meister, D. C., Bambha, R. P., Kliner, D. A., and Soderlund, M. J., Gamma radiation effects in Yb-doped optical fiber, *Proc. SPIE*, 6453, 645328, 2007.

Fukuda, M., *Reliability and Degradation of Semiconductor Lasers and LEDs*, Artech House, Boston, MA, 1991.

G. R. Hopkinson, Radiation effects in a CMOS active pixel sensor, *IEEE Trans. Nucl. Sci.*, 47(6), 2000.

Galvanauskas, A. and Samson, B., Large-mode-area designs enhance fiber laser performance, *SPIE's OE Magazine*, 15–17, July 2004.

Gusarov, A., Starodubov, D. S., Berghmans, F., Deparis, O., Defosse, Y., Fernandez, A. F., Decreton, M. C., Patrice, M., and Michel, B., Design of a radiation-hard optical fiber Bragg grating temperature sensor, *SPIE Conf. proc.*, 3872, 1999.

Hemdow, S., Falvey, S., Nelson, B., Thienel, L., and Drape, T., Overview of qualification protocol of fiber lasers for space applications, *SPIE Conf. Proc.*, 6100, 61001, 2006.

Hemmati, H., JPL internal memo, 2002.

Hemmati, H., *Deep-Space Optical Communications*, Wiley, Hoboken, NJ, 2006.

Hemmati, H. and Lesh, J. R., Environmental testing of a diode-pumped Nd: YAG laser and a set of diode laser arrays, *SPIE Conf. Proc.*, 1059, 146–153, 1989.

Hendow, S., Falvey, S., Nelson, B., Thienel, L., and Drape, T., Overview of qualification protocol of fiber lasers for space applications, *SPIE Conf. Proc.*, 6100, 1Y1–1Y13, 2006.

Henschel, H., Kohn, O., Schmidt, H. U., Kirchhof, J., and Unger, S., Radiation-induced loss of rare-earth doped silica fibers, *IEEE Trans. Nucl. Sci.*, 45, 1552–1557, 1998.

Hillard, G. B. and Ferguson, D. C., Measured rate of arcing from an anodized sample on the SAMPIE flight experiment, *33rd Aerospace Sciences Meeting & Exhibit*, AIAA, Reno, NV, 95–0487, 1995.

Hovis, F. E., Qualification of the laser transmitter for the CALIPSO aerosol Lidar mission, *SPIE Conf. Proc.*, 6100, 1X1–1X10, 2006.

Howard Jr., J. W. and Hardage, D. M., Spacecraft environments interactions: Space radiation and its effects on electronic systems, NASA Publication # NASA/TP-1999-209373, 1999.

Huston, S. L., NASA publication # NASA/CR-2002–211785, Space environments and effects, 2002.

Johnson, A., Radiation degradation mechanisms in laser diodes, *IEEE Trans. Nucl. Sci.*, 51, 3564–3571, 2004.

Kim, Q., Wrigely, C. J., Cunningham, T. J., and Pain, B., *Space Qualification of Photonic Devices*, SPIE Conf. Proc., 4640, 15–21, 2002.

Lalucaa, V., Goiffon, V., Magnan, P., Rolland, G., and Petit, S., Single-Event effects in CMOS image sensors, *IEEE Trans. Nucl. Sci.*, 60(4), 2013.

Le Kernec, A., Sotom, M., Bénazet, B., Barbero, J., Peñate, L., Maignan, M., Esquivias, I., Lopez, F., and Karafolas, N. "Space evaluation of optical modulators for microwave photonic on-board applications." International Conference on Space Optics—ICSO 2010 (Vol. 10565, p. 105652I), SPIE Conf. Proc., November 2017.

Tan, L-Y, Li, F-J, Xie X-L., Zhou, Y-P., Ma J., Proton Radiation Effects on GaAs/AlGaAs Core-Shell Ensemble Nanowires Photo-detector, *Chinese Physics B*, 26, Num. 8, 2017.

Limpert, J., Schreiber, T., Liem, A., Nolte, S., Zellmer, H., Peschel, T., Guyenot, V., and Tünnermann, A., Thermo-optical properties of air-clad photonic crystal fibers in high-power operation, *Opt. Exp.*, 11, 2982–2990, 2003.

Marshall, P. W., Dale, C. J., and LaBel, K. A., Space radiation effects in high performance fiberoptic data links for satellite data management, *IEEE Trans. Nucl. Sci.*, 43, 645–653, 1996.

Milanowski, R., Aniceto, R., Hardy, F., Vermeire, B., Jacox, M., Moro, S., and Cahoy, K. (2017, July). Proton Radiation Effects Assessment of a Commercial 12-Megapixel CMOS Imager. In Radiation Effects Data Workshop (REDW), 2017 IEEE (pp. 1–7), 2017.

Miroshnichenko, Leonty I. Radiation conditions in space, *Radiation Hazard in Space*, Springer, the Netherlands, 23–46, 2003.

Moyer, R. S., Grencavich, R., Smith, R. W., and Minford, W. J., Design and qualification of hermetically packaged lithium niobate optical modulator, *Proceedings of the IEEE Electronic Components and Technology Conference*, San Jose, CA, 425–429, 1997.

NASA Goddard Space Flight Center, Vibroacoustic qualification testing of payloads, subsystems, and components, Preferred Reliability Practices, Practice # PT-TE-1419, NASA Goddard Space Flight Center, Greenbelt, MD, p. 8, 2000.

Ott, M. N., Fiber laser components technology readiness review, chrome-extension://oemmndcbldboiebfnladdacbdfmadadm/https://photonics.gsfc.nasa.gov/tva/meldoc/fiberlaserradiationeffects.pdf, 2004.

Ott, M. N., Coyle, D. B., Canham, S., and Leidecker, H. W., Qualification and issues with space flight laser systems and components, *SPIE Conf. proc.*, 6100, 1V-1–1V-15, 2006a.

Ott, M. N., Jin, X. L., Chuska, R., Friedberg, P., Malenab, M., and Matuszeski, A., Space flight requirements for fiberoptic components, qualification testing and lessons learned, *SPIE*, 6193, 619309, 2006b.

Ott, M., Vela, J., Magee, C., and Shaw, H., Reliability of optical fiber modulators for space flight environments. NASA Parts and Packaging Program Report, IPPAQ Task Report, NASA GSFC greenbelt, Maryland, 1–17, 2002.

Pease, R. L., Johnston, A. H., and Azarewicz, J. L., Radiation testing of semiconductor devices for space electronics, *Proc. IEEE*, 1501–1526, 1988.

Peri, F., Heaps, W. S., and Singh, U. S., Laser risk reduction technology program for NASA's earth science enterprise, *SPIE Conf. Proc.*, 4893, 166–175, 2003.

Robinson, B. S., Boroson, D. M., Schieler, C. M., Khatri, F. I., Guldner, O., Constantine, S., and Garg, A., TeraByte InfraRed Delivery (TBIRD): A demonstration of large-volume direct-to-Earth data transfer from low-Earth orbit. *Free-Space Laser Communication and Atmospheric Propagation 1024*, SPIE Conf. Proc., 10524, 105240V, February 2018.

Rose, T., Gunn, D., and Valley, G.C., Gamma and proton radiation effects in erbium doped fiber amplifiers: Active and passive measurements, *J. Lightwave Tech.*, 19, 1918–1923, 2001.

Samaras, Anne. Single Event Effects and Rate Calculation. TRAD Short Course, 2014.

Sherry, W. M., Gaeb, C., Miller, T. J., and Schweizer, R. C., High performance opto-electronic packaging for 2.5 and 10 Gb/s laser modules, *Proceedings of the Electronic Components and Technology Conference*, Orlando, FL, 620–627, 1996.

Shoone, H., Carson, R. F., Paxton, A. H., and Taylor, E. W., AlGaAs vertical cavity surface emitting laser responses to 4.5-MeV proton irradiation, *IEEE Photonics Tech. Lett.*, 9, 1552–1557, 1997.

Stark, John P.W., The spacecraft environment and its effect on design. *Spacecraft Systems Engineering.* Fortescue, P., Swinerd, G., and Peter W. Stark, (Eds.), Chapter 2, Wiley, Hoboken, NJ, 2011.

Stueber, T.J. and Mundson, C., Evaluation of kapton pyrolysis, arc tracking, and flashover on SiOx-coated polyimide insulated samples of flat flexible current carriers for SSF, NASA report #CR-191106, 1993, chrome-extension://oemmndcbldboiebfnladdacbdf-madadm/https://ntrs.nasa.gov/archive/nasa/casi.ntrs.nasa.gov/19930014241.pdf

Suhir, E., Accelerated life testing (ALT) in microelectronics and photonics, its role, attributes, challenges, pitfalls, and its interaction with qualification tests, *J. Electron. Packaging*, 124, 281–291, 2002.

Suparta, Wayan, The variability of space radiation hazards towards LEO spacecraft, *J. Phys. Conf. Ser.*, 539(1), 2014 (IOP Publishing).

Taylor, E.W., Space and enhanced radiation induced effects in key photonic technologies, *Proceedings of the IEEE Aerospace Conference*, Snowmass, CO, 307–316, 1999.

Thomes, W. J., LaRocca, F. V., Ott, M. N., Jin, X. L., Chuska, R. F., MacMurphy, S. L., and Jamison, T. L. Investigation of hermetically sealed commercial LiNbO3 optical modulator for use in laser/LIDAR space-flight applications. In Nanophotonics and Macrophotonics for Space Environments (Vol. 6713, p. 67130T). International Society for Optics and Photonics, September 2007.

Troska, Jan, Gill, Karl, Grabit, Robert, and Vasey, Françoi, "Neutron, proton and gamma radiation effects in candidate InGaAs pin photodiodes for the CMS tracker optical links." No. CERN-CMS-NOTE-1997-102, 1997.

Troupaki, E., Vasilyev, A., Kashem, N., Allan, G.R., and Stephen, M.A., Space qualification and environmental testing of quasi-continuous wave laser diode arrays, *J. Appl. Phys.*, 100, 063109, 2006.

Webb, R. C. and Greenwell, R. A., Photonics projections and needs for survivable space systems in the 21st century, *SPIE*, 3440, 1998.

Williams, G. M., Putnam, M. A., and Friebele, E. J., Space radiation effects on erbium-doped fibers, *SPIE Conference Proc.*, 2811, 30–37, 1996.

Wright, M. W., Franzen, D., Hemmati, H., Becker, H., and Sandor, M., Qualification and reliability testing of a commercial high-power fiber-coupled semiconductor laser for space applications, *Opt. Eng.*, 44, 054204, 2005.

www.hq.nasa.gov/office/codeq/87395c14.pdf

www.nasa.gov/offices/oce/llis/0817.html

Zellmer, H., Riedel, P., and Tünnermann, A., Visible upconversion lasers in praseodymium ytterbium-doped fibers, *Appl. Phy. B: Lasers Opt.*, 69, 417–429, 1999.

11 Optical Satellite Networking

The Concept of a Global Satellite Optical Transport Network

Nikos Karafolas

CONTENTS

11.1 PREFACE TO THE NEW EDITION

Since the first edition of this book, Satellite Networks have witnessed a new era of development with the emergence of several proposals for Non-Geosynchronous Orbits (NGSO) constellations. To a large extent, this era resembles the surge of similar proposals for narrowband (telephony) and broadband (Internet) constellations in the late 1990s, which eventually led to the implementation of IRIDIUM and GLOBALSTAR. The difference today is that the experience of the mishaps in the constellation business plans of the late '90s offers a solid basis to consider the new opportunities with careful analysis. In parallel, the relevant technologies have advanced markedly, and optical inter-satellite links (ISLs) are now operational in the first operational system, that of the European Data Relay System (EDRS). Relevant photonic components and equipment such as the Photonic Switch have also advanced, primarily driven by their introduction in Microwave Photonic Communication Satellite (COMSAT) payloads. Optical-Fiber Transport Networks have also advanced remarkably offering the necessary experience to leverage to space concepts for "fiber in the sky" solutions seamlessly linked to the terrestrial fiber network [Hauschildt 2018]. Hence, it is appropriate to consider the "Optical Satellite Networking" concepts in the light of all these developments and especially of the new broadband satellite constellation proposals.

11.2 INTRODUCTION

Recently, a number of broadband satellite networks have been initiated seeking financing and some of them proceeding through the initial study phase for the network and satellite design. All proposals target to offer Internet services globally and/or high bit rate, low-latency communications to time-sensitive applications. Most of the constellations make use of optical inter-satellite links, as shown in Table 11.1 **[Leosat 2015, Laserlight, 2014; Telesat 2015 Space-X 2016, Handley 2018, Del Portillo 2018]**. They are based primarily in NGSO sometimes also including a Geosynchronous Earth Orbit (GEO) component part.

These proposals are similar to the ones proposed in the 1990s, which gained frequency use license, and two of them, Iridium and Globalstar, were launched and have been operating since then **[Evans, 1997; Freidell, 1998; Pratt et al., 1999]**. Iridium and Globalstar have been an outstanding technological leap forward demonstrating for over 15 years:

- The capability of using a system of tens of satellites (66 for Iridium and 48 for Globalstar) for global communications via space.

TABLE 11.1
Proposed Broadband Constellations with Optical ISLs

Project	Country	Number of Satellites	Frequency Band	Orbit	Optical ISL
Laser Light Communications	USA	12	Optical	MEO	YES
Leosat	USA	108	Ka	LEO	YES
SpaceX	USA	4425	Ka	LEO	YES
Telesat	Canada	117	Ka	LEO	YES
Theia	USA	112	Ku, Ka	LEO	YES

- The capability of having complete (Iridium) or almost complete (Globalstar) coverage of the Earth by a single communication system.
- The use of inter-satellite links at a massive scale (in the Ka band in Iridium with 121 bidirectional ISLs) without interruption for many years.
- The use of onboard traffic routing, which brings the network layer of the communication protocol suit to satellites.
- The formation of a complete autonomous system by a satellite network.

The main reasons for the present broadband satellite networks to consider the more complex than GEO solutions at NGSO are the potential for extensive frequency reuse using multi-spot beams and very low latency due to the small (when compared to GEO) propagation distances. It is actually worth noticing that propagating through space is faster than through a terrestrial fiber-optic links due to the small propagation speed in the fiber medium (2/3 of the speed of light in vacuum). This can be a decisive factor for some business cases that are time-sensitive.

Since the first edition of this book, ISLs have progressed markedly. The EDRS system demonstrates the use of optical inter-satellite link in an operational environment following 15 years of in orbit trials n various satellites [**Benzi, 2016**].

Furthermore, the uplink/downlink capacity of Communications Satellites has reached 1 Tbps through the use of sophisticated beam-forming and antennas that allow hundreds or even thousands of narrow spot beams to reuse the BW very efficiently [**VIASAT 2017**]. It is within reason to foresee that if such capacities are to be handled by an ISL network, then the ISLs not only need to be optical, but they will also need to reach a capacity in the range of Tbps through the use of Wavelength Division Multiplexing (WDM), and the satellites will need to be equipped with efficient onboard optical Wavelength Routing (WR).

WDM links is the technology adopted by terrestrial fiber-optic networks as a means to increase the link capacity by hundreds of Gbps as well as to offer a means of routing optical signals end-to-end without intermediate regeneration through a WDM-router composed of optical WDM multiplexer/demultiplexer (MUX/DEMUX) and an Optical Switch/Router [**Mukherjee, 2006**].

The technologies and equipment involved in fiber-optic WDM networks such as WDM sources, WDM MUX/DEMUX, and Optical Space Switches are now being investigated for COMSAT Microwave Photonic Payloads [**Anzalchi et al., 2014; Sotom et al., 2016**]. Together with appropriate Photonic Frequency Generation Units and Frequency Converters for a Photonic implementation of the classical in a Microwave Payload where the RF carriers carrying the value-added data from ground are imposed on WDM carriers to be frequency converted and routed onboard the PL by photonic means. The first demonstrators have been build and the first IoDs are underway [Sotom et. al, 2016, Piqueras et.al. 2018].

All the ingredients required to form a Satellite "Optical Transport Network" based on WDM ISLs and onboard Wavelength Routing (WDM/WR S-OTN) are in place [**Karafolas et al., 2000; Chan, 2003**]. This is the topic of this chapter. In order to study the characteristics of a WDM/WR-based S-OTN both at the network as well as at the technology level, a reference scenario is used. This is the Celestri LEO system proposed by Motorola and is shown in Figure 11.1 [**Kennedy et al, 1997**]. Celestri was the first

FIGURE 11.1   The Celestri constellation proposal.

system to specify laser terminals as the baseline ISL technology. The analysis provided in the following chapters apply equally to Medium Earth Orbit (MEO) and GEO OTNs. The same analysis tools for capacity dimensioning and path routing as in fiber OTNs are used. The power requirement (and to a lesser extend the mass and footprint) is however a differentiating a factor when considering the satellite OTN, given that a spacecraft is a minimalistic system in all aspects and probably most importantly in power. So, the power management is taken into account when considering ways to optimize the link capacity.

The access from ground users is done by state-of-the-art high-capacity RF links in a multi-spot beam scenario. One could consider to have optical terminals for access to airborne users such as airplanes, Unmanned Aerial Vehicles (UAVs), and stratospheric platforms [**Chan, 2003**]. Optical links between an airplane and a GEO satellite have been already demonstrated with the LOLA experiment using the ESA ARTEMIS satellite [**Vaillon, 2008**]. With the proliferation of UAVs and stratospheric platforms, such optical links that are not strongly impaired by the presence of atmosphere would be more frequently used as they offer very high bandwidth and secure links without the need of frequency coordination and licensing. Of course, optical links can be performed with satellites orbiting in different orbits (or even in deep space) such as the optical links between Low Earth Orbit (LEO) satellites and the EDRS GEO satellites [**Benzi, 2016**]. As far as the satellite OTN is concerned, the optical inputs/outputs are treated as any input/output to the satellite network node making no difference on the overall network design methodology.

Furthermore, the potential of using multigigabit optical "feeder" uplinks/downlinks from ground to satellite and satellite to ground is under consideration for use in a mess- or ring-based satellite OTNs especially in MEO. LASERLIGHT was the first proposal to incorporate optical up/downlinks in its design to provide a satellite optical network that will integrate seamlessly with the terrestrial and submarine optical networks [**LASERLIGHT, 2014**]. Although optical up/downlinks have been demonstrated in very long distances including the 622-Mbps Moon-Earth link [**NASA 2013**], the challenges of achieving such Earth-LEO/MEO/GEO links at the rates of several hundreds of Gbps, as it is required by backbone optical networks, are nontrivial and at the moment unresolved. It is not only the issue of availability of optical ground stations (OGS) that to a large extent can be addressed by having several such OGS positions in suitable geographic locations and using spatial diversity to access the space segment. The atmospheric turbulence distorts the optical signal impairing its total throughput. The present, second, edition of this book will hence not cover this part. As substantial work is under way (ESA works toward the concept of a High-Throughput Satellite Optical Transport Network including optical feeder links, onboard optical processing, and WDM ISLs) the hope is that the third edition of this book will report the operation of such satellite networks using optical feeder links.

In the following chapters, the network properties and the technology requirements of the satellite OTNs are analyzed.

11.3 ORBITAL GEOMETRY AND NETWORK PHYSICAL TOPOLOGY OF SATELLITE CONSTELLATIONS

The orbital geometry of NGSO networks is characterized by a number of parameters most notably:

- The total number of satellites
- The number of satellites per orbital plane
- The phasing factor (relative spacing between satellites in adjacent planes)

Two main approaches in the constellation design have been followed [**Walker, 1984; Ballard, 1980; Wood, 2001**]:

- π constellations (or Walker star)
- 2π constellations (or Walker delta, or Ballard rosette)

In π constellations, satellites are launched from a single hemisphere, and hence the two edge-orbits appear counter-rotating with no ISLs established between them forming a *network seam*; Iridium is such a constellation. In 2π constellations, satellites are launched from an almost-360° region, and from a ground user perspective appear as an ascending and a descending surface. The effective general network topology is that of *Manhattan Street Network* (MSN), which in the two cases takes the specific form of [**Maxemchuk, 1987**]:

- Cylindrical mesh topology (π constellations), consisting of orbital rings and interorbital chords. Iridium is a π constellation of 9 orbits each with 11 satellites.

- Toroidal mesh topology (2π constellations), consisting of orbital rings and interorbital chords, which depending on the relative spacing between satellites in adjacent orbits (phasing factor) can form a ring (phasing factor $0°$), or one or more spirals (phasing factor other than $0°$). Celestri is a 2π constellation with 7 orbits each, with 9 satellites.

The two network topologies are depicted in Figure 11.2.

π constellation

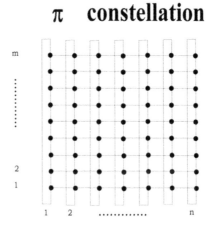

2π constellation

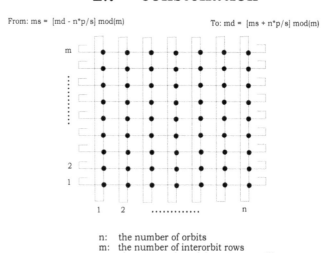

From: ms = [md - n*p/s] mod(m)   To: md = [ms + n*p/s] mod(m)

n: the number of orbits
m: the number of interorbit rows
ms: the row number of the source satellite
md: the row number of the destination satellite
p: the inter-orbit phasing factor
s: the intra-orbit spacing

FIGURE 11.2   The π and 2π constellation network physical topology is that of a bidirectional Manhattan Street Network (cylindrical and toroidal respectively for the two cases).

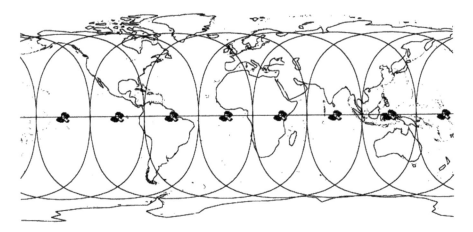

FIGURE 11.3 An equatorial satellite network with two ISL terminals per S/C in a global ring physical topology.

In the case of equatorial networks, the network can acquire the topology of a single ring, as shown in Figures 11.3 and 11.4. Such networks have been proposed for operation in LEO to serve the specific region around the equator. However, if this network is in GEO, it can have an almost full coverage of the Earth by just three satellites as suggested by Clarke **[1945]**.

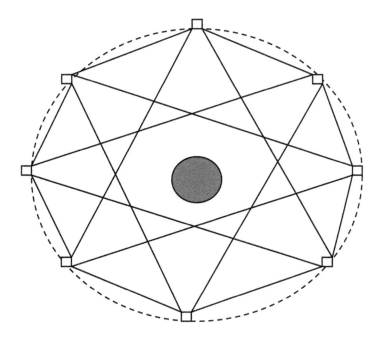

FIGURE 11.4 A GEO satellite network with four ISL terminals per S/C in a mesh topology.

The traffic in the space segment served by such a network would be highly vola-tile and it is described by:

- The *satellite-sourced/sunk traffic*, which varies continuously as a func-tion of geographic longitude and latitude of the satellite position and time at the area of coverage. For most of the time, the NGSO satellites cover oceans or sparsely populated areas. Therefore, the average add/drop traffic is expected to be significantly smaller than the peak value.
- The *source-to-destination traffic matrix*, which in NGSO changes con-tinuously as a function of geographic longitude and latitude of source and destination, and time in source and in destination. Similar to the previous discussion, the probability to have a particular connection heavily loaded is significantly small, as it requires that both satellites be above areas of high traffic. Consequently, the average bandwidth requirement of a connection is expected to be much smaller than the corresponding peak traffic. Moreover, the traffic in the connections is expected to be *asymmetric*.

A large part of the traffic served by a satellite system will be *transit traffic* that has to be processed by the onboard processor in each one of the satellites involved along the path from the source to the destination satellite. The denser (higher number of satellites in the network) a constellation, the higher the transit traffic.

The *latency* in data transmission involves propagation and processing delays. Although the first can be predicted with satisfactory accuracy calculating the source to destination satellite propagation path, the second type of delays depend on a large number of factors and primarily on the traffic conditions (for which no reliable data exist) and the available resources of the network. In any case, the delays that may be encountered by intermediate processing can impair the quality of service offered as expressed, e.g., in *cell transfer delay* and *cell delay variation* to use the Asynchronous Transfer Mode (ATM) terminology. This is one of the reasons to con-sider a technique that can allow end-to-end paths to be established without interme-diate processing of data [**Karafolas et al., 2000**].

11.4   OPTICAL NETWORKING TECHNIQUE

A satellite network is essentially a global access network relying on a global trans-port network that is formed in space by the ISLs. The designs of the proposed NGSO broadband constellations have been based on the use of a single-wavelength ISL, electronic onboard processing with dynamic routing reflecting the changes in the satellites geographic positions, inter-satellite link state (range, utilization), and On Board Processor (OBP) state (queuing delays).

Optical WDM networking with wavelength routing is based on the concept of *lightpath*, i.e., a continuous optical connection that is established between a source node and a destination node (satellites in this case) via a number of links (ISLs in this case) and intermediate nodes without requiring processing of traffic in the intermedi-ate nodes [**Baroni et al., 1997**]. Several lightpaths can use the same ISL by being at different wavelength through WDM. Wavelength routing or, in other words, optical

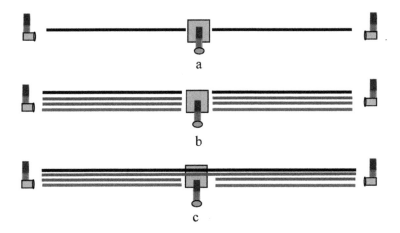

FIGURE 11.5 (a) Single-wavelength optical ISL, (b) WDM optical ISL, (c) WDM ISL with wavelength routing.

circuit switching, is then used onboard each satellite to properly direct the lightpaths from an input ISL to an output one or to the downlink based on a fixed traffic matrix (Figure 11.5). Optical networks based on these principles are currently deployed in terrestrial fiber networks throughout the world [**Mukherjee, 2006**].

The main benefits in adopting optical networking for the ISL segment is

- The simplification in data routing (the space segment appears to be almost static hiding the satellite movement).
- The elimination or minimization of processing delays, which contribute to degradation of the Quality of Service (QoS) in the system.

Potential benefits also include savings in OBP power consumption and ISL utilization improvement through statistical multiplexing. Although these are generally applied characteristics, the actual impact of adopting optical networking principles in the ISL segment depends on the design of each specific network.

From a network perspective the system is modeled as a bidirectional MSN, where:

- The nodes (satellites) are of equal importance
- The maximum satellite source capacity is fixed
- The ISL capacity is fixed
- The source/destination capacity is highly volatile
- The connection traffic matrix changes continuously

In designing the optical transport network (OTN), only *fixed topologies* are considered for several reasons and primarily to avoid frequent lightpath handover that can degrade the performance of the system while increasing the OBP and ISL transceiver hardware requirements significantly. *Reconfiguration* of the OTN is only considered in specific designs and for failure restoration purposes. The main challenge of the

OTN design for a satellite constellation is to balance the desired improvement in the communication system performance measured in:

- End-to-end or cell (in general packet) transfer delay
- Cell-delay variation
- Percentage of packet rejection (related to delay)
- Overhead (related to delay and packet rejection for load balancing)

to the required satellite resources primarily measured in power consumption. The reliability of the unit that is a function of its complexity is a paramount issue in any space system needs to be proved before considering in launching these technologies. Developments in optical device manufacturing will eventually address the reliability issue allowing the manipulation of tens of lightpaths. The required power remains the real limiting factor affecting the total ISL capacity that is the product of:

- Number of lightpaths per ISL
- Lightpath bandwidth

There are two main approaches in the OTN design: single-hop and multi-hop.

11.4.1 SINGLE-HOP OPTICAL NETWORKING

Single-hop OTNs are based on establishing end-to-end lightpaths for all the end-node connections that in a network with N satellites are $N(N-1)/2$. Lightpaths (wavelength and propagation path) are fixed and determined following a certain algorithm that first defines the physical path allocation, then the wavelength allocation, and finally it allocates a restoration path for the case of each individual ISL failure **[Baroni et al., 1997]**. The full optical connectivity provides the maximum benefit in terms of the communications parameters performance, since absolutely no intermediate processing is required under nominal operational condition. This is paid in increased terminal complexity since tens of WDM lightpaths would be required along with a large optical switch. Unless dynamic power management and/or variable transmission rate is employed (increasing the complexity), the ISLs would be underutilized due to the volatile nature of the end-node connection traffic.

A first estimation on the *minimum number of wavelength channels per ISL, Wc,* can be derived by the *network limiting cut* approach in a uniform all-to-all traffic case which provides that at minimum $Wc_{min} = max \{Wc_{min}\} = max \{K(N-K)/C\}$ lightpaths would be required per ISL, where: **[Baroni et al., 1997]**

- N is the number of satellites in the network
- K is a disjoint segment of the network disjoint by the rest $(N-K)$ by interrupting
- C ISLs.

This number allows conclusions to be drawn regarding the feasibility of establishing an OTN for the constellation. As an example, the Celestri 2π constellation of 63 satellites and 6 ISLs per satellite requires at least 45 wavelengths per ISL to form a single-hop OTN across the ISL network. The Iridium π constellation of 66 satellites and 4 ISLs per satellite requires in excess of 90 wavelength channels per ISL. The comparison of these two networks, which have a similar number of satellites, reveals the impact of the number of ISLs and the type of constellation (with or without network seam) in the network connectivity. In general, the denser a network is the less efficient the utilization of the lightpaths would be. This conclusion points toward adoption of the single-hop approach only by networks of a modest amount of satellites (10–20). Such networks can be formed at higher orbits, MEO and GEO. Hence, a MEO or GEO in a ring physical topology of 9 satellites (such as the GEO component of the Celestri proposal) would be served in single-hop connectivity by 10 lightpaths per ISL.

11.4.2 MULTI-HOP OPTICAL NETWORKING

Multi-hop networking is the alternative approach for the dense constellations. In multi-hop, the end-to-end connection may be served not only by a single lightpath but also by two (double-hop) or even more (multi-hop) lightpaths if the network is very dense **[Acampora et al., 1992]**. The target is to bring the number of lightpaths per ISL to a more acceptable level of 8 to 16 channels. The use of double-hop means that some end-to-end connections have to be switched electronically in one (double-hop) or more (multi-hop) intermediate satellites. The designing challenge is to define the criteria for deciding which connections will be served by single-hop and which by double or more hops. Again, network performance parameters have to be balanced with the network (satellites) resources utilization. In comparison with the performance of the single-hop networks, the multi-hop OTN design would target to smoothen the extreme peaks in the ISL traffic load (and hence improve the link utilization) while using the minimum number of ISL hops with the shortest range inter-orbit ISLs. Some technological facts that have to be taken into account include the higher reliability of the intra-orbit ISLs (due to fixed pointing), while in terms of the traffic characteristics one has to consider the high ratio of peak-to-average source traffic and of satellite-to-satellite connection traffic. Several design approaches can be conceived and applied to a constellation depending its size and type (π or 2π). An approach that can in general be applied is the *Matrix* approach **[Gisper et al., 1996]**. Matrix is based on the bidirectional MSN form of the constellation networks and applies a simple rule of defining lightpaths for all the intra-orbit connections as well as for all the inter-orbit connections. For connections that link satellites of different orbit and inter-orbit planes, a hop in the common node (the satellite in the junction of orbit and inter-orbit plane) is required. Such a lightpath allocation principle results in an OTN that has only single-hop and double-hop connections **[Karafolas et al., 2000]**. Using the Matrix approach:

- Celestri would have 32.2% of its connections served by single-hop, whereas 67.8% by double. The maximum lightpath load for the ISL is 10 for the intra-orbit rings and 6 for the inter-orbit ISLs.
- Iridium would have 23% of its connections served by single-hop and 77% by double-hop. The lightpath load for the ISLs is 15 for the intra-orbit ISLs and 5 for the inter-orbital ISLs.

- Double-hop logical connectivity represents an efficient solution for constellations of less than 100 satellites, like most of the proposed ones. However, in the case of networks with a very large number of satellites, *multi-hop optical connectivity* is the only solution that ensures a realistic number of lightpaths per ISL and efficient resources utilization. Teledesic was an interesting case of a constellation that requires multi-hop OTN [**Liron, 1995**]. In the 288-satellites version of Teledesic which is a π constellation (12 orbits by 24 satellites each) with 8 ISLs per S/C accessing the adjacent and second adjacent nodes in the 4 directions, two single-hop rings consisting of the odd and the even numbered nodes in the orbital ring can be configured. This would require that each intra-orbit ISL in the rings supports $W = 18$ lightpaths and would enable any intra-ring connection to be established with at most 2 lightpath hops. Similarly, by using linear chord topologies in the inter-orbit chord any inter-orbit chord connection can be performed with at most 2 lightpath hops. Overall, any node-pair connection in Teledesic can be established with at most 4 optical hops.

Another approach in allocating lightpaths is the *Gradual expansion* where the optical connectivity expands progressively from the adjacent satellite to the second one. In constellations like Celestri and Iridium, each satellite communicate only with the adjacent satellites so they have "1st degree connectivity". On the other hand, Teledesic has links also with the second adjacent satellites having "1st and 2nd-degree of connectivity" [**Wood, 2001**]. However, the Teledesic system uses 4 additional ISL terminals to achieve 2nd-degree connectivity (total 8 ISL terminals) (Figure 11.6). In the wavelength routing approach, 4 terminals employing 2 wavelength channels with wavelength routing and in-line amplification in the adjacent satellites could achieve the same connectivity. Looking it from another point of view one can double or triple the connectivity of a satellite by adding a second or a third wavelength channel at each ISL terminal and employing wavelength routing in the intermediate satellites. The gradual expansion progresses by accessing satellites according to the

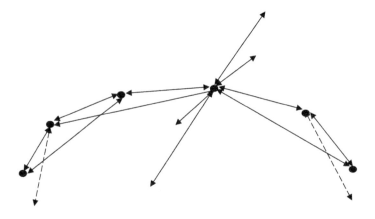

FIGURE 11.6   Example of 2nd-degree ISL connectivity with 8 ISLs per S/C.

TABLE 11.2
Expansion in the Celestri Connectivity

Degree of Expansion	Number of Connections	
1	189	10%
2	567	29%
3	1134	58%
4	1600	82%
5	1827	93%
6	1953	100%

required number of ISL hops. Beyond the 2nd-degree of connectivity, the issues of path/wavelength allocation arise. When the connectivity degree reaches the network diameter (max-min number of ISLs between any two satellites in the constellation), full mesh connectivity is achieved. The connectivity in Celestri progresses is shown in Table 11.2.

It has to be mentioned that the Japanese "Next Generation LEO Satellite" proposal was the first proposal to adopt wavelength routing allocating wavelengths based on the gradual expansion approach to the four directions (each satellites is equipped with 4 ISLs) up to the third-degree of connectivity **[Suzuki, 1999]**.

Evaluating the *capacity of the OTN* for a constellation, one has to know the ISL and lightpath capacities and estimate the average number of ISL hops, h_{av}, that a message travels across the constellation. h_{av} will depend on the type of OTN, whereas the links and lightpath capacities can be shaped accordingly to the calculated traffic demand for each constellation. As an example, for Celestri, h_{av} is 3.38 when no wavelength routing is applied, 1 when single-hop OTN is applied, and 1.41 when double-hop is applied. Hence, the capacity of the OTN is estimated to be:

1. $[N\ K\ C_{ISL}]\ /3.38$
2. $[1953\ C_{lp}]/1$
3. $[252\ C_{clp}\ +\ 378\ C_{rlp}]/1.41$

where C_{ISL} is the ISL capacity, C_{lp} is the lightpath capacity (assumed the same for all lightpaths in the single-hop scenario), C_{clp} is the column (intra-orbit) lightpath capacity in the column of the matrix (252 such lightpaths exist), and C_{rlp} is the lightpath capacity in the rows (inter-orbit) of the matrix (378 such lightpaths exist). The lightpath capacity dimensioning will be performed taken into account the specific traffic demand for each constellation.

11.4.3 TRAFFIC ROUTING AND LIGHTPATH/ISL DIMENSIONING

Traffic routing decisions in the OTN are made on the requirement to access the destination satellite with the minimum number of lightpath and ISL hops spanning the shortest possible total range. In single-hop OTNs, there is a fixed lightpath to be followed so there is not a real issue of deciding how to route traffic under nominal

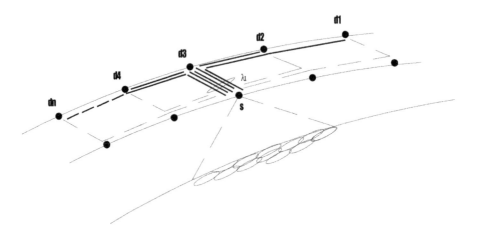

FIGURE 11.7 Statistical multiplexing of connections s-d$_i$ over the lightpath s-d$_3$.

operational conditions. In the double-hop matrix OTN case and for the connections requiring double hop, there are always two paths to be followed using first the intra-orbit lightpath followed by the inter-orbit one and vice versa. It is important to notice that the range of two inter-orbit paths changes in a deterministic and therefore pre-dictable manner according to the geographic latitude, as shown in Figure 11.7.

In dimensioning the lightpath capacity, it is important to consider the vola-tile changes on both the source/sunk traffic and of the connection traffic matrix as well as the complementary traffic properties among the end-node connections **[Perdigues et al., 2001]**. Based on this fact, several end-to-end connections share the same lightpath. The actual lightpath capacity will be determined by the traf-fic (i.e., the connections) statistics, and expected traffic. Several connections can be *statistically multiplexed* over the same lightpath determining, hence, the required lightpath capacity. As an illustration, using the ATM/SONET-SDH/WDM layer suit a lightpath would carry a higher order SONET/SDH signal (e.g., OC48/STM12 at 2.5 Gbps), which can consist of concatenated lower order signals (e.g., OC12/STM4 at 622 Mbps or OC4/STM1 at 155 Mbps) serving different end-node connections **[Werner et al., 1997]**. Statistical multiplexing takes place at the source satellite since the higher order SONET/SDH signal accommodates traffic for a large number of connections which do not take simultaneously their peak values (Figure 11.8). Virtual Path Identifier (VPI) switching would take place at the common node of matrix for the connection employing two lightpaths.

Since the first edition of this book, a number of papers have been published dealing with the optimization of the process of allocating wavelengths in the ISLs **[Werner, 2001; Li, 2010; Tan, 2010; Yang, 2010, 2011; Guo, 2014; Zhe, 2016]**. The physical limitations of possible Doppler shift and interrupted links are analyzed. What one must always consider is that the satellite OTN design is optimized taking into consideration the performance of the current technologies. The possible Doppler shift is known a priori; hence, the OTN design foresees an appropriate wavelength spacing. If a large number of wavelengths are needed in multi-hop schemes, then

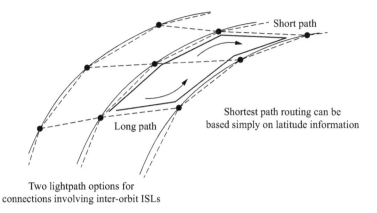

Short path

Shortest path routing can be
based simply on latitude information

Long path

Two lightpath options for
connections involving inter-orbit ISLs

FIGURE 11.8   Traffic can be routed via the two shortest paths in the bidirectional MSN (BMSN)-type of constellation topology that can be selected in an alternative fashion based on predefined geographic latitude information.

different optical amplifiers are used and are optimized for different wavelength sub-bands within the 1500–1620 band. And of course, if a critical number of wavelengths is reached, then the number of hops is reduced. Similarly, in the 2π constellation, all links are maintained uninterrupted, whereas in the π constellations, the network designer may opt for interrupted links in the "network seam," which is known a priori based on the geographical coordinates, and the traffic may be balanced by the onboard processor.

11.5   OPTICAL NETWORKING TECHNOLOGIES

The general layout of the equipment required onboard the satellite is depicted in the block diagram of Figure 11.9. It consists of the following main parts:

- The ISL terminal including:
  - The ISL telescopes (shown here is the case of ISLs in the four directions)
  - The PAT (pointing, acquisition, and tracking) subsystem responsible for each ISL telescope
- The optical lightpath routing unit including:
  - The optoelectronic transceivers
  - The lightpath router (DEMUX, OXC, MUX)
  - The electronic processor linked to the optical router
- The harness connecting the ISL terminals to the optical router

11.5.1   THE ISL TERMINAL

The developments in the design of optical ISL terminals have led to a modular approach that keeps the "Communication Unit" separate from the Telescope and PAT. Hence, the ISL terminal involves only the parts necessary to acquire and

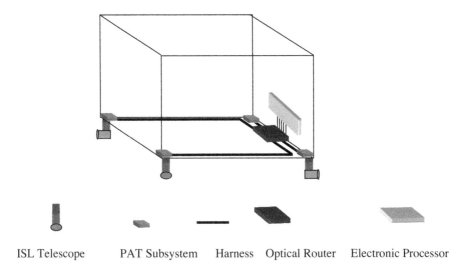

ISL Telescope   PAT Subsystem   Harness   Optical Router   Electronic Processor

FIGURE 11.9   A simplified schematic of the OTN devices layout onboard the S/C bus.

maintain the inter-satellite link irrespective of the type of the optical signal to be carried (number of WDM channels, modulation format, reception scheme, etc.). All functions related to the reception, processing, and transmission of the communication signals are integrated to the onboard (electronic and optical) router as part of the satellite payload. Link budget calculations will indicate the required telescope aperture, since the ISL bit error rate (BER) depends on the gain of the telescopes, which in turn is a function of the wavelength of the transmitted signal and the aperture of the transmit and receive telescopes. Neither the telescope design (separate or common Tx/Rx telescopes) nor the type of the PAT to be employed (using either a beacon or scanning beam techniques) influence the design of the OTN. The harness consists of the single mode optical fibers for the signal transfer from/to the telescopes to/from the Rx/Tx communications electronics and the optical router. The design of the transmitter side has to assess the potential appearance of nonlinear phenomena due to the high optical power traveling along the short length of the fiber **[Jaouen et al., 2000]**.

11.5.2   THE LIGHTPATH ROUTER

The layout of the lightpath router including the ISL communication part and the optical router is depicted in the box of block diagram of Figure 11.10.

The optoelectronic transceivers consist of the following elements:

- The transmitters (transmission electronics, optical sources, transmission filters)
- The optical booster amplifiers at the ISL transmit side
- The optical preamplifiers at the ISL receive side
- The receivers (filters, photodetectors, reception electronics)

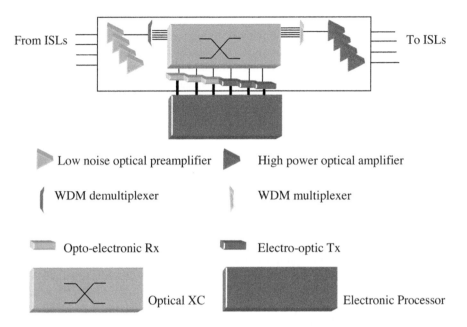

FIGURE 11.10 The Optical receiving, routing, and transmitting unit linked with four ISLs and with the electronic processor.

While the lightpath router consists of:

- WDM demultiplexers
- Optical add/drop crossconnect
- WDM multiplexers

11.5.3 WAVELENGTH BAND AND COMMUNICATION SCHEME

The use of WDM and the requirement of high-reliability components point toward the use of the *1550-nm optical band* that has been so vastly developed over for the fiber-optic communications industry. All the required components exist in this band including very accurately defined and stable in the frequency domain optical sources to produce the WDM signal, optical processing elements such as filters, multiplexers, and demultiplexers with very accurately and stable frequency response, and active elements such as optical amplifiers (doped fiber amplifiers as well as semiconductor ones). The multiwavelength transmission also points toward the adoption of direct detection or self-homodyne schemes due to simplicity as opposed to the homodyne reception using a distinct local oscillator produced at the receiver. Homodyne reception of the PSK-modulated signals is well known to offer much higher sensitivity reception [**Czichy, 1997; Lange, 2007**]. For the case of a single-wavelength signal, the added complexity of the receiver is justified; however, in a WDM reception of 8 to 16 signals, a homodyne reception scheme would result in a rather unacceptable complexity. The use of direct detection or self-homodyne detection dictates the modulation/reception formats to be employed. In the general case of *direct detection*, *ASK modulation* would be employed. Pulse coding has been shown theoretically that can increase the sensitivity of the system **[Wizner et al., 1999]**. If the more sensitive *self-homodyne detection* is used, the signal would be modulated in a *Differential PSK* format [**Suzuki, 1999**]. It is again a trade-off between receiver sensitivity and complexity.

11.5.4 THE TRANSCEIVER

The lightpath router receives the lightpaths from the N ISLs as well as the ones generated locally by the satellite and destined to other satellites via the ISL network. It has as output the same number of ISLs plus the downlink of the satellite. The signal sources can be distributed feedback (DFB) lasers that are used as a standard optical frequency source in terrestrial WDM systems. The optical frequency reference after modulation would be fed to a Doped Fiber Amplifier to be boosted at the required output power. Such amplifiers can now provide output powers of several watts, enough to satisfy most link requirements for Earth-orbiting satellites. The selection of the amplifier has to be tested for its radiation tolerance, which has been proved to depend on several parameters including the doping elements [**Williams et al., 1998; Ott, 2005; Girard S 2018**]. Similar devices with 40-dB amplification but with lower output power and near the theoretical limit Noise Figure will be used for pre-amplification at the receiver side.

11.5.5 THE WAVELENGTH MUX/DEMUX

The wavelength multiplexed lightpaths will first have to pass through a demultiplexer to provide the individual lightpath inputs to the Optical Switch. Similarly, at the output of the switch, the individual lightpaths will have to be multiplexed into a WDM signal to be fed to the ISLs. Multiplexing is achieved by a device reciprocal to the demultiplexer. The MUX/DEMUX must provide very good isolation between the various channels to avoid crosstalk. The resolution requirements depend on the wavelength channel spacing. Assuming a number of channels between 8 and 16 (taking advantage of the multiplexing while not adding high complexity on the system) would mean that 1 nm to 2 nm is a reasonable spacing. This requirement is easily met by current WDM demultiplexing technologies that are designed to meet the very strict specification of fiber-optic WDM networks based on International Telecommunication Union (ITU) regulations. WDM demultiplexers can be based on a number of principles and a variety of technologies. Examples are the Thin Film technology and the Arrayed Waveguide Gratings (AWGs) that can be implemented in the microphotonic form (Figure 11.11).

11.5.6 THE SWITCH

The core of the lightpath router is an *Optical Switch* that has to perform the function of switching the lightpaths from any input to any output including inputs/outputs from the ISL terminals and the Electronic Processor that handles the satellite uplink/

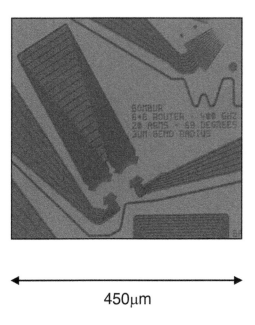

450μm

FIGURE 11.11   (a). Type of Arrayed Waveguide Gratings (AWGs) developed by the University of Gent for ESA for onboard routing and wavelength multiplexing/demultiplexing applications. Notice the 6 × 6 router with a 20-branch array and the I/Ps – O/Ps waveguides to the AWG.

downlink. This optical switch principle is served by an *optical crossconnect* (OXC) unit with add/drop functionality **[Jackman et al., 1999; Neilson, 2006]**. The OXC must be fully reconfigurable and adjusted as desired. This can be of crucial importance as lightpaths may have to be rearranged from time to time to accommodate, for example, the case of a component or even a S/C failure in the network. They also have to exhibit a very low crosstalk level and low throughput losses. Optical crossconnect technologies for space applications have been investigated since the late '80s. Several technologies have been demonstrated including integrated optic matrices using electro-optic switches and Spatial Light Modulators for satellite switched Time Division Multiple Access TDMA systems **[Baister et al., 1994]**. Currently, the focus is on the more sophisticated optical MEMS (micro-electro-mechanical systems)-based fabrics, which carry the potentials to implement large, switch matrices. MEMS have emerged over the last years as the main candidate to build such OXC from the WDM fiber networks. MEMS are configured in two main types using either a two-dimensional (2-D) matrix or a combination of two such matrices that create a three-dimensional (3-D) structure. The 3-D MEMS technology is the one that allows the large scaling required in future optical networks (with hundreds of input/output ports), whereas the 2-D OXC is adequate to handle a lower number of channels and would be preferred primarily for reliability reasons [**Dobbelaere et al., 2003**]. A figure of a 2-D MEMS OXC is shown in Figure 11.12a. ESA works with optical switch manufacturers in order to space qualify their products and produce Photonic Routers. Such routers

(a)

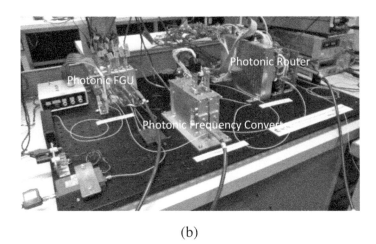

(b)

FIGURE 11.12   (a) A graphic of a 2-D optical MEMS OXC (Courtesy "FibreSystems".)
(b) A photonic microwave payload demonstrator including a 3-D Photonic Router from an
ESA-funded activity on Photonic Microwave Payloads (From **Sotom et al., 2016**.)

have been a part of ESA-funded demonstrators of Photonic Microwave Payloads as
the one depicted in **Figure 11.12b [Sotom et al., 2016]**. Such demonstrators help
to advance the idea of a fully photonic PL as the one to be used in a satellite OTN.
Looking further to the future, developments of microphotonic routers with hundreds
of distinct elements will be targeted in the coming years.

11.5.7   CROSSTALK AND NOISE

The nature of the OTN, based on the use of wavelength multiplexed channels that
cross several nonlinear optical elements (such as the Doped Fiber Amplifier DFAs)
and many wavelength selective devices (filters, mux/demux, etc.), requires the

examination of the potential appearance of crosstalk. The amplified spontaneous emission (ASE) produced at each DFA accumulates with the signal (the part that falls within the signal optical BW) along the optical path and it impairs the signal-to-noise ratio (SNR) by creating interferometric noise. In this way, it imposes an upper limit of the number of ISLs that a lightpath can propagate along before the SNR becomes prohibitively low and signal regeneration is required (estimating that at least a booster DFA and a preamplifier are involved at each ISL stage with a possible inline amplifier to compensate for the losses in the router) [**Wilson, 1997**]. Therefore, for the design of the OTN, careful estimation of the ASE must be performed to determine the lightpath feasibility. Given the maturity of optical components that now provide isolation better than 30 dB, no crosstalk from adjacent channels should be expected in a channel spacing of 1 nm to 3 nm. Also, due to the very high optical power levels involved in the DFAs, nonlinear phenomena such as four-wave mixing may appear despite the very short distance (of some meters) the light propagates in a single-mode fiber. The importance of these phenomena has to be evaluated for each particular layout of the onboard hardware (optical power levels and fiber lengths) and especially with the reference to the WDM transmission [**Agrawal, 2012**].

11.6   OTN POWER CONSIDERATIONS

The power consumption is a main concern in any space application, and therefore when designing the OTN, one also has to budget the required electrical power. Certain OTN strategies will probably be ruled out just on the power consumption basis. The main power consumption parts of the OTN are:

- Free-space ISL transmission (which depends on ISL range, transmission rate, and required BER and is a function of the telescope aperture). Power is primarily required to drive the pump lasers at the booster amplifiers.
- OBP throughput loss compensation (which depends primarily on the size of the switch, i.e., on the number of lightpaths to be switched). Again, DFA amplifiers will most likely be used.

The main parameter of importance is the so call *wall-plug efficiency* of the DFAs that reflects the electrical power required to produce an optical output of a certain power. Currently, the type of optical amplifiers designed for ISLs operate at a wall plug efficiency of less than 10%. The required ISL power can be calculated based on the link requirements for a certain data rate. The ISL range is either fixed (intra-orbit) or changes in a slow (in comparison to a communication session), deterministic and therefore predictable fashion (inter-orbit). In the latter case, *dynamic power management* can be employed in order to save power when the link becomes shorter. For example, in the Celestri network, the ISL range varies between a minimum of about 2900 km and a maximum of about 5860 km. This leads to a power requirement variation of 6 dB. The principles of dynamic power management can be extended to include the case of variable ISL data rates, although this would imply dynamic modifications at the transmitter and the receiver. From the OTN design viewpoint in the general case of fixed lightpaths (and therefore ISL data transmission rates) the critical parameter to consider is the aggregate data rate of all lightpaths that form the ISL beam. *Statistical multiplexing* is therefore important in order to maximize the utilization of the lightpaths effectively measured as W/Gbps.

11.7   OTN MANAGEMENT

The OTN needs to be managed as any telecommunications system. The same principles shall be adopted as in terrestrial fiber OTNs with adaptations to take into account all the new elements introduced by a space application (including the distance involved between the nodes, the free-space nature of the inter-satellite links, the traffic characteristics, the reliability issue with the inability to intervene physically to a malfunctioning hardware). As an example of the difference between the space and the terrestrial OTNs, the standard lightpath *ring protection* usually implemented in terrestrial fiber ring networks (using the reverse path in the ring) is not automatically applicable in space, given the extremely large ring circumference (42,000 km around the globe) that introduces unacceptable propagation delays.

The management of the ISL segment in a satellite OTN involves monitoring and management of the optical wavelength channels and of the individual circuits. Parameters that have to be monitored include:

- Lightpath BER
- Lightpath power level
- Optical amplifier requirements (input channels of certain power level)
- Crossconnect/add/drop multiplexer configuration

Automated control or via telemetry has to be exercised to hardware such as:

- Laser source drivers
- Photodetector amplifiers
- Optical amplifiers pump sources
- Multiplexers
- Demultiplexers
- Optical filters
- Attenuators

A three-layered structure is defined by the ITU for the WDM OTN Management. Each layer serves a higher electrical/optical client layer. The optical network layer stack is as follows with the meaning in a satellite OTN context indicated in brackets:

- *Optical Channel* (**OCh**) layer (which refers to the individual lightpaths)
- *Optical Multiplex Section* (**OMS**) layer (which refers to the WDM ISLs)
- *Optical Transmission Section* (**OTS**) layer (which is limited for connections on board the satellite optical module)

The relevant information on the health of each part in the network can be carried across the constellation either by a dedicated (1-ISL hop) wavelength channel, or by using a (low-rate) TDM, or Subcarrier Multiplexing (SCM) channel incorporated in the 1-hop lightpaths. This is called optical *Supervisory Channel*. Each node uses this channel to reach the constellation's Management Network Centre on ground. The case of failure at various stages of the communication system has to be foreseen and addressed. There are three main categories of failures:

- *Single optical channel lightpath failure:* This reflects the case where a laser source is out of order or the case of an optical modulator failure.
- *Link failure:* This is the case where the whole ISL has failed (e.g., due to amplifier failure, massive laser sources/transmitters failure, pointing errors due to PAT problems, etc.
- *Node failure:* This is the case of a whole onboard processor or even the whole satellite being broken down (e.g., power problems, attitude and orbital control problem, etc.)

Two main approaches for solving failures in a network exist: protection and restoration. *Protection* is distinguished from restoration in that protection systems reserve unused resources (e.g., lightpaths) for automatic engagement once a failure is detected. It is common to apply 1 + 1 protection policies reserving one physical path for each operating one. *Restoration* on the other hand uses existing resources to accommodate possible failures and the task is to optimize the use of the resources even when failures occur. The protection/restoration procedure involves three steps:

1. Fault detection
2. Fault notification message
3. Initiation of protection/restoration action

In the case of the satellite OTN, the effort is to provide solutions to failure restoration (instead of protection) through network reconfigurability at the minimum cost of additional hardware. Satellite constellations can be seen as *Dynamic Path Rearrangeable Mesh Architectures* due to the orbital nature of the satellite constellation. Both lightpath and IS link failure restoration can be provided by rerouting data over other paths/ISLs using spare capacity that is reserved for this purpose (and also for covering traffic demands in cases of overflow of a particular lightpath). This is similar to *SONET/ATM mesh path restoration* and requires that extra lightpath capacity is reserved for restoration. This approach also takes advantage of the statistical multiplexing policy followed in the dimensioning of the lightpath capacity and the redundancy inherent in the network. Restoration can in principle be achieved without significant impact in delay if the alternative Shortest Path is in service (no extra ISL hops) or with at most one or two extra ISLs. Let's recall that the path and wavelength allocation algorithm includes a last step of allocating a restoration path for the case of each individual ISL failure. In the case of a *satellite node failure*, the constellation will be equivalent to a network of N-1 nodes and then it has to be assessed whether the network has the coverage redundancy to allow it to operate with a satellite less or it has to recruit a spare satellite.

11.8   INTEROPERABILITY

The concept of the OTN was proposed for the scenario of autonomous high-capacity commercial satellite constellation networks designed to provide broadband telecommunication services. However, the same principle can be extended to include interoperation of various satellite networks as well as with a look to the future other non-satellite space systems such as stratospheric networks consisting of tens of platforms kept geostationary at a height between 20 km and 25 km [**Karafolas, 2001**]. In terms of hardware, interoperability requires proper design of the PAT subsystem of the ISL terminals (wide view angle, fast acquisition) and the coordination of the operational wavelength bands with the subsequent specifications to the onboard wavelength selective elements. Once the onboard hardware is designed in such a flexible fashion, proper software can allow the OTN elements to interoperate with similar optical networks of other systems dynamically during the lifetime of the constellation. One can envisage the optical interoperation of a large number of systems with space infrastructure at different altitudes spanning all the way from the cloud limit altitude of about 20 kms to GEO. Furthermore, proposals have been made for the fiber-optic network to be linked with the satellite OTN through multigigabit optical uplink/downlink from a ground station [**Chan, 2003; LASERLIGHT, 2014**]. ESA has demonstrated the feasibility of such links using the Optical Ground Station at Tenerife, Canary Islands, Spain, with the ARTEMIS satellite in GEO. In an operational scenario, it is of paramount importance to apply spatial diversity in the distribution of the ground stations in order to achieve a high link-availability and adopt sophisticated transmission techniques that will counteract the signal disturbances caused by the atmospheric turbulence.

11.9   SUMMARY

The space segment of the upcoming broadband satellite constellation networks with ISLs will form an optically linked Transport Network with onboard electronic routing. Further extension including Optical Routers can form a full OTN that improves the data traffic in the ISL segment by:

- Eliminating queuing delays in intermediate nodes and hence also improving the delay variation in communications.
- Using simplified fixed routing and reducing the amount of overhead information while reducing the processing load for the onboard processor with potentially significant savings in power.

The analysis and management of the OTN follows the same treatment as a terrestrial fiber transport network adapted to the specific characteristics of a global space system. The required technologies have been developed for applications in fiber networks, most components have been space qualified for other missions, and adaptation of some new devices such as the optical switches employed in the core of the optical network is needed to confirm their use under the conditions of a many years of operation in the near-Earth environment.

Careful traffic engineering involving statistical multiplexing proves to be of key importance from a resource utilization perspective, especially with reference to the electrical power requirement. The latter determines the applicability of the OTN in the power-limited system of the spacecraft.

Interoperability expressed both in hardware adaptation and signal format and frequency coordination can allow the dynamic link of the optical infrastructure of various satellites and stratospheric systems operating at various altitudes spanning from the cloud limit to GEO. Furthermore, using a spatially diverse network of optical Ground Stations, the satellite OTN can be seen as an extension element of the terrestrial and submarine OTN.

REFERENCES

Acampora A. S., Shah S., (June 1992), "Multihop lightwave networks: A comparison of store-and-forward and hot-potato routing," IEEE Transactions on Communications, vol. 40, no. 6, pp. 1082–1090.

Agrawal G., (2012), Nonlinear Fiber Optics, Academic Press, Cambridge, MA, eISBN 9780123973078.

Anzalchi J., et al., (2014), Application of photonics in next generation telecommunication satellite payloads, International Conference on Space Optics 2014, SPIE volume 10563.

Del Portillo I., Cameron B.G., Crawley E.F., "A technical comparison of three low earth orbit satellite constellation systems to provide global broadband', 69th International Astronautical Conference, Bremen, Germany, 1-5 October 2018.

Baister G. C., Gatenby B. V., (1994), "The optical crossbar switch for signal routing on board communication satellites," International Journal of Satellite Communications, vol. 12, pp. 135–145.

Ballard A. H., (September 1980), "Rosette constellations of Earth satellites," IEEE Transactions on Aerospace and Electronic Systems, vol. AES-16, no. 5, pp. 656–673.

Baroni S., Bayvel P., (February 1997), "Wavelength requirements in arbitrarily connected wavelength-routed optical networks," IEEE/OSA Journal of Lightwave Technology, vol. 15, no. 2, pp. 242–251.

Benzi E., (May 2016), Optical Inter-Satellite Communication: The Alphasat and Sentinel-1 A in orbit experience, 14th International Conference on Space Operations Conferences.

Chan V., (November 2003), "Optical Satellite networks," IEEE Journal of Lightwave Technology, vol. 21, no. 11, pp. 2811–2827.

Clarke A. C., (October 1945), "Extra-terrestrial relays," Wireless World, pp. 305–308.

Czichy R. H., (March 10–11, 1997), "Miniature Optical Terminals," CRL International Workshop on Space Laser Communications, CRL, Tokyo, Japan, pp. 105–112.

Dobbelaere P. D., Falta K., Gloeckner S., (2003), "Advances in Integrated 2D MEMS-based solutions for optical network applications," IEEE Optical Communications, vol. 1, no. 2, pp. S16–S23, (supplement to the IEEE Communications Magazine, vol. 41, no. 5).

Evans J. V., (1997), "Personal communications satellite systems," Space Communications, vol. 14, pp. 243–260.

Freidell J. E., (1998), "Why commercial broadband satellites absolutely must have laser ISLs and how the free-space laser communications community could let them down," SPIE Proceedings, vol. 3266, pp. 99–110.

Guo Y., et al., (2014), Research on Routing and Wavelength/Subcarrier Assignment Algorithm based on Layered-Graph Model in Optical Satellite Network, International Conference on Optical communication Systems, 2014.

Gisper T., Kao M., **(May 1996)**, "An all-optical network architecture," IEEE Journal of Lightwave Technology, vol. 14, no. 5, pp. 693–702.

Girard S., Morana A., Ladaci A., Robin T., Mescia L., Bonnefois JJ., Boutillier M., Mekki J., Paveau A., Cadier B., Marin A., Querdane Y., Boujenter A., "recent advances in radiation-herdened fiber-basaed technologies for space applications", Journal of Optics, 20, 2018

Handley M (2018), Delay is not an option: Low Latency Routing in Space", HotNets-XVII,

Hauschildt H., Elia C., Jones A., Moeller H.L., Perdiques J.M., "ESA's Scylight Programms, Activities and Status of the High Throughput Optical Network "Hydron" ", International Conference on Space Optics **2018**, SPIE volume 11180.

Jackman N. A., Patel S. H., Mikkelsen B. P., Korotky S. K., **(January–March 1999)**, "Optical Cross Connects for Optical Networking," Bell Labs Technical Journal, vol. 4, no. 1, pp. 262–281.

Jaouen Y., Bouzinac J.-P., Delavaux J.-M. P., Chabran C., Flohic M. Le, **(2000)**, "Generation of four wave mixing products inside wdm c-band 1 W Er3+/Y b3+ amplifier," Electronics Letters, vol. 36, no. 3 pp 233–235.

Karafolas N., Baroni S., **(December 2000)**, "Optical Satellite networks," IEEE Journal of Lightwave Technology, special issue on "Optical Networks," vol. 18, no. 12, pp. 1792–1806.

Karafolas N., **(April 17–20, 2001)**, "The case of Optical Networking using Stratospheric Platforms," 19th AIAA International Communications Satellite Systems conference, Paper no. 39. Toulouse, France.

Kennedy M. D., Malet P. L., **(June 1997)**, "Application for authority to construct, launch and operate the Celestri multimedia LEO system," technical report, Motorola Global Communications Inc. filed before FCC.

Lange R., **(2007)**, Homodyne BPSK-based optical inter-satellite communication links, Free-Space laser Communication Technologies XIX February 2007, SPIE volume 6457.

Laserlight, **(January 21, 2014)**, Laser Light Executive Summary Presentation, Pacific Telecommunications Conference PTC-14, Honolulu, Hawaii.

Leosat **(2015)**, LEOSAT patent "System and Method for Satellite Routing of Data", International Publication Number WO2015/175958 A2

Li Y. J., et al., **(June 2010)**, "A novel two-layered optical satellite network of LEO/MEO with zero phase factor," Science China (Information Sciences), vol. 53, no. 6, pp. 1261–2716.

Liron M. L., **(June 7, 1995)**, "Traffic routing for satellite communication system," application for a US Patent, no. 481,573.

Maxemchuk N., **(May 1987)**, "Routing in the Manhattan Street Network," IEEE Transactions on Communications, vol. 35, no. 5, pp. 503–512.

Mukherjee B., **(2006)**, Optical WDM Networks, Springer, New York, ISBN 0-387-29188-1.

NASA (2013), **Lunar Laser Communications (LLCD)**, https://www.nasa.gov/sites/default/files/llcdfactsheet.final_.web_.pdf

Neilson D. T., **(July/August 2006)**, "Photonics for switching and routing," IEEE Journal of Selected Topics in Quantum Electronics, vol. 12, no. 4, pp. 669–678.

Ott M., **(March 2005)**, "Validation of Commercial Fiber Optic Components for Aerospace Environments," SPIE Conference on Smart Structures and Materials, Smart Sensor Technology and Measurement Systems, vol. 5758.

Perdigues J., Werner M., Karafolas N., **(April 17–20, 2001)**, "Methodology for traffic analysis and ISL dimensioning in broadband constellations using Optical WDM networking," 19th AIAA International Communications Satellite Systems conference, paper no. 509, Toulouse, France.

Piqueras M., Marti J., Delgado S., Singh R., Beeder D., Parish R., Turgeon G., Castells A., Roux L., "A flight demonstration photonic payload for up Q/V-band implemented in a satellite ka-band hosted payload aimed at broadband high throughput satellites", International Conference on Space Optics **2018**, SPIE volume 11180.

Pratt S., Richard A. Raines, Carl E. Fossa Jr., Michael A. Temple, (1999), "An operational and performance overview of the Iridium low earth orbit satellite system," IEEE Communications Surveys, Second Quarter'99, http://www.comsoc.org/pubs/surveys

Sotom M., et al., (2016), "Flexible photonic payload for broadband telecom satellites: from concepts to system demonstrator," International Conference on Space Optics, SPIE volume 10562.

Space-X, (2016) https://licensing.fcc.gov/myibfs/download.do?attachment_key=1158350

Suzuki R., (1999), "A study of next generation LEO system for global multimedia mobile satellite communications," Proceedings of the 5th European Conference on Satellite Communications, 3-5/11 '99, Toulouse, France.

Tan L., et al., (April 2010), "Wavelength dimensioning of optical transport networks over nongeosynchronous satellite constellations," Journal of Optical Communications and Networks, vol. 2, no. 4, pp. 166–174.

Telesat (2015), Telesat Ka-band NGSO constellation FCC filling SAT-PDR-20161115-00108

Vaillon L., et al., (2008), Optical Communications Between an Aircraft and a GEO Relay Satellite: Design & Flight Results of the LOLA Demonstrator, International Conference on Space Optics **2008**, SPIE volume 10566.

Viasat (2017), https://www.viasat.com/products/high-capacity-satellites

Walker J. C., (1984), "Satellite constellations," Journal of the British Interplanetary Society, vol. 37, pp. 559–571.

Werner M., Delucchi C., Vogel H.-J., Maral G., De Ridder J. J., (January 1997), "ATM-based routing in LEO/MEO satellite networks with intersatellite links," IEEE Journal on Selected Areas in Communications, vol. 15, no. 1, pp. 69–82.

Werner M. (2001), Optical Inter-satellite Link Networks: design and Dimensioning Issues, EU COST Action 272.

Williams G. M., Friebele E. J., (June 1998), "Space radiation effects on Erbium-Doped Fiber Devices: Sources, Amplifiers and passive measurements," IEEE Transaction on Nuclear Science, vol. 45, no. 3, pp. 1531–1536.

Wilson K. E., (1997), "Link engineering design methodology for tandem free space optical communication links using all optical architecture," SPIE Proceedings, vol. 2990, pp. 215–229.

Wizner P. J., Kalmar A., (1999), "Sensitivity enhancement of optical receivers by impulsive coding," Journal of Lightwave Technology, vol. 17, no. 2, pp. 171–177.

Wood L., (2001), "Internetworking with satellite constellations," PHD Thesis, University of Surrey, 2001 http://www.ee.surrey.ac.uk/Personal/L.Wood

Yang Q., et al., (March 2010), "Analysis of crosstalk in Optical Satellite Networks with wavelength division multiplexing architectures," Journal of Lightwave Technology, vol. 28, no. 6, pp. 931–938.

Yang Q., et al., (February 2011), "An analytic method of dimensioning required wavelengths for optical WDM satellite networks," IEEE Communications Letter, vol. 15, no. 2, pp. 247–249.

Zhe L., et al., (2016), "Wavelength dimensioning for wavelength-touted WDM satellite network," Chinese Journal of Aeronautics, vol. 29, no. 3, pp. 763–771.

12 Future Directions

Hamid Hemmati

CONTENTS

12.1 INTRODUCTION

The laser communications technology has moved beyond mostly technology demonstration to full operational use and ambitious in-space applications at a large scale. In the past two decades the primary objective was successful lasercom link technology validation with aircraft and spacecraft while minimizing flight transceiver's mass, power-consumption, size and cost was a secondary priority.

Transceiver maturity, the advent of highly compact and efficient coherent fiber-optic transceivers that scaled quickly to greater than 1 Tbps along with more efficient ground receivers is expected to result in the development of highly compact laser communication transmitters for flight. These developments should shift much of the communications link burden to the ground transceiver and away from the flight transceiver.

A recent publication presents pathways whereby ongoing optical technology advances for planetary lasercom would allow 30 dB of increased data rate relative to the current state of the art in technology [1]. In other words, technological advances should allow deep space data-rate returns of 1 Gbps from the maximum Mars range, 100 Mb/s from Jupiter distances, and 10 Mbps from Uranus. Many orders of magnitude greater than the present radio frequency (RF) capabilities.

The enhanced link margin afforded by these improvements can be traded for higher data rate, smaller optical aperture diameter/size, lower mass, and lower power consumption of the flight transceiver hardware.

Some of the recent and future lasercom developments worldwide are summarized below.

12.2 INTER-SATELLITE LINKS

Substantial interest in satellite constellations located in the low Earth orbit (LEO) covering the globe has created the best opportunity in the history of laser communications operational use of the technology at large transceiver quantity [2].

Several such satellite telecom operators (e.g., SpaceX, Telesat, and LeoSat) have publicly announced their intention to use LEO-LEO inter-satellite links in orbit to minimize the quantity of ground-based gateway stations that are required to establish robust and seamless links, or decrease latency relative to terrestrial (fiber of wireless) links [3–5].

Laser communications is well suited to this application due to small size, mass, power, and particularly a vacuum channel free of adverse atmospheric effects. At least four flight transceivers per satellite (2 for in-plane transceivers and 2 for out-of-plane links) and a few hundred to a few thousand satellite are required in each constellation. This requirement presents manufacturing-challenges for the fabrication of high-tolerance optical transceivers in large quantities and at a cost that is a fraction of the current cost of flight lasercom transceivers. Large scale manufacturing at low cost is now being actively addressed by certain manufacturers globally.

12.3   US, EUROPEAN, AND ASIAN LASERCOM EFFORTS

Airplane Links: Several research groups, e.g., Facebook Inc. and Mynaric AG have recently demonstrated optical bidirectional links with airplanes at the rate of up to 40 Gbps [6, 7]. Future capacity scaling of duplex links with airplanes to 100s of Gbps is now a real possibility.

LEO Links: Research groups in the United States, Europe, Japan, and China have demonstrated and are planning to demonstrate by 2020 high-rate links from and to LEO satellites [8, 9]. Again, today the technology is at hand to attain lasercom links at >1 Tbps capacity. Quantum-key-distribution (QKD) demonstrations have been and remain to be the by-product of LEO-ground optical communication links [10, 11].

The Aerospace Corporation recently demonstrated lasercom links from a microsatellite (CubeSat) at the rate of 200 Mbps, and Tesat Spacecom/DLR and other organizations are also planning for lasercom links from CubeSat with flight transceivers as small as 0.3U [12, 13].

GEO Links: Tbps-scale link to and from geosynchronous orbit (GEO) and between LEO and GEO have also been the focus of attention in recent years [14]. With Ka-band satellite connections from GEO reaching 1 Tbps (through 100s of beams at 1–2 Gbps each) [15, 16] and the appetite to scale such capacity per satellite, undoubtedly, laser communications feeder link (uplink) to GEO satellite will play a major role to make this quest a reality [17, 18]. Ground-based station diversity for optical uplink to GEO is expected to make possible high-link availability to the satellite, a major requirement for operational satellites [19].

Lunar Links: Lincoln Laboratory along with the team-mates in a NASA-funded project called Lunar Laser Communications Demonstrator (LLCD) successfully demonstrated the first sub-Gbps laser communications from the Moon, setting the stage for optical links beyond the Earth orbit [20].

Planetary Links: The first-ever planetary laser communication link to an asteroid is by NASA as part of the Psyche robotic spacecraft mission is planned for the 2022/2023 time frame launch and 2026 operations [21]. The 5-meter aperture diameter Palomar telescope will be used as the ground station.

12.4   COMBINED RF AND OPTICAL LINK EXPERIMENTS

DARPA's (Defense Advanced Research Projects Agency's) Optical-Radiofrequency combined link experiment communication adjunct (ORCA) developed and tested a secure hybrid electro-optical, RF, Internet protocol-based communication system for tactical reach-back applications [22]. This development was in response to the requirements of high-reliability and high-capacity links for tactical applications that are placing increasing demands on the throughput of current RF systems and RF satellite links. Moreover, the incorporation of lasercom capability is expected to relieve RF congestion and allow for a more efficient use of allocated RF capacity. The objective of this successful technology demonstration campaign was to provide links between a ground-based moving node and an aircraft at a distance of approximately 50 km, flying 8 km above the mean sea level. Also, links between two aircraft approximately 200 km apart.

As an example of a hybrid RF/Optical link, the National Institute for Information and Communications Technology (NICT, Japan) has embarked on a 10-Gbps optical uplink to a GEO satellite followed by high-rate Ka-band downlink from the ETS-9 satellite in the 2022 timeframe [23].

12.5   TECHNOLOGY ADVANCEMENT OPPORTUNITIES

Atmospheric effects on the laser beam, communicating within a few degrees of the Sun-angle, more efficient power amplifiers, simplified and significantly lower-cost flight lasercom transceivers, and insufficient ground infrastructure remain as engineering challenges.

Mitigation of Atmospheric Effects: Alleviation of adverse atmospheric effects, such as scintillation and attenuation, is typically required to enhance the robustness of a lasercom link. Multi-beam uplink, spoiling of spatial coherence of the transmit beam, and the use of interleaver and erasure codes have been used to partially mitigated scintillation effects [24, 25]. The advent of adaptive optics (AO) systems that operate not only in weak turbulence, but also in moderate to strong turbulence will help to substantially mitigate atmospheric scintillation effects. An AO system helps with fiber-out, fiber-in type links where one can effectively leverage major developments in the fiberoptics transceiver industry.

While future Earth optical receivers may be located on airborne or Earth-orbiting platforms to circumvent the atmospheric issues, technological and operational demonstrations for the first decade or so of space laser communications are most likely to rely on lower risk and less expensive ground-based receivers. Ground station diversity can adequately mitigate cloud cover. Large diameter membrane filters located at the entrance aperture of the ground telescope will aid in communications at small Sun angles. The advent of affordable, large (1–2 m) diameter, near-diffraction-limited ground-based telescopes should make the infrastructure development realizable.

Navigation and Ranging: Future spacecraft are expected to use the higher frequency of optical beams for improved spacecraft navigation and precision ranging, since this data is provided routinely by the spacecraft's telecommunications subsystem [26]. Doppler information is gleaned from the recovered pulse-position-modulation (PPM) slot clock. With sub-millimeter precision, spacecraft range data can be inferred from coding frame boundaries. Analysis and simulations indicate that, depending on how the code word slot clock is phase-locked, Doppler and range data two to ten times better than the Ka-band can be obtained using PPM and direct detection. Ranging via an active laser transponder orbiting or landing on Moon will provide multiple science data with accuracy unattainable from ranging to passive retro-reflectors on the Moon.

Multifunctional Transceivers: Passive and active electro-optic systems (e.g., imagers, spectrometers, laser ranging, and laser remote sensing) typically comprise the same subsystems (e.g., lasers, imager(s), detectors, and beam directors) as a lasercom instrument. This feature provides a unique opportunity for the telecommunications transceiver to be combined with other electro-optic instruments. Undoubtedly, some compromises (e.g., on field-of-view of sensors or characteristics of the laser used) are necessary for a multifunctional instrument [27].

Networking and Multiple Access: Optical satellite networks are expected to become a key aspect of satellite networking. Chan provides suggestions for new dimensions of space system architectures enabled by such a network [28]. Point-to-multipoint laser communications using a single transceiver is of interest, but as yet an immature technology. The design of optical systems with wide enough field of view without sacrificing performance is a prime enabler of this technology.

Light Science: The "light science" (akin and complementary to radio science measurements) should be possible and enabled by lasercom, but it remains a largely unexplored field. Examples include occultation measurements to probe ionospheres, magnetic fields, gravitational field measurements, and tests of the fundamental theories of relativity [29].

Key Advancements: Lasercom technology is expected to revolutionize future air and space system architectures by delivering multi-Tbps links, while at the same time reducing mass, power, and size. Implementation of standard telecommunications interoperability protocols is expected to drive down the cost of operations. For many of the near-Earth communications link scenarios, lasercom technology is ready for deployment. The span of potential link margin (or capacity) improvements over the state of the art exceed 30 dB, and space flight reliability is projected to significantly improve with time, exemplified by:

Laser transmitter / fiber optical amplifiers	Improving the overall efficiency to >>20% and a longer lifetime.
Flight optics	Higher throughput efficiency, optimized laser coupling.
Photo-detectors	Higher quantum efficiency and lower noise relative to what is available today. Photon counters with simultaneously optimized parameters.
Receiver architecture	Realizing several dB enhancement as identified by analysis.
Optical receiver	High-throughput low-cost apertures with 0.6-1.2 m diameter
Modulation and coding	More efficient means of mitigating atmospheric effects, particularly on the uplink.
Atmospheric propagation data	Comprehensive quantitative data taken over extended period of time in satellite-ground is lacking in the literature, but sorely needed for proper link analysis.
Uplink adaptive optics	The advent of an efficient scheme that also compensates for the isoplanetic angle.
Reliability	Data accumulation for better reliability assessment, and redundancy implementation without major mass penalty.

REFERENCES

1. H. Hemmati, A. Biswas, and D. Boroson, "Prospects for improvement of interplanetary laser-communication data rates by 30 dB," Proc. IEEE, V. 95, 10, pp. 2082–2092 (2007).
2. http://www.losangeles.af.mil/news/story.asp?id=123115329
3. https://www.universetoday.com/140539/spacex-gives-more-details-on-how-their-starlink-internet-service-will-work-less-satellites-lower-orbit-shorter-transmission-times-shorter-lifespans/
4. https://www.telesat.com/services/leo/what-makes-it-work
5. http://leosat.com/technology/
6. C. Chen et al., "Demonstration of bi-directional coherent air-to-ground optical link," SPIE Photonics West Conf. Proc., V. 10524 (2018).
7. J. Horwath et al., "Test results of error-free bidirectional 10 Gbps link for air-to-ground optical communications," SPIE Proc., V. 10524 (2018).
8. A. Carrasco-Casado, H. Kunimori, H. Takenaka, T. Kubo-Oka, M. Akioka, T. Fuse, Y. Koyama, D. Kolev, Y. Munemasa, and M. Toyoshima, "LEO-to-ground polarization measurements aiming for space QKD using Small Optical TrAnsponder (SOTA)," Opt. Express, V. 24, pp. 12254–12266 (2016).
9. W. Chen et al., 5.12 Gbps optical communication link between LEO satellite and ground station," 2017 IEEE International Conf. on Space Optical Systems and Applications (ICSOS), Naha, Japan, 14–16 Nov (2017).
10. R. Hughes and J. Nordholt, "Quantum space race heats up," Nature Photonics, V. 11, 456–458 (2017).
11. L. Calderaro, C. Agnesi, D. Dequal, F. Vedovato, M. Schiavon, A. Santamato, V. Luceri, G. Bianco, G. Vallone, and P. Villoresi, "Space quantum communication with higher orbits," Quantum Information and Measurement (QIM), V: Quantum Technologies, Italy, 4–6 April (2019).
12. T. Rose et al., "Optical communication downlink from a 1.5U CubeSat: OCSD program," SPIE Proc., V. 10910 (2019).
13. J. Crabb, G. Stevens, C. Michie, W. Johnstone, and E. Kehayes, "Laser transmitter for CubeSat-class applications," SPIE Proc. V. 10910 (2019).
14. J. Poliak, D. Giggenbach, F. Moll, F. Rein, C. Fuchs, and R. Mata Calvo, "Terabyte-throughput GEO satellite optical feeder link testbed," 2015 13th International Conf. on Telecom. (ConTel), Graz, Austria, 13–15 July (2015).
15. H. Zech, F. Heine, D. Troendle, P. M. Pimentel, K. Panzlaff, M. Motzigemba, R. Meyer, and S. Phillip-MayR , "LCTS on ALPHASAT and Sentinel 1a: In orbit status of the LEO TO GEO data relay system," International Conf. on Space Optics (ICSO), Tenerife, Canary Islands (2014).
16. C. Miller, "How and why commercial high-capacity satellites offer superior performance and survivability in the future space threat continuum," 32ns Space Symposium, Colorado Springs (2016).
17. H. Hauschildt, C. Elia, H. L. Moeller, J. M. Perigues Armengol, "HydRON: High throughput optical network," SPIE Proc., V. 10910 (2019).
18. C. Volans, Z. Sodnik, J. M. Perdigues-Armengol, D. Alauf, N. Vedrenne, and K. Nicklaus, "Towards optical data highways through the atmosphere," SPIE Proc., V. 10910 (2019).
19. R. Alliss, "Optimizing the performance of space and ground optical communications," SPIE Proc., V. 10910 (2019).
20. F. I. Khatri, B. S. Robinson, M. D. Semprucci, and D. M. Boroson, "Lunar laser communications demonstration architecture," Acta Astronautica, V. 111, pp. 77–83 (2015).
21. A. Biswas, M. Srinivasa, S. Piazzolla, and D. Hoppe, "Deep space optical communications," SPIE Proc., V. 10524 (2018).

22. J. C. Juarez, A. J. Goers, J. E. Maloicki, R. J. Dimeo, and V. Bedi, "Evaluation of curvature adaptive optics for airborne laser communication systems," SPIE Proc., V. 10770 (2018).

23. Y. Munemasa et al., "Design status of the development of a GEO-to-ground optical feeder link," HICALI, SPIE Proc., V. 10524 (2018).

24. R. Pernice, A. Ando, A. Parisi, A. C. Cino, and A. C. Busacca, "Moderate-to-strong turbulence generation in a laboratory indoor free space optics link and error mitigation via RaptorQ codes," 18th IEEE Inter. Conf. on Transparent Optical Networks (ICTON), Trento, Italy, 10–14 July (2016).

25. S. Yamamoto, H. Takahira, and M. Tanaka, "5 Gbit/s optical transmission terminal equipment using forward error correcting code and optical amplifier," Electronics Let., V. 30, 254–255 (1994).

26. Y. Chen, K. Birnbaum, and H. Hemmati, "Active laser ranging over planetary distances with millimeter accuracy," Appl. Phys. Lett., V. 102, 241107 (2013).

27. H. Hemmati and J. R. Lesh, "A combined laser-communication and imager for micro spacecraft (ACLAIM)," SPIE Proc., V. 3266, 165–170 (1988).

28. V. W. S. Chan, "Optical satellite networks", J. Lightwave Tech., V. 21, 311 (2003).

29. H. Hemmati, "Deep-Space Optical Communications," Chapter 7.3, ISBN: 978-0-470-04240-3, Wiley Intercedence, Hoboken, NJ 2006.

Index

Note: Page numbers in *italics* indicate figures and **bold** indicate tables in the text.

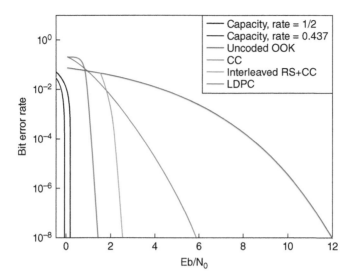

FIGURE 1.8 Performance of different modulation schemes, indicating that the recently developed codes for optical detection are within <1 dB of capacity limit. (OOK, on-off keying; CC, convolutional code; RS, Reed-Solomon; LDPC, low-density parity check) (see Chapter 6).

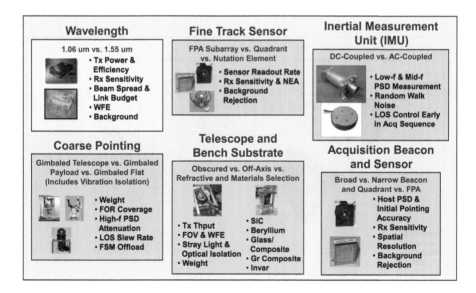

FIGURE 3.12 Key laser terminal trades optimize the design.

Characteristic	Aluminum	Beryllium	Composite M55J-954	Silicon	Silicon Carbide	ULE	Zerodur
E (GPa) Modulus of Elasticity	68	303	50 (z axis) 110 (x,y axes)	131	331	67	90
ρ (g/cc) Density	2.7	1.85	1.62	2.33	2.7	2.2	2.5
E/ρ Specific Stiffness	25	164	31 (z axis) 68 (x,y axes)	56	123	30.5	36
κ (W/mK) Thermal Conductivity	237	210	20 (longitudinal) 0.4 (transverse)	156	165	1.3	1.6
α (ppm/K) Thermal Expansion	11	11.3	+/- 0.1	2.6	2.0	0.03	0.02
κ/α Thermal Stability (Steady State)	21.5	18.6	40-200	60	82.5	43	80

FIGURE 3.15 Mechanical and thermal properties influence optimum design choices.

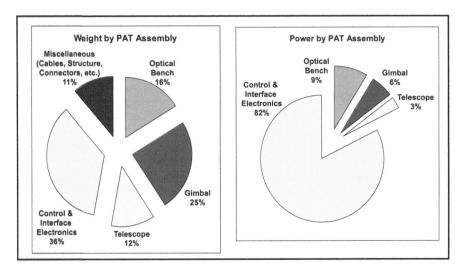

FIGURE 3.23 PAT weight and power contributors by key assemblies.

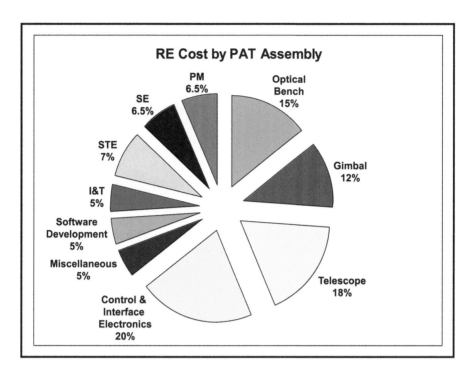

FIGURE 3.24 PAT recurring cost contributors by assembly.

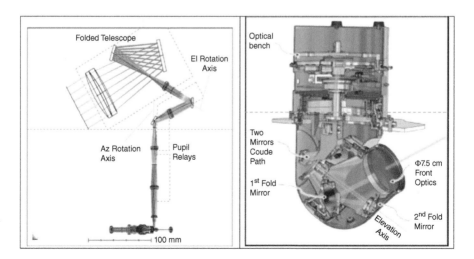

FIGURE 5.2 The optics layout and schematic of a laser communications transceiver optics developed for 100 Gbps bidirectional coherent links between airplanes and atmosphere. (Credit Facebook, Inc.)

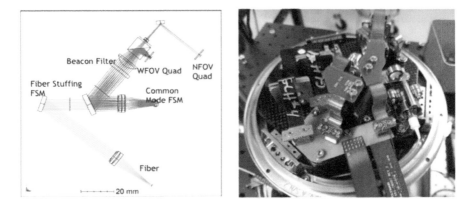

FIGURE 5.3 The small optics bench for the transceiver shown in Figure 5.2 incorporating a large number of components into a highly compact area. (Credit: Facebook, Inc.)

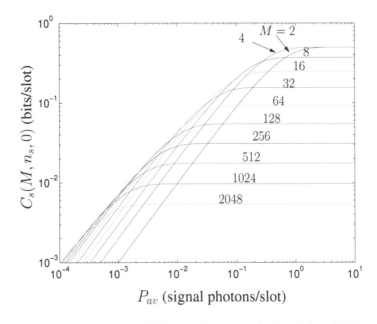

FIGURE 6.3 Capacity for Poisson PPM channel, $C_{h,e}$, $n_b = 0$, $M \in \{2, 4, \ldots, 2048\}$.

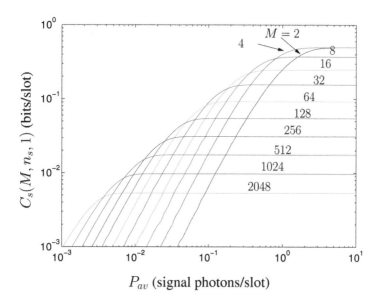

FIGURE 6.4 Capacity for Poisson PPM channel, $C_s(M, n_s, n_b)$, $n_b = 1.0$, $M \in \{2, 4, \ldots, 2048\}$.

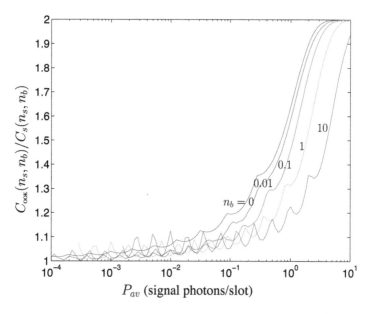

FIGURE 6.9 Relative loss due to using PPM, $C_{OOK}(n_s, n_b) / C_s(n_s, n_b)$, $n_b \in \{0, 0.01, 0.1, 1.0, 10\}$.

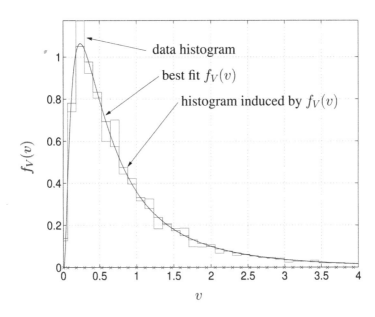

FIGURE 6.11 Histogram of fading samples and a lognormal probability density function (PDF) fit $f_V(v)$: $\hat{\sigma}_l^2 = 0.98$.

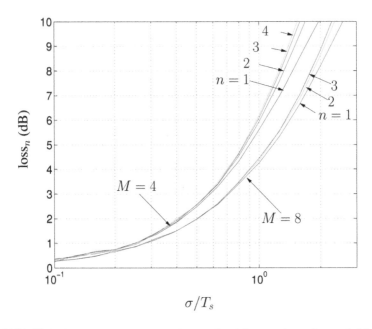

FIGURE 6.16 Loss as a function of normalized jitter variance for $n = 1, 2, 3, 4$, $M = 4$ and $M = 8$. Gaussian jitter, $\sigma = 1.0$ nsec, $\lambda_b = 1$ photon/nsec, $R = 1/2$ ECC, $\beta = 1.0$.

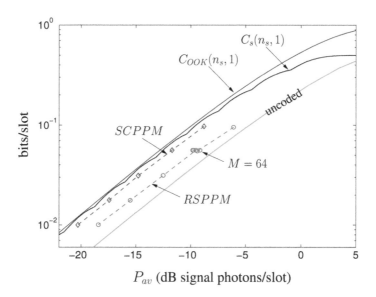

FIGURE 6.21   Sample operating points, $n_b = 1$.

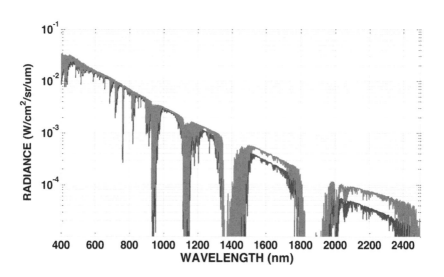

FIGURE 8.16   Daytime sky radiance at 2 km above sea level. The Sun zenith angle is 45°. Two cases (radiance curves) are shown: (1) the observer zenith angle on the ground is at 40° (higher radiance curve) and (2) the observer zenith angle on the ground is at 70° (lower radiance curve). The rural aerosol model with a visibility of 23 km at sea level was used. Data obtained after MODTRAN simulation.

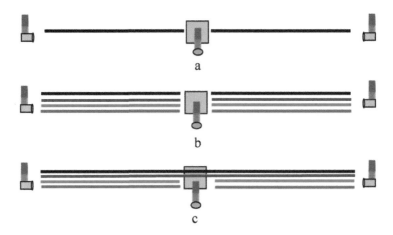

FIGURE 11.5 (a) Single-wavelength optical ISL, (b) WDM optical ISL, (c) WDM ISL with wavelength routing.

Printed and bound by CPI Group (UK) Ltd, Croydon, CR0 4YY

17/10/2024

01775660-0015